METHODS IN MOLECULAR BIOLOGY™

Series Editor
John M. Walker
School of Life Sciences
University of Hertfordshire
Hatfield, Hertfordshire, AL10 9AB, UK

For other titles published in this series, go to
www.springer.com/series/7651

Protocols for In Vitro Propagation of Ornamental Plants

Edited by

S. Mohan Jain and Sergio J. Ochatt

University of Helsinki, Helsinki, Finland
and
Institut National de la Recherche Agronomique, UMRLEG, PCMV, Dijon, France

Humana Press

Editors
S. Mohan Jain, Ph.D.
University of Helsinki
Helsinki
Finland
mohan.jain@helsinki.fi

Sergio J. Ochatt
Institut National de la
Recherche Agronomique
UMRLEG, PCMV, Dijon
France
ochatt@epoisses.inra.fr

ISSN 1064-3745 e-ISSN 1940-6029
ISBN 978-1-60327-390-9 e-ISBN 978-1-60327-114-1
DOI 10.1007/978-1-60327-114-1
Springer New York Dordrecht Heidelberg London

Library of Congress Control Number: 2009939425

Printed on acid-free paper

Humana Press is part of Springer Science+Business Media (www.springer.com)

Foreword

It was around 1970, I had just completed a 5-year breeding project aiming at fixing flower colour in gerbera progenies: white, yellow, pink, and red; colour homogeneity was sound, but size and shape still required some improvement. The problem was definitely resolved by Murashige and Skoog, USA who published a reliable protocol for gerbera micropropagation. In short, my gerbera seed lines were immediately rendered obsolete by this efficient cloning system, able to produce millions of plants of a matchless and previously unknown homogeneity, the uniformity of flower shape and colour being the basic requirements for the market. The success of micropropagation resulted in a tremendous growth in gerbera flower production worldwide, and this species conquered a leading place in the floriculture industry.

This personal experience stresses the impact of micropropagation on the genetic improvement research strategies in ornamentals. Micropropagation has become "invasive", especially in ornamental plant material issues. Today, hundreds of protocols exist; however, only a modest percentage of them are exploited economically. Thus, only micropropagation of plants with a high market price range, like orchids for instance, has proved cost-effective and achieved great success.

Micropropagation is a labour-intensive system: hand-power is estimated to represent 60–70% of total costs. This explains the outsourcing of the major labs in developing countries where labour is cheaper. Nevertheless, certain industrial protocols remain a proprietary technology of leading labs, mostly western, with the exception of Japan and Taiwan.

To gain cost-effectiveness, some micropropagation systems were developed recently, applying sophisticated in vitro manipulation methods involving somatic embryogenesis and organogenesis protocols.

Early on, researchers faced problems in the application of a particular protocol, genetic instability, and/or low multiplication rates, all negative factors for industrial production. Moreover, nowadays, some ornamental species still remain elusive to micropropagation. Breeders found that the fitness of in vitro multiplication in many species is a genotype-specific feature and therefore the competence for micropropagation should be considered in selection programs as a new "agronomic" trait.

The economical value of the clone fostered the application of micropropagation worldwide, and in vitro culture gained increasing popularity not only where the cost was recovered by the added value of plants in the market, but also in applied research where test tube plants represent a novel subject of study per se: no biotech lab may today disregard in vitro multiplication as an activity following mutant recovery, for embryo rescue, for pathogen-free plant production, for the selection of sporophytes and gametophytes and, finally, for the application of genetic transformation.

In short, selection methods have migrated from the field to the test tube, where the efficiency may attain results hitherto never thought of. In effect, new protocols for species, considered until yesterday to be recalcitrant, continuously increase the list of micropropagated plants, and this fact fully justifies an updated handbook. Such reliable and tested protocols of micropropagation represent, in my view, a new step forward in the development of proprietary technology, the only strategy to be competitive in today's world market.

Sanremo, Italy

Tito Schiva
Former Director CRA-ISF
t.schiva@alice.it

Preface

Ornamental plants are produced mainly for their aesthetic value; thus, the propagation and improvement of quality attributes such as leaf types, flower colour and fragrance, longevity and form, plant shape and architecture, and the creation of novel variation are important economic goals for the ornamental industry. The economic value of ornamental plants has increased significantly worldwide, and is increasing annually by 8–10% especially in the developing countries, due to low labour cost. In the developed world, the labour cost is extremely high, resulting in high cost of propagules and leaving a low-profit margin to the companies. Micropropagation of ornamental plants is labour-oriented, and for that reason, outsourcing of plant multiplication activities is shifted to the countries having low labour cost. In spite of all these problems, in vitro propagation of plants is still an efficient system and is being used by many commercial laboratories and national institutes worldwide for rapid plant multiplication, germplasm conservation, pathogen elimination, genetic manipulations, and for secondary metabolite production. Annually, millions of ornamental plants are routinely produced in vitro. The great potential of micropropagation for large-scale plant multiplication can be tapped by cutting down the cost of production per plant by applying low-cost tissue culture, adopting practices and optimizing the use of equipment and resources to reduce the unit cost of micropropagule and plant production without compromising the quality. Furthermore, the development and rapid multiplication of new ornamental cultivars are required to meet the demand of consumers all year round. The combination of mutagenesis and in vitro culture would be an ideal approach for the development of new ornamental cultivars. The existing and refined protocols for in vitro culture and transgenesis and their direct applications in improving and developing new ornamental cultivars would immensely benefit both the industry and academic institutes. Therefore, the need for a book describing step-wise protocols for rapid plant multiplication and in vitro storage of major commercially viable ornamental plants seems evident.

This book has a total of 32 chapters and is divided into two major sections, A and B. Section A contains 26 chapters mainly on micropropagation of cut and pot flowers, and Section B has six review chapters including in vitro production of sweet peas, the status of transgenics in ornamental plants, in vitro conservation, the status of floriculture in Europe, azalea phylogeny, and thin cell layers. Each chapter in Section A contains a step-wise description of in vitro culture explants, medium preparation, detailed medium table, shoot initiation and proliferation, root induction, in vitro plant hardening, and field transfer. All submitted manuscripts have been peer reviewed and revised accordingly.

The readership of the book will be floriculturists, researchers, commercial companies, biotechnologists, and students.

Helsinki, Finland — *S. Mohan Jain*
Dijon, France — *S.J. Ochatt*

Contents

SECTION B PROTOCOLS FOR IN VITRO PROPAGATION OF ORNAMENTAL PLANTS

Contributors

PARAMVIR SINGH AHUJA • *Division of Biotechnology, Institute of Himalayan Bioresource Technology (CSIR), Palampur, India*

M. FIROZ ALAM • *Department of Botany, Rajshahi University, Rajshahi, Bangladesh*

MEJÍAS ALEXANDER • *Laboratorio de Biotecnología Vegetal, Centro de Botánica Tropical, Instituto de Biología Experimental, Facultad de Ciencias, Universidad Central de Venezuela, Los Chaguaramos, Caracas, Venezuela*

MARGHERITA BERUTO • *Istituto Regionale per la Floricoltura, Sanremo, Italy*

PREM L. BHALLA • *Plant M olecular Biology and Biotechnology Laboratory, Australian Research Centre of Excellence for Integrative Legume Research, Faculty of Land and Food Resources, The University of Melbourne, Parkville, VIC, Australia*

SABITA BHATTACHARYA • *Department of Botany, Bose Institute, Kolkata, India*

SANGHAMITRA BHATTACHARYYA • *Department of Molecular Cardiology, Cleveland Clinic, Cleveland, OH, USA*

LUCA BRAGLIA • *CRA-FSO Research Unit for Floriculture and Ornamental Species, Sanremo, Imperia, Italy*

JOSE L. CASAS • *Instituto Universitario de Investigación CIBIO (Centro Iberoamericano de la Biodiversidad), Universidad de Alicante, San Vicente del Raspeig, Alicante, Spain*

MARCOS CASTELLANOS • *Plant and Crop Sciences Division, School of Biosciences, University of Nottingham, Loughborough, UK*

C. CONREUX • *INRA, CR Dijon, UMR LEG, Dijon Cedex, France*

MICHAEL R. DAVEY • *Plant and Crop Sciences Division, School of Biosciences, University of Nottingham, Loughborough, UK*

NOËLLE DORION • *Unité Mixte de Recherche GenHort (INRA, INH, University of Angers), Institut National d'Horticulture, INH, Angers, France*

TOM EECKHAUT • *Plant Unit, Applied Genetics and Breeding, Institute for Agricultural and Fisheries Research (ILVO), Melle, Belgium*

GIANCARLO FASCELLA • *CRA-SFM Research Unit for the Recovery and Improvement of Mediterranean Ornamental Plants S.S. 113-km 245,500, 1-90011 Bagheria, Palermo, Italy*

SEIICHI FUKAI • *Department of Horticulture Science, Faculty of Agriculture, Kagawa University, Kagawa, Japan*

ANTHONY GALLARD • *Unité Mixte de Recherche GenHort (INRA, INH, University of Angers), Institut National d'Horticulture, INH, Angers, France*

ANNALISA GIOVANNINI • *CRA-FSO Research Unit for Floriculture and Ornamental Species, Sanremo, Imperia, Italy*

TOSHINARI GODO • *Botanic Gardens of Toyama, Fuchu-machi, Toyama, Japan*

AGNÈS GRAPIN • *Unité Mixte de Recherche GenHort (INRA, INH, University of Angers), Institut National d'Horticulture, INH, Angers, France*

Miguel Pedro Guerra • *Laboratory of Plant Developmental Physiology and Genetics, Graduate Program in Plant Genetic Resources, Federal University of Santa Catarina, Florianópolis, SC, Brazil*
S. Dutta Gupta • *Agricultural and Food Engineering Department, Indian Institute of Technology Kharagpur, Kharagpur, India*
Nguyen Thanh Hai • *Plant Molecular Biology and Plant Breeding Laboratory, Tay Nguyen Institute of Biology, Dalat, Lam Dong, Vietnam*
M. Lokman Hakim • *Institute of Food and Radiation Biology, Atomic Energy Research Establishment, Ganakbari, Savar, Dhaka, Bangladesh*
Takashi Handa • *Laboratory of Ornamental Plants, School of Agriculture, Meiji University, Higashi-mita, Tama-ku, Kawasaki, Japan*
Anber Hassanein • *Unité Mixte de Recherche GenHort (INRA, INH, University of Angers), Institut National d'Horticulture, INH, Angers, France*
Johan Van Huylenbroeck • *Plant Unit, Institute for Agricultural and Fisheries Research, Melle, Belgium*
L. Jacas • *INRA, CR Dijon, UMR LEG, Dijon Cedex, France*
Kristien Janssens • *Plant Unit, Applied Genetics and Breeding, Institute for Agricultural and Fisheries Research (ILVO), Melle, Belgium*
Hatem Ben Jouira • *Unité Mixte de Recherche GenHort (INRA, INH, University of Angers), Institut National d'Horticulture, INH, Angers, France*
Navtej Kaur • *Division of Biotechnology, Institute of Himalayan Bioresource Technology (CSIR), Palampur, India*
Ellen De Keyser • *Plant Unit, Applied Genetics and Breeding, Institute for Agricultural and Fisheries Research (ILVO), Melle, Belgium*
Nobuo Kobayashi • *Laboratory of Plant Breeding, Faculty of Life and Environmental Science, Shimane University, Nishikawatsu, Matsue, Shimane, Japan*
S.L. Kothari • *Department of Botany, University of Rajasthan, Jaipur, India*
Maurizio Lambardi • *IVALSA/Istituto per la Valorizzazione del Legno e delle Specie Arboree, National Research Council (CNR), Sesto, Fiorentino, Firenze, Italy*
Yuanxue Lu • *Kunming Botanical Garden, Kunming Institute of Botany, Chinese Academy of Sciences, Heilongtan, Kunming, Yunnan, China*
Oropeza Maira • *Laboratorio de Mejoramiento Vegetal, Centro de Botánica Tropical, Instituto de Biología Experimental, Facultad de Ciencias, Universidad Central de Venezuela, Los Chaguaramos, Caracas, Venezuela*
Carlo Mascarello • *CRA-FSO Research Unit for Floriculture and Ornamental Plants, Sanremo, Italy*
M.L. Mayor • *Cátedra de Genética, Fac. De Cs. Agrarias, Universidad Nacional de Rosario, Zavalla, Argentina*
Antonio Mercuri • *CRA-FSO Research Unit for Floriculture and Ornamental Species, Sanremo, Imperia, Italy*
Masahiro Mii • *Laboratory of Plant Cell Technology, Graduate School of Horticulture, Chiba University, Matsudo, Chiba, Japan*
Takashi Nakatsuka • *Iwate Biotechnology Research Center, Narita, Kitakami, Iwate, Japan*

MAZEN NASSOUR • *Unité Mixte de Recherche GenHort (INRA, INH, University of Angers), Institut National d'Horticulture, INH, Angers, France*
DIANA NENCHEVA • *Institute of Ornamental Plants, Agricultural Academy, Sofia, Bulgaria*
G. NESTARES • *Cátedra de Genética, Fac. De Cs. Agrarias, Universidad Nacional de Rosario, Zavalla, Argentina*
DUONG TAN NHUT • *Plant Molecular Biology and Plant Breeding Laboratory, Tay Nguyen Institute of Biology, Dalat, Lam Dong, Vietnam*
MASAHIRO NISHIHARA • *Iwate Biotechnology Research Center, Narita, Kitakami, Iwate, Japan*
CHALERMSRI NONTASWATSRI • *Department of Horticulture, Faculty of Agricultural Production, Maejo University, Chiangmai, Thailand*
SERGIO J. OCHATT • *INRA, CR Dijon, UMR LEG, Dijon Cedex, France*
ENRIQUE OLMOS • *Departamento de Estrés Abiótico y Patología Vegetal, Centro de Edafología y Biología Aplicada del Segura, Espinardo, Murcia, Spain*
ELIF AYLIN OZUDOGRU • *Department of Biology, Laboratory of Plant Tissue Culture, Gebze Institute of Technology, Kocaeli, Turkey*
OCTAVIO PAREDES-LÓPEZ • *Depto. Biotecnología y Bioquímica, CINVESTAV-IPN, Irapuato, México*
PRATAP KUMAR PATI • *Department of Biotechnology, Guru Nanak Dev University, Amritsar, India*
MAI XUAN PHAN • *Plant Molecular Biology and Plant Breeding Laboratory, Tay Nguyen Institute of Biology, Dalat, Lam Dong, Vietnam*
L.A. PICARDI • *Cátedra de Genética, Fac. De Cs. Agrarias, Universidad Nacional de Rosario, Zavalla, Argentina*
ABEL PIQUERAS • *Departamento de Mejora Vegetal, Centro de Edafología y Biología Aplicada del Segura, Espinardo, Murcia, Spain*
LUCA PIPINO • *CRA-FSO Research Unit for Floriculture and Ornamental Species, Sanremo, Imperia, Italy*
MAŁGORZATA PODWYSZY SKA • *Research Institute of Pomology and Floriculture, Skierniewice, Poland*
J. BRIAN POWER • *Plant and Crop Sciences Division, School of Biosciences, University of Nottingham, Loughborough, UK*
V.S.S. PRASAD • *Agricultural and Food Engineering Department, Indian Institute of Technology Kharagpur, Kharagpur, India*
ALBERTO PREVIATI • *Centro Sperimentale Ortofloricolo "Po di Tramontana", Veneto Agricoltura, Rosolina, Rovigo, Italy*
AGATA PTAK • *Department of Plant Breeding and Seed Science, Agricultural University of Kraków, Kraków, Poland*
SILVIA RADICE • *Centro de Estudios Farmacológicos y Botánicos (CEFYBO – UBA – CONICET), Ciudad Autónoma de Buenos Aires, Argentina*
JAN DE RIEK • *Plant Unit, Applied Genetics and Breeding, Institute for Agricultural and Fisheries Research (ILVO), Melle, Belgium*
BARBARA RUFFONI • *CRA-FSO Research Unit for Floriculture and Ornamental Plants, Sanremo, Italy*

MARCO SAVONA • *CRA-FSO Research Unit for Floriculture and Ornamental Plants, Sanremo, Italy*
VALENTINA SCARIOT • *Department of Agronomy, Forest and Land Management, University of Turin, Grugliasco, Italy*
MADHU SHARMA • *Division of Biotechnology, Institute of Himalayan Bioresource Technology (CSIR), Palampur, India*
P.K. SHARMA • *Department of Botany, University of Rajasthan, Jaipur, India*
PINAKI SINHA • *Department of Botany, Rajshahi University, Rajshahi, Bangladesh*
DARIUSZ SOCHACKI • *Research Institute of Pomology and Floriculture, Skierniewice, Poland*
JAIME A. TEIXEIRA DA SILVA • *Faculty of Agriculture and Graduate School of Agriculture, Kagawa University, Kagawa-ken, Japan*
PURNIMA TYAGI • *Department of Botany, University of Rajasthan, Jaipur, India*
PABLO EMILIO VANEGAS-ESPINOZA • *Depto. de Biotecnología, Centro de Desarrollo de Productos Bióticos, Instituto Politécnico Nacional, Yautepec, México*
TERESA EDITH VARGAS • *Laboratorio de Biotecnología Vegetal, Centro de Botánica Tropical, Instituto de Biología Experimental, Facultad de Ciencias, Universidad Central de Venezuela, Los Chaguaramos, Caracas, Venezuela*
T. VEGA • *Cátedra de Genética, Fac. De Cs. Agrarias, Universidad Nacional de Rosario, Zavalla, Argentina*
LIRIO LUIZ DAL VESCO • *Laboratory of Plant Developmental Physiology and Genetics, Graduate Program in Plant Genetic Resources, Federal University of Santa Catarina, Florianópolis, SC, Brazil*
ALMA ANGÉLICA DEL VILLAR-MARTÍNEZ • *Depto. de Biotecnología, Centro de Desarrollo de Productos Bióticos, Instituto Politécnico Nacional, Yautepec, México*
TRAUD WINKELMANN • *Institute of Floriculture and Woody Plant Science, Tree Nursery Science Section, Leibniz University Hannover, Hannover, Germany*
CHUI ENG WONG • *Plant Molecular Biology and Biotechnology laboratory, Australian Research Centre of Excellence for Integrative Legume Research, Faculty of Land and Food Resources, The University of Melbourne, Parkville, VIC, Australia*
R. ZORZOLI • *Cátedra de Genética, Fac. De Cs. Agrarias, Universidad Nacional de Rosario, Zavalla, Argentina*

Section A

Protocols for In Vitro Propagation of Ornamental Plants

Chapter 1

Micropropagation and Organogenesis of *Anthurium andreanum* Lind cv Rubrun

Oropeza Maira, Mejías Alexander, and Teresa Edith Vargas

Abstract

Tissue culture techniques are routinely used for mass propagation and the establishment of disease free stock material. Virtually all pot type Anthuriums available in the market today are produced by tissue culture. In this chapter, we describe an efficient protocol to obtain *Anthurium andreanum* cv Rubrun vitro plants through micropropagation and organogenesis. Seeds from plant spadixes were germinated on MS medium supplemented with 0.5 mg/L BA. Micro-cuttings from in vitro germinated seedlings were subcultured on MS medium containing 2 mg/L BA and 0.5 mg/L NAA. Four-week-old in vitro plants obtained from microcuttings, showed callus proliferation at the stem base. The development of shoots and plantlets was observed from callus tissue. We also describe a detailed method for the histological analysis of callus tissue and a vitro plants acclimatization protocol.

Key words: Micropropagation, Organogenesis, Histological analysis, Acclimatization, Anthuriums

1. Introduction

Anthuriums have been cultivated for many decades as cut flowers. They are relatively easy to grow and have attractive foliage. Under the ideal environment, Anthuriums produce long lasting flowers all year round. Anthuriums are grown worldwide with the highest production in the United States and The Netherlands.

Tissue culture has become a routine technique for plant propagation, which has revolutionized the ornamental industry. This technique is most suitable for mass propagation and the establishment of disease free stock material. Today, virtually all pot type Anthuriums available in the market are produced by various tissue culture techniques.

S.M. Jain and S.J. Ochatt (eds.), *Protocols for In Vitro Propagation of Ornamental Plants*, Methods in Molecular Biology, vol. 589
DOI 10.1007/978-1-60327-114-1_1, © Humana Press, a part of Springer Science+Business Media, LLC 2010

Plant regeneration of *Anthurium andreanum* has been achieved through adventitious shoot formation from callus (1–4), axillary buds (5), root explants (6), and direct shoot regeneration from lamina explants (7). Teng (8) established that in liquid or raft cultures, most adventitious *Anthurium* shoots regenerated singly or in loose aggregates, which is an advantage over solid cultures. Geier (9) concluded that plant age and plant genotype influence plant regeneration of *Anthurium andreanum*, as well as NH_4NO_3 on callus and shoot formation from young leaf tissues.

This paper describes a detailed protocol for in vitro regeneration of *Anthurium andreanum* cv Rubrun plants from microcuttings and callus tissue through organogenesis.

2. Materials

2.1. Plant Material

Commercially obtained *A. andreanum* cv Rubrun plants and in vitro plants of *A. andreanum* cv Rubrun germinated from seeds separated from spadixes of potted *A. andreanum* plants (*see* Note 1).

2.2. Surface Sterilization of Source Material

1. Tap water.
2. Ethanol 70% (v:v).
3. Commercial bleach solution [e.g., "Nevex" bleach; (Procter and Gamble, Venezuela) 5% (v/v) NaClO], diluted 1:10 (v:v) with tap water.
4. Autoclaved water; 250 mL aliquots in 500 mL screw capped bottles.
5. Whatman No. 1 filter paper.
6. Magnetic stirrer, magnetic bar, 500 mL beaker (autoclaved).
7. Tissue culture facilities – Instruments (scalpel, spatula, forceps, knife, spirit burner to flame sterilize instruments), laminar flow bench, culture room.
8. 9 cm diam. Petri dishes (HYSIL, England).

2.3. Medium for Seed Germination

MS medium (6) (Table 1.1).

2.4. Media for Micropropagation and Organogenesis

MS medium (10) supplemented with 2 mg/L BA (Sigma) (add 2 mL stock solution at 1 mg/L), and 0.5 mg/L naphthalene acetic acid (NAA) (Sigma) (add 0.5 mL stock solution at 1 mg/L).

Table 1.1
Murashige and Skoog (1962) medium

Solution	Components	Stock concentration (g/L)	Volume (mL) for 1 L medium	Final Concentration (mg/L)
A	NH_4NO_3	82.5	20	1,650
B	KNO_3	95.0	20	1,900
C	H_3BO_3 KH_2PO_4 KI $Na_2MoO_4{\cdot}2H_2O$ $CoCl_2{\cdot}6H_2O$	1.24 34.00 0.166 0.050 0.005	5	6.2 170 0.83 0.25 0.025
D	$CaCl_2{\cdot}2H_2O$	88.0	5	440
E	$MgSO_4{\cdot}7H_2O$ $MnSO_4{\cdot}4H_2O$ $ZnSO_4{\cdot}7H_2O$ $CuSO_4{\cdot}5H_2O$	74.0 4.46 1.72 0.005	5	370 22.3 8.6 0.025
F	Na_2EDTA $FeSO_4{\cdot}7H_2O$	3.72 2.78	10	37.2 27.8
G	Thiamine-HCl	1	4.2	1
	Myo-Inositol			100
	Nicotinic acid Pyridoxin Glycine	1 1 1	0.5 0.5 2	0.5 0.5 2.0
	Sucrose		30 g/L	

2.5. Histological Analysis

1. Tertiary butyl alcohol.
2. Paraplast (SPI supplies).
3. Ethanol.
4. FAA (Formaldehyde: acetic acid: ethanol).
5. Picric acid (Sigma) saturated in 95% ethanol solution.
6. Safranine 1,125 % (Sigma).
7. Fast Green FCF 0.1% (Sigma).
8. Orange G 1% (Methyl Orange BDH) (Sigma).
9. Xylol (Sigma).
10. Egg white.
11. Canadian balsam (BIORAD).
12. Leica microtome, Germany, RM 2135.
13. Wild Heerbrugg optical microscope model M20EB, equipped with a camera.
14. Brushes, water bath, slides, microscope cover glasses, gauze.

3. Methods

3.1. Preparation and Sterilization of Culture Media

1. Mix all the components as they appear in Table 1.1.
2. Initially raise volume to 800 mL with distilled H_2O.
3. Adjust medium to pH 5.8 using NaOH, HCl, or KOH.
4. Add 2 g gelrite (Sigma).
5. By adding distilled water, raise final volume up to 1 L.
6. Heat until solution is clear and autoclave at 121°C, 1.05 kg/cm^2 (15–20 psi) for 20 min.
7. Dispense 30 mL medium into the petri dishes.

3.2. Surface Sterilization and In vitro Germination of A. andreanum Seeds

1. Separate fruits from spadixes and sterilize for 15 min in 3% NaOCl (Fig. 1.1).
2. Rinse three times with sterile distilled water (5 min each).
3. Isolate seeds and sterilize for 20 min in 1% NaOCl.
4. Rinse two times with sterile distilled water (10 min each).
5. Cultivate 10 seeds per petri dish on MS medium (6).
6. Incubate under continuous fluorescent light 50 µmol/m^2/s at 25°C for 2 weeks. *See* Note 2.
7. Transfer plants to appropriate culture flasks and incubate for 2 weeks.

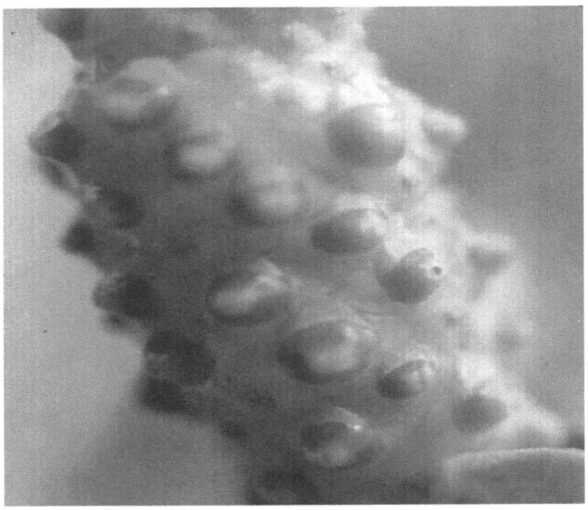

Fig. 1.1. Mature spadix of *A. andreanum*.

Fig. 1.2. *A. andreanum* plantlets from germinated seeds under continuous light conditions.

3.3. Micropropagation of Anthurium andreanum

1. Use in vitro germinated seedlings as a source of explants (Fig. 1.2).
2. Culture four micro-cuttings per flask on MS medium supplemented with 2 mg/L BA and 0.5 mg/L NAA.
3. Incubate under continuous fluorescent light 50 μmol/m^2/s at 25°C for 8 weeks.
4. Isolate micro- cuttings from 8-week-old plants. *See* Note 3.
5. Incubate under the same conditions as described earlier. Shoot development must be observed 6 weeks later (Fig. 1.3).

3.4. Indirect Organogenesis of A. andreanum

At the stem bases of the 8-week-old plants originated from micro-cuttings, a proliferation of callus tissue is observed (Fig. 1.4).

1. Take segments of approximately 1 × 1 cm of these callus tissues.
2. Sub culture on the same MS basal medium containing 2 mg/L BA and 0.5 mg/L NAA.
3. Incubate under continuous fluorescent light 50 μmol/m^2/s at 25°C for 4 months.
4. Subculture on the fresh medium at every 30-day interval (Fig. 1.5).

Fig. 1.3. Shoot development from micro cuttings 6 weeks after induction.

Fig. 1.4. Callus proliferation on *A. andreanum* plantlets obtained from micro cuttings 5 months after inoculation.

3.5. Acclimatization

1. Transfer vitroplants to MS medium without hormones and culture them for 3 months.
2. Remove them from the culture tubes carefully to avoid any damage to the roots.
3. Remove culture medium sticking to the roots by washing in running tap water.

Fig. 1.5. *A. andreanum* callus showing shoot proliferation 4 months after subculture.

Fig. 1.6. *A. andreanum* plants under greenhouse conditions.

4. Transfer plants to pots containing a mixture of soil and organic humus (1:1).
5. Keep plants in chambers with 95% relative humidity (RH) and low light intensity 700 μmol/m^2/s.
6. After 1 month, transfer plants to the greenhouse (Fig. 1.6).

3.6. Histological Analysis

For histological observations, the organogenic callus derived from micro-cutting explants was submitted to the following protocol:

3.6.1. Fixing and Dehydration (see Note 4)

1. Fix callus fragments on FAA (40% formaldehyde, 10% glacial acetic acid, 50% ethanol, store at room temperature) for a minimum of 2 h. *See* Note 5.

Table 1.2
Alcohol dehydrations series

	Percentage of alcohol				
Mixture	50	70	85	95	100
dH_2O	50	30	15	–	–
95% Ethyl alcohol	40	50	50	45	–
Tertiary butyl alcohol (TBA)	10	20	35	55	75
100% ethyl alcohol	–	–	–	–	25

2. Place the sample, previously identified (e.g., *A. Andreanum* callus fragment N° 1), on an adapted syringe to allow easy sample manipulation.
3. Treat the sample with 70% ethanol for 2–3 h.
4. Treat the sample with a series of alcohol dehydrations for a day at each concentration (Table 1.2) *See* Note 6.

3.6.2. Paraplast Inclusion

1. Place approximately 1 g paraplast chips on the bottom of a vial.
2. Fill 2/3 of the vial with TBA (Tertiary butyl alcohol).
3. Place the samples with their proper identification.
4. Add more paraplast chips, till they reach 1 cm over the TBA.
5. Incubate at 60°C. Add more paraplast and agitate briefly and gently every 30 min until the paraplast is dissolved and the smell of TBA is gone. *See* Note 7.
6. Leave at 60°C for 2–3 days until the mixture is odorless.

3.6.3. Block Preparation

1. Make small cardboard boxes (3 cm each side and 1 cm high).
2. Place the samples in Petri dishes and heat for 2 min at 60°C.
3. Heat two forceps.
4. Melt paraplast, and fill the cardboard boxes with it.
5. Transfer the samples with hot forceps to the cardboard box. *See* Note 8.
6. Place the boxes on water and when an opaque film is formed, stick the identification with paraplast.
7. Once the paraplast has solidified, remove the cardboard.
8. Store blocks at 4°C for at least 2 days before cutting.

3.6.4. Paraplast Block Cutting

1. Make a cut with a knife where the samples are placed.
2. Trim the paraplast with the knife.

3. Fill the support mold to be used with melted paraplast.
4. Stick the sample block on the support mold and trim with a hot spatula.
5. Set the block on the microtome (Leica microtome, Germany, RM 2135) and orientate the sample so that it would be perpendicular to the microtome blade. *See* Note 9.
6. Make thick cuts (18–20 μ) until the sample is reached, then set the microtome to 12 μ.
7. Pick up the cut sections with a brush and place on top of a white paper in a box.

3.6.5. Setting of Cuts on Slides

1. Heat a water bath to 50°C.
2. Put the cut sections on the warm water bath until they stretch (opaque side facing up/bright side facing down).
3. Clean the slides with water and soap, and then put them on 95% ethanol.
4. Dry the slides.
5. Treat the slides with 50% glycerine and 50% egg white solution. *See* **Note 10**.
6. Dry slides in a vertical position for at least an hour.
7. Put the cut sections over the slides.

3.6.6. Fast Green or Safrnine Staining See *Note 11*

1. Treat the slides with xylol to remove the paraplast.
2. Treat with a rehydration series (2 min each):
   Absolute Ethanol
   95% Ethanol
   70% Ethanol
   50% Ethanol
3. Prepare the Safranine 1,125% solution: 2,25 g Safranine O (Sigma) or 3 g Safranine T (Sigma). Add 225 mL 95% ethanol and 225 mL H_2O, store at room temperature.
4. Incubate with Safranine 1,125% for 1 min.
5. Dry the excess safranine and dehydrate (2 min each):
   50% Ethanol
   70% Ethanol
   95% Ethanol
6. Treat with drops of Picric Acid saturated in 95% ethanol solution (add picric acid to 100 mL 95% ethanol until the picric acid begins to precipitate, store at room temperature) until there is no more red observed.
7. Wash with 95% ethanol.

8. Add Orange G 1% (1 g Orange G, 100 mL 95% Ethanol, filter through Whatman No. 1 filter paper and store at room temperature) and Fast Green 0,1% (100 mg Fast Green, 100 mL 95% Ethanol, store at room temperature) to the slide, mix until the color is uniform, and leave for 5 min.
9. Wash with 95% ethanol-100% ethanol-xylol (10 s each).
10. Eliminate the excess xylol and make sure there is no water left.
11. Mix Canadian balsam with xylol until it is in liquid form (it should not be too viscous or too watery).
12. Add some drops of the balsam over the sample and cover with a microscope cover glass.
13. Dry for several days.
14. Remove the excess balm with a blade.
15. The observation and photomicrograph can be done with a Wild Heerbrugg optical microscope model M20EB, equipped with a camera (Fig. 1.7).

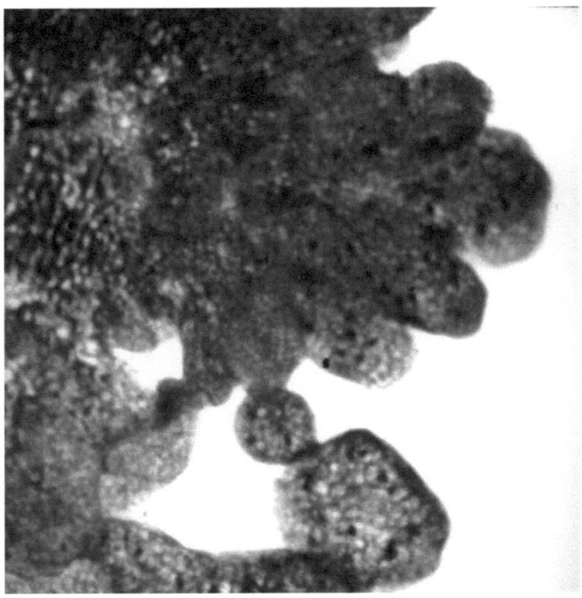

Fig. 1.7. Three-month-old organogenic callus showing shoot primordia (Scale bar = 310 μm).

4. Notes

1. To establish *Anthurium andreanum* regeneration systems using explants obtained from plants maintained under greenhouse conditions is a difficult task. We tested many disinfection protocols and obtained from 0 to 60% healthy explants. In this protocol, we propose the use of explants obtained from in vitro plants. The best protocol to obtain in vitro plants in our experience was from in vitro germinated seeds. Micro cuttings can be obtained from these plants and the micropropagation protocol established, to avoid genetic variability between the regenerated plants.
2. *A. andreanum* seeds can also be germinated in the darkness; however, after 1 week, we observed that 74% seeds germinated under continuous light—much higher than 30% seed germination in the darkness on MS medium containing 2 mg/L BA. One week later, the radicle emerged, and the shoot developed under continuous light; it failed to do so in the darkness.
3. We recommend the culture of at least 50 micro cuttings or callus fragments (5 micro cuttings or callus fragments per flask) to guarantee an appropriate number of replications for statistical analysis.
4. We show a detailed protocol for the histological analysis of callus tissue. It is important to observe the development of the organogenic process from callus tissue at the microscopic level to determine the origin and development of meristemoids, shoot primordia, or roots from these callus tissues.
5. Callus fragments no bigger than 1 cm in diameter should be used.
6. Some safranine pellets can be added to the series to have an appropriate orientation to cut the callus fragments.
7. On the first day, during paraplast inclusion, it is better to stop agitating and to add more paraplast before leaving the mixture overnight because the alcohol might evaporate completely and the samples may be lost.
8. The samples must be orientated correctly, and the formation of bubbles must be avoided.
9. The blade must be cleaned with xylol frequently.
10. The 50% glycerol: 50% egg white solution must be used fresh and filtered through gauze. Some phenol crystals must be added to this solution to avoid contamination.
11. A slide jar with a glass cover should be used to make slide transfer easy during the stain protocol.

Acknowledgments

This research was supported by the Scientific and Humanistic Council of the "Universidad Central de Venezuela," Caracas, Venezuela.

References

1. Kuehnle AR, Sugii N (1991) Callus induction and plantlet regeneration in tissue cultures of Hawaiian *anthuriums*. Hortsci 26:919–921
2. Pierik RLM, Steegmans HHM (1976) Vegetative propagation of *Anthurium scherzerianum* Schott through callus cultures. Scientia Hortic 4:291–292
3. Pierik RLM, Steegmans HHM, Van der Meys JAJ (1974) Plantlet formation in callus tissues of *Anthurium andreanum* Lind. Scientia Hortic 2:193–198
4. Vargas TE, Mejías A, Oropeza M, de García E (2004) Plant regeneration of *Anthurium andreanum* cv Rubrun. Electr J Biotech 7:282–286
5. Kunisaki JT (1980) In vitro propagation of *Anthurium andreanum* Lind. Hortsci 15:508–509
6. Chen FC, Kuehnle AR, Sugii N (1997) *Anthurium* roots for micropropagation and *Agrobacterium tumefaciens*-mediated gene transfer. Plant Cell Tiss Org Cult 49:71–74
7. Martin KP, Joseph D, Madassery J, Philip VJ (2003) Direct shoot regeneration from lamina explants of two commercial cut flower cultivars of *Anthurium andreanum* Hort. In Vitro Cell Dev Biol Plant 39:500–504
8. Teng WL (1997) Regeneration of anthurium adventitious shoots using liquid or raft culture. Plant Cell Tiss Org Cult 49:153–156
9. Geier T (1986) Factors affecting plant regeneration from leaf segments of *Anthurium scherzerianum* Schott cultured in vitro. Plant Cell Tiss Org Cult 6:115–125
10. Murashige T, Skoog F (1962) A revised medium for rapid growth and bio-assays with tobacco tissue cultures. Physiol Plant 15:473–497

Chapter 2

A Highly Efficient Protocol for Micropropagation of *Begonia tuberous*

Duong Tan Nhut, Nguyen Thanh Hai, and Mai Xuan Phan

Abstract

A protocol for micropropagation of begonia was established utilizing a thin cell layer (TCL) system. This system has been employed to produce several thousand shoots per sample. Explant size and position, and plant growth regulators (PGRs) contribute to the tissue morphogenesis. By optimizing the size of the tissue and applying an improved selection procedure, shoots were elongated in 8 weeks of culture, with an average number of 210 ± 9.7 shoots per segment. This system has facilitated a number of studies using TCL as a model for micropropagation and will enable the large-scale production of begonia. On an average, the best treatment would allow production of about 10,000 plantlets by the micropropagation of the axillary buds of one plant with five petioles, within a period of 8 months.

Key words: Begonia, Thin cell layer, Micropropagation, TDZ, Petiole, Stem, Floral stalk

1. Introduction

Begonia spp. are dispersed throughout the tropical and subtropical regions in Central and South America, Asia, and Africa, with over 2,000 species recorded (1). Begonias are primarily perennial herbs with thick rhizomes or tubers (2) and are important ornamental plants throughout the world, used as garden plants, pot plants, in hanging baskets, and as plants for indoor decoration. Over 200 species (e.g. *Begonia semperflorens, B. tuberhybrida, B.* × *hiemalis, B.* × *elatior, B.* × *cheimantha,* and *B. socotrana*) have been introduced by commercial growers. Begonias are easy to propagate vegetatively by means of stem, leaf or top cuttings, or by splitting. Among these conventional propagation methods, leaf cuttings are most commonly used in *Begonia*; they are also used in a number of other species with the ability to form both adventitious buds and roots on detached leaves (3). Sometimes,

S.M. Jain and S.J. Ochatt (eds.), *Protocols for In Vitro Propagation of Ornamental Plants*, Methods in Molecular Biology, vol. 589
DOI 10.1007/978-1-60327-114-1_2, © Humana Press, a part of Springer Science+Business Media, LLC 2010

Begonia leaves are highly regenerative. Undetached leaves can bear many leaf-like appendages on their surface, as in the case of *B. hispida* Schott. var. *cucullifera* Irmsch (4). The use of phytohormones, especially cytokinins, enhances the regeneration of buds on undetached leaves of *B. rex* (5).

The thin cell layer (TCL) system, first described in tobacco, is an efficient model to study the mechanisms of control morphogenesis and transformation (6). This method has proven to be very efficient in many dicotyledonous and monocotyledonous species (7). Nowadays, it has become an ideal method for the rapid micropropagation of many favorite plants, and especially for recalcitrant to in vitro regeneration, such as some (a) legumes: *Psophocarpus tetragonolobus* (8), *Soja biloxi* (9); (b) woody dicotyledonous species: *Hardwickia binata* Roxb (10), *Citrus* (11); (c) woody monocotyledonous species: *Bambusa glaucescens* (12), *Cocos nucifera* (13); and (d) herbaceous monocot species: *Oryza sativa* (14), *Rhynchostylis giganteum* (15). A great number of shoots or plantlets can easily be obtained through TCL culture technology within a relatively short time that allows for further research in areas such as DNA transformation and breeding. The TCL system was utilized to study *Begonia rex* petioles (16) by using 6-benzyladenine (BA) and 1-naphthaleneacetic acid (NAA). Shoot regeneration was also obtained from petioles on MS medium containing NAA and BA via organogenesis (17, 18). However, using TDZ for direct shoot formation of *Begonia* petiole transverse-TCls (tTCLs) has not previously been described. To establish reliable, practical gene transfer and propagation systems, it is crucial to increase the efficiency of shoot regeneration, such as in *Begonia*.

The success of the micropropagation procedure depends on several factors, which should be observed during the process. Some researchers have reported procedures for the micropropagation of begonia. This protocol, starting from petiole TCLs, has been proved to be efficient for regenerating plantlets of begonia and can be effectively utilized in genetic transformation studies in the future. With the availability of high-quality shoots, this method can also be standardized for large-scale propagation of other species and further extended to other economically important flower species.

2. Materials

2.1. Explants

1. Petiole.
2. Stem.
3. Floral stalks.

They were obtained from 2-year-old *Begonia tuberous* plants, grown in the greenhouse (25° C, 70–85% relative humidity, natural light).

2.2. Culture Media and Conditions

1. The culture media employed in this procedure are basal MS medium (19). The media are adjusted to pH 5.8 with 1 M KOH and solidified with 2.5 g/L Phytagel (Sigma) unless otherwise stated. The media are pH adjusted before autoclaving at 121° C, 1 atm for 20 min.
2. The cultures are incubated in growth chambers whose environmental conditions are adjusted to 25 ± 2° C and 75–80% relative humidity, under a photosynthetic photon density flux (PPDF) 45 μmol/m^2/s with a photoperiod of 10 h per day.
3. The shoots are cultured in 100 mm × 20 mm petri dishes containing 40 mL semi-solid medium.

3. Methods

3.1. Sterilization

1. The petiole, stem and floral stalk of *Begonia tuberous* are washed thoroughly under running tap water for 30 min, soaked in detergent for 10 min, and rinsed six times with distilled water and once with 70% ethanol for 60 s.
2. After being washed again with distilled water, the segments are disinfected with a 0.1% $HgCl_2$ aqueous solution for 6 min then rinsed six times in sterile distilled water.

3.2. Thin Cell Layer Cutting

1. The Begonia floral stalks are segmented as shown in Fig. 2.1.
2. The petiole and floral stalk are cut into transverse thin cell layers (tTCL), 0.2–1.0 mm thick (Fig. 2.2, *see* Note 1).
3. The Begonia main stem is cut transversely into several sections 2 mm thick and cultured on MS medium in combination of BA and NAA (*see* Note 2).

3.3. Culture

3.3.1. Petiole TCLs

1. Culture petiole TCLs on MS medium supplemented with 0.2 mg/L TDZ (*see* Note 3). Shoots initiate around the epidermal cells of TCLs after 3 weeks of culture.
2. After 8 weeks of culture initiation, transfer the explants onto MS medium without PGRs for shoot elongation.
3. Excise the in vitro shoots derived from the tTCLs and transfer onto the media without PGRs for further elongation.
4. After 2 weeks, the elongated shoots are cultured on MS medium supplemented with 1.0 mg/L BA alone (*see* Note 4).

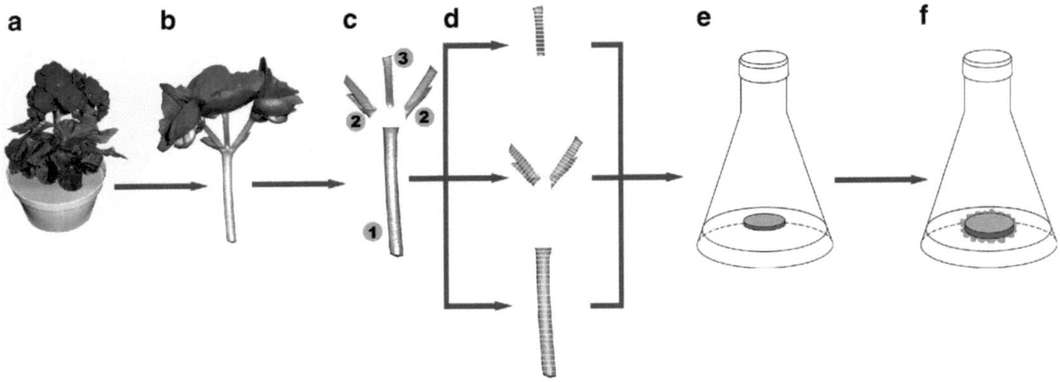

Fig. 2.1. Explanation of steps of *Begonia* micropropagation from floral stalk tTCLs.

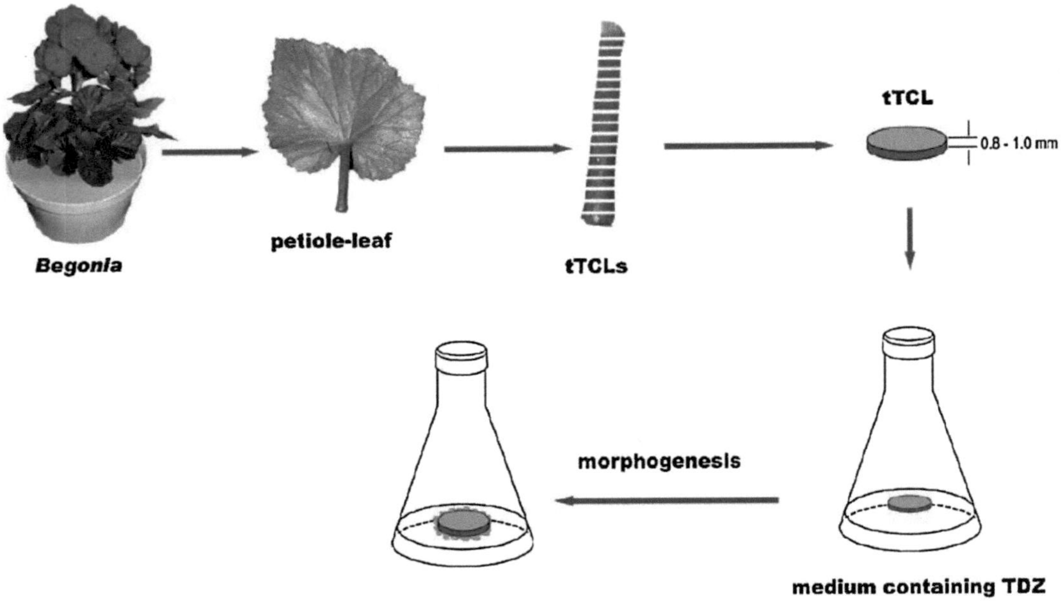

Fig. 2.2. Scheme of *Begonia* micropropagation from petiole tTCLs.

3.3.2. Stem TCLs

1. Culture petiole TCLs on MS medium supplemented with 0.2 mg/L BA and 0.2 mg/L NAA.
2. The highest shoot regeneration rate can be obtained after 4 weeks (*see* Note 5).
3. Then culture these shoots on MS liquid medium supplemented with 0.2 mg/L BA and 0.2 mg/L NAA.

3.3.3. Floral Stalk TCLs

1. Culture TCLs derived from position 2 on MS medium supplemented with 0.2 mg/L TDZ alone.
2. The highest regeneration rate can be obtained after 3 weeks.

3.4. Rooting

1. Adventitious shoots about 1–1.5 cm high with 3–4 leaves are subjected to rooting medium.
2. These shoots can be rooted on MS medium supplemented with 0.5 mg/L BA, 0.1 mg/L NAA and 1 g/L activated charcoal after 45 days of culture, and then these plantlets are transferred to the greenhouse.

3.5. Acclimatization

1. After the induction of the roots, the plantlets are transferred for hardening to sterile moistened sand:soil mixture in the ratio 1:1, pH 5.8, in jam bottles.
2. The plantlets are transferred to polybags containing the same potting mixture after 5–6 weeks; they are kept in the greenhouse for acclimatization at 25° C and 80% humidity.
3. The plantlets are expected to adapt well and flower after 6 months.

4. Notes

1. TCL thickness also significantly affects the shoot regeneration rate. Three mm thickness of TCL is recommended for obtaining high shoot formation frequency and uniform shoots.
2. When only BA is used for petiole TCLs, the caulogenesis rate might be lower than that in media containing BA and NAA, and the shoot formation rate might be less than 60% (20). Caulogenesis is recorded after 8 weeks.
3. Low concentration of TDZ, less than 0.2 mg/L, is effective for caulogenesis from *Begonia* petiole tTCLs.
4. The highest shoot formation (56.67%) was recorded with the use of 1.0 mg/L BA (20).
5. When both BA and NAA are used for petiole TCLs at low concentrations, shoot formation is over 75% (20).

References

1. Inoue Y (1983) Encyclopedia of horticulture, vol 2B. Seibundo-Shinkosha, Tokyo
2. Willis JC (1973) A dictionary of the flowering plants and ferns, 8th edn. Cambridge University Press, Cambridge, England Revised by HK Airy Sahw
3. Heide OM (1964) Effects of light and temperature on the regeneration ability of *Begonia* leaf cuttings. Plant Physiol 17: 789–804
4. Maier U, Sattler R (1977) The structure of the epiphyllous appendages *Begonia hispida* var. *cucullifera*. Can J Bot 55:264–280
5. Chlyah A, Van Tran Thanh K (1968) Budding capacity of undetached *Begonia rex* leaves. Nature 218:493
6. Van Tran Thanh M (1973) In vitro control of de novo flower, bud, root and callus differentiation from excised epidermal tissues. Nature 246:44–45

7. Nhut DT, Aswath CR, da Silva JAT, Le VL, Thorpe T, Tran Thanh Van K (2003) Tobacco thin cell layer morphogenesis. In: Nhut DT, Le BV, Tran Thanh Van K, Thorpe T (eds) Thin cell layer culture system: regeneration and transformation applications. Kluwer Academic Publishers, Dordrecht, The Netherlands, pp 285–342
8. Tran Thanh Van K, Lie-Schricke H, Marcotte JL, Trinh TH (1986) Winged bean (*Psophocarpus tetragonolobus* L. DC. In: Bajaj YPS (ed) Biotechnology in agriculture and forestry. Crop I, vol 2. Springer, Berlin Heidelberg, New York, USA, pp 556–567
9. Juillien F, Wyndaele R (1992) Precocious in vitro flowering of soybean cotyledonary node. J Plant Physiol 140:251–253
10. Das AB, Rout GR, Das P (1995) In vitro somatic embryogenesis from callus culture of the timber yielding tree *Hardwickia binata* Roxb. Plant Cell Rep 15:147–149
11. Carimi F, de Pasquale F, Crescimanno FG (1999) Somatic embryogenesis and plant regeneration from pistil thin cell layers of *Citrus*. Plant Cell Rep 18:935–940
12. Juillien F, Van Tran Thanh K (1994) Micropropagation and embryoid formation from young leaves of *Bambusa glaucescens* 'Golden Goddess'. Plant Sci 98:199–207
13. Sugimura Y, Salvana MJ (1989) Induction and growth of callus derived from rachilla explants of young inflorescences of coconut palm. Can J Bot 67:272–274
14. Nhut DT, Bui VL, Van Tran Thanh K (2000) Somatic embryogenesis and direct shoot regeneration of rice (*Oryza sativa* L.) using thin cell layer culture of apical meristematic tissue. J Plant Physiol 157:559–565
15. Bui VL, Hang Phuong NT, Anh Hong LT, Van Tran Thanh K (1999) High frequency shoot regeneration from *Rhynchostylis gigantea* (Orchidaceae) using thin cell layers. Plant Growth Regul 28:179–185
16. Chlyah A, Van Tran Thanh K (1975) Different reactivity in epidermal cells of *Begonia rex* excised and grown in vitro. Plant Physiol 35:16–20
17. Mikkelsen EP, Sink KC (1978) Histology of adventitious shoot and root formation on leaf-petiole cuttings of *Begonia* × *hiemalis* Fotsch Aphrodite Peach. Sci Hortic 8:179–192
18. Takayama S, Misawa M (1981) Mass propagation of *Begonia* × *hiemalis* plantlets by shake culture. Plant Cell Physiol 22:461–467
19. Murashige T, Skoog F (1962) A revised medium for rapid growth and bioassays with tobacco tissue cultures. Physiol Plant 15:473–479
20. Nhut DT, Hai NT, Huong DTQ, Hang NTT, da Silva JAT (2005) Thidiazuron induces high frequency shoot regeneration in begonia petiole-leaf transverse thin cell layer culture. Prop Ornament Plants 5:149–155

Chapter 3

Micropropagation of *Vanda teres* (Roxb.) Lindle

M. Firoz Alam, Pinaki Sinha, and M. Lokman Hakim

Abstract

For high frequency clonal propagation of *Vanda teres*, nodal segments are cultured on VW medium supplemented with 2% sucrose, 2 mg/L Kinetin, 0.5 mg/L NAA, 2 g/L peptone, 1 g/L activated charcoal and 2.2 g/L gelrite. The cultures are incubated at 24 ± 2° C under fluorescent light 50 μmol/m^2/s for a 16 h photoperiod per day. The PLBs (protocorm like bodies) are induced within 12 weeks of culture and are subcultured to proliferate on the fresh nutrient culture medium for 8 weeks. The clumps of the PLBs are dissected and cultured on VW medium containing 2% sucrose, 15% coconut water (CW), 2 g/L peptone, 150 mg/L l-glutamine and 1 g/L activated charcoal. The PLB sections elongate to form shoots, and new PLBs are induced from the base within 8 weeks of culture. For plantlets formation, the shoots are cultured on VW medium amended with 2% sucrose, 15% CW, 2 g/L peptone, 1 g/L activated charcoal, 50 g/L banana pulp and 1 mg/L Indole-3-butyric acid (IBA). The regenerated plants are acclimatized and cultivated in the nursery, where they bloom within 3 years.

Key words: *Vanda teres*, Nodal segment, In vitro culture, Clonal propagation, Acclimatization, Blossom

1. Introduction

The genus *Vanda* consists of orchids that are found throughout tropical South and South East Asia to New Guinea and North-Eastern Australia and is widely grown as a cut-flower in Thailand and Singapore. A large export industry has developed using a handful of species in an extensive hybridization program, both within *Vanda* and in combination with related genera.

The first experiments on *Vanda* tissue culture were carried out at the University of Singapore with callus masses derived from seedlings (1) in a medium containing 2,4-Dichlorophenoxy acetic acid (2,4-D) and tomato juice (2). Various protocols developed for *Vanda* microcloning show considerable variation with respect to explants, media composition and environmental conditions (3).

S.M. Jain and S.J. Ochatt (eds.), *Protocols for In Vitro Propagation of Ornamental Plants*, Methods in Molecular Biology, vol. 589
DOI 10.1007/978-1-60327-114-1_3, © Humana Press, a part of Springer Science+Business Media, LLC 2010

Haiyan and Kondo (4) cultured shoot apices of *Vanda pumila* in Gamborg's B_5 (5) liquid media supplemented with ∞-naphthaleneacetic acid (NAA) and 6-benzylamini purine (BA) at various concentrations, in which shoot primordia were induced. Plantlets were developed 120 days after subculture of shoot primordia in B_5 liquid medium supplemented with 0.02 mg/L BA. Seeni and Latha (6) developed a reliable technique for the rapid multiplication of *Vanda coerulea* through culture of shoot tips and leaf bases of mature plants on Mitra et al. (7) medium supplemented with 10% CW and 0.5–1.0 g/L casein hydrolysate (CH). Explants with proliferating buds/PLBs were subcultured on medium enriched with 35 g/L banana pulp, 30% CW and 1.08 μM NAA. The highest number of shoots – 70 and 100 – were obtained from a single shoot/leaf base explant after 30 weeks of culture. Rooting was achieved in medium containing 30 g/L Banana pulp and 1.08 μM NAA within 3–4 weeks. Aggarwal (8) developed a micropropagation system based on foliar explants of *Vanda coerulea* cultured on VW medium. A combined treatment with NAA and BA was usefully employed to induce PLB mediated regeneration in 100% explants. The replacement of BA with kinetin (Kn) promoted direct development of shoot buds.

Vanda teres has high commercial value and is collected indiscriminately from its natural habitat; that is why it has become vulnerable to extinction. This reason warrants the development and establishment of an efficient protocol for mass clonal propagation and consequently, commercial exploitation and conservation of this important indigenous orchid of tropical Asia.

The present investigation was, therefore, accomplished with the goal of developing a protocol for high frequency regeneration of *Vanda tere*.

2. Materials

2.1. Plant Materials

1. Mature shoots are collected from the plants naturally grown on the trees in the forests of Sylhet district in Bangladesh and are used as a source of explants. Shoots may also be collected from nursery grown plants.

2.2. Surface Sterilization of Plant Materials

1. Tap water.
2. Ethanol 70% (v/v).
3. Tween (Loba, India) 80 5% (v/v).
4. Mercuric chloride (Loba, India) 2% (w/v).
5. Autoclaved water 150 mL aliquots in 250 mL cotton plugged conical flask.

6. Magnetic stirrer, magnetic bar, 500 mL beaker (autoclaved).
7. Tissue culture facilities – laminar air flow bench, instruments, culture room.

2.3. Culture Media

Macro- and micro salts of VW (9) medium, with different vitamins and hormones as detailed below:

1. *Macronutrients:* 200 mg/L $Ca_3(PO_4)_2$, 28 mg/L ferric citrate, 525 mg/L KNO_3, 250 mg/L KH_2PO_4, 250 mg/L $MgSO_4$,$7H_2O$, 500 mg/L $(NH_4)_2SO_4$. (To prepare the stock solution, tenfolds of each compound are dissolved in 1 L of distilled water and stored at 10° C.) 100 mL stock solution is used to prepare a total volume of 1 L of medium (*see* Note 1).
2. *Micronutrients:* 7 mg/L $MnSO_4$. (700 mg $MnSO_4$ is dissolved in 100 mL distilled water and stored at 10° C. 1 mL stock solution is used to prepare a total volume of 1 L of medium).
3. *Vitamins:* 0.5 mg/L nicotinic acid, 0.5 mg/L pyridoxine HCl, 0.1 mg/L thiamine HCl, 2.0 mg/L glycine and 100 mg/L myoinositol. (To prepare the stock solution, 100-folds of each compound are dissolved in 100 mL distilled water and stored at –20° C up to 4 months (*see* Note 2). 1 mL stock solution is used to prepare a total volume of 1 L of medium).
4. *Organic supplements*: Coconut water – 15%; it is taken from green fruit and stored at –20° C (*see* Note 2). Banana pulp – 50 g/L; ripened banana is peeled and mashed using a mortar and pestle, and used fresh in medium preparation. Peptone – 2 g/L; it is weighed and added directly to the medium preparation. Activated charcoal – 1 g/L, it is weighed and added directly in medium preparation.
5. Sucrose (20 g/L) is added directly to the medium. pH of the medium is adjusted to 5.6 with 1 N NaOH, and 2.2 g/L gelrite is added; then the medium is autoclaved at 121° C for 20 min.

2.4. Acclimation of Regenerated Plants to Ex vitro Conditions

1. Tap water.
2. Plastic basket (25 cm diameter).
3. Clay pots (10 cm diameter).
4. Charcoal chips – Wood-charcoal is obtained by burning tree-wood, and 10 mm^3 chips are prepared.
5. Coconut husk – Husk of the ripe coconut fruit is extracted, and 20 mm^3 sized husks are prepared by cutting.
6. Fertilizer – 30N-10P-10K and 20N-20P-20K mixtures are collected from PROSHIKA, Bangladesh. Three g of mixture are dissolved in 1 L of water.

3. Methods

3.1. Preparation and Sterilization of Culture Media

1. Prepare culture media for (a) induction of PLBs from nodal segment explants, (b) proliferation of PLBs and development of shoots and (c) root induction in regenerated shoots according to the formulation given in Table 3.1. The medium for the

Table 3.1
Composition of media used for the different stages of plantlets formation from nodal segment explants of *Vanda teres*

Compounds[a] (mg/L)	Medium 1 PLBs induction (mg/L)	Medium 2 PLBs proliferation and development of shoots (mg/L)	Medium 3 Formation of plantlets (mg/L)
$Ca_3(PO_4)_2$	200	200	200
Ferric citrate	28	28	28
KNO_3	525	525	525
KH_2PO4	250	250	250
$MgSO_4$ $7H_2O$	250	250	250
$(NH_4)_2SO_4$	500	500	500
$MnSO_4$	7	7	7
Myoinisitol	100	100	100
Nicotinic acid	0.5	0.5	0.5
Pyridoxine HCl	0.5	0.5	0.5
Thiamine HCl	0.1	0.1	0.1
Glycine	2.0	2.0	2.0
Sucrose	20,000	20,000	20,000
Peptone	2,000	2,000	2,000
Kn	2.0	–	–
NAA	0.5	–	–
IBA	–	–	1.0
Coconut water	–	15%	15%
L-glutamine	–	150	–
Activated charcoal	1,000	1,000	1,000
Banana pulp	–	–	50,000
pH	5.6	5.6	5.6

[a]All macro- and micro elements are the ingredients of VW medium

induction of PLBs contains 2.0 mg/L Kn and 0.5 mg/L NAA as growth regulators and 1 g/L of activated charcoal.

2. Adjust the pH of the media to 5.6 using 1 M HCl or 1 M KOH.
3. Add 2.2 g/L gelrite (Dachifa), and boil the medium for melting the gelrite.
4. Dispense the medium into suitable glass vessels, e.g., 60 mL aliquots into 250 mL capacity conical flasks or 250 mL round glass bottles.
5. Sterilize the media by autoclaving at 121° C for 20 min.
6. Store the autoclaved media at room temperature in the dark for a maximum of 2 weeks.

3.2. Surface Sterilization and Preparation of Explant

1. Take young shoot segments from collected mature plants.
2. Wash them in running tap water and then in 5% (v/v) Tween 80 for 5 min.
3. Take the explants in a laminar air flow cabinet and transfer to a sterile conical flask.
4. Wash the explants in double-distilled water, rinse in 70% ethanol for 30 s, and surface sterilize them by rinsing in 0.2% (w/v) mercuric chloride solution for 5 min.
5. Wash the explants thoroughly with sterile double-distilled water 3–4 times.
6. Prepare the surface sterilized explants for inoculation. Cut 5.0–10.0 mm long nodal segments with a single node.

3.3. Explants Inoculation and PLBs Induction

1. Inoculate the prepared explants (nodal segments) aseptically on the surface of the prepared culture medium 1 (Table 3.1) containing 2% sucrose, 2.0 mg/L Kn, 0.5 mg/L NAA, 2 g/L peptone, and 1 g/L activated charcoal.
2. Incubate the cultures at 24 ± 2° C under white fluorescent light (Philips Electronics Ltd., Dhaka, Bangladesh) at 50 μmol/m^2/s for a 16 h photoperiod per day.
3. Subculture the explants on freshly prepared medium of the same composition every 4weeks to maintain active growth.
4. At the end of the second subculture, 12weeks after inoculation, small PLBs should be formed on at least 95% of the cultured explants, with 25–30 PLBs per culture. The PLBs are globular in shape and somewhat elongated (Fig. 3.1b, c).

3.4. Proliferation of PLBs and Their Development Into Shoots

1. Subculture the initially induced PLB clumps in the same nutrient medium, as mentioned in Subheading 3.3, step 1, for a further 8 weeks. During this subculture, the number of PLBs should have increased to an average of 150 (Fig. 3.1d).

Fig. 3.1. (**a–h**) Regeneration of *Vanda teres* through in vitro culture of nodal segment. (**a**). Flower of *Vanda teres*, plant parts of which were used as explants. (**b**). Induction of PLBs from explant cultured on VW + 2% sucrose + 2.0 mg/L Kn + 0.5 mgL NAA + 15% CW + 2 g/L peptone + 1 g/L AC. (**c**). Second subculture on the same medium, (**d**). Eight-week-subculture of initially induced PLBs on the same medium. (**e**). Development of shoots from the old PLBs and proliferation of new PLBs from the base of the old ones, cultured on VW + 2% sucrose + 150 mg/L l-glutamine + 15% CW + 2 g/L peptone + 1 g/L AC. (**f**). Eight-week-culture of regenerated shoots with induction of roots as well as shoot growth on VW + 2% sucrose + 15% CW + 2 g/L peptone + 1 g/L AC + 50 g/L banana pulp + 1.0 mg/L IBA. (**g**). Two-year-old regenerated plants. (**h**). Flowering phase of 3-year-old regenerated plants.

2. Dissect the clumps of the PLBs longitudinally so that each section contains an average of 15 PLBs.
3. Transfer each section of the clumps on to gelrite-gelled VW medium supplemented with 2% sucrose + 15% CW + 2 g/L peptone + 1.0 g/L AC + 150 mg/L L-glutamine (Table 3.1, medium 2). After 8 weeks of culture in this medium, the PLBs should have developed into shoots and elongated properly. The mean height of the shoots (including leaves) should be 48.8 mm. In addition, a large number of secondary PLBs should have been induced from the base of the developing shoots (Fig. 3.1e).

3.5. Formation of Plantlets

1. Subculture the well developed shoots (4.5–7 cm long) on gelrite-gelled VW medium supplemented with 2% sucrose, 15% CW, 2 g/L peptone, 1 g/L AC, 50 g/L banana pulp and 1.0 mg/L IBA (Table 3.1, medium 3).
2. Incubate the cultures at 24 ± 2° C with a 16 h photoperiod (50 mol/m^2/s^2; white fluorescent light). In 8 weeks, 5–6 roots should have been induced from the base of the shoots (Fig. 3.1f).

Within the first 36 weeks of initiation of the culture, 150 plants and a huge amount of PLBs and secondary PLBs should be produced from a single explant of nodal segment. Through repeated subculturing of shoots on rooting medium, and PLBs on proliferation medium, regeneration of plantlets may be continued, and 150,000 plantlets could be produced every 32 weeks.

3.6. Acclimatization and Potting of the Regenerated Plants and Their Establishment in the Nursery

1. Transfer the culture vessels with the rooted plantlets from the growth room to an open room and keep them there for 7 days. When the culture vessels with the plantlets are kept at normal room temperature (30 ± 2° C) and indirect sunlight, the plantlets should be acclimatized to some extent.
2. Remove the rooted shoots from the culture medium and wash the roots carefully with warm (approx. 35° C) tap water to remove the gel adhered to the roots.
3. Transfer the plantlets to a plastic basket of 25 cm diameter (to hold 50 plantlets in a basket) containing coconut husk (20 mm^3) and charcoal chips (10 mm^3) (2:1).
4. Keep the baskets containing the plantlets in indirect daylight and mist the plantlets at 6-h intervals for 30 days; at the end of this period, 80–85% plants should have survived.
5. Transfer the plants individually to 10 cm diameter clay pots containing coconut husk (20 mm^3) and charcoal chips (10 mm^3) (2:1) (*see* Note 3).
6. Rear the plants for up to 12 months in clay pots (Fig. 3.1g), spraying 3 g/L 30N-10P-10K solution at 10-day intervals.

7. After 12 months, tie plants on the tree logs (*see* Note 4), water them by spraying every day and fertilize them by spraying with 30N-10P-10K solution (3 g of fertilizer dissolved in 1 L of water) at 10-day intervals for 12 months.
8. After 12 months, fertilize the plants by spraying with 20N-20P-20K solution (3 g of fertilizer dissolved in 1 liter of water) at 10-day intervals. The regenerated 3-year old plants should have started flowering (Fig. 3.1h).

4. Notes

1. Stock solutions of VW macro- and micro salts are stored at 10° C up to 4 months.
2. Stock solutions of auxins, cytokinins and vitamins are stored at -20° C; this ensures the solution is completely thawed before taking any aliquots.
3. Before use, the potting mix (coconut husk and charcoal chips) should be autoclaved, to kill bacteria and fungi.
4. In *Vanda teres,* roots emerge from each node, and for this reason the plants survive very well, if they are tied to a tree log allowing the roots to cling to it.

References

1. Rao AN (1963) Organogenesis in callus cultures of orchid seeds. In: Maheshwari P, Rangaswami NS (eds) A Symposium. International Society of Plant Morphologists., University of Delhi, Delhi, India pp 332–344
2. Rao AN (1977) Tissue culture in the orchid industry. In: Reinert J, Bajaj YPS (eds) Applied and fundamental aspects of plant cell, tissue, and organ culture. Springer, Berlin, Heidelberg, pp 44–69
3. Arditti J, Ernst R (1993) Micropropagation of orchids. Wiley, New York 649 pp
4. Haiyan NA, Kondo K (1995) Conservation of gene resources in epiphytic *Vanda pumola* (Orchidaceae) by tissue culture shoot primordia. Plant Tissue Cult Lett 12:273–279
5. Gamborg OL, Miller RA, Ojima K (1968) Nutrient requirements of suspension cultures of soybean root cells. Exp Cell Res 50:151–158
6. Seeni S, Latha PG (2000) In vitro multiplication and ecorehabilitation of the endangered Blue *Vanda.* Plant Cell Tissue Organ Cult 61: 1–8
7. Mitra GC, Prasad RN, Roychoudhury A (1976) Inorganic salts and differentiation of protocorms in seed callus of an orchid and correlated changes in its amino acid content. J Exp Biol 14:350–351
8. Aggarwal S (2001) Micropropagation of *Vanda coerulea* Griff through leaf culture. Sixth National Seminar on Orchid Diversity in India, Science and Commerce and Orchid Show, October 11–13, 2001, Institute of Himalayan Bioresource Technology, Palampur, Abstract No. 49, pp 84–85
9. Vacin E, Went F (1949) Some pH changes in nutrient solutions. Bot Gaz 110:605–613

Chapter 4

In vitro Propagation Through Axillary Shoot Culture of *Ranunculus asiaticus* L.

Margherita Beruto

Abstract

Ranunculus asiaticus is an important ornamental species mainly cultivated in the countries surrounding the Mediterranean sea. So far, the multiplication of this plant has been mainly carried out by seed and tuberous root division; however, these systems present many drawbacks. Tissue culture is an attractive alternative for accelerated propagation of selected and indexed genotypes. In this chapter, we present a flow chart for the commercial production of *Ranunculus* clones by using *in vitro* axillary budding. Although the price of micropropagated plants is higher compared to traditional material (seedlings and tuberous roots from seed populations), we need to consider that micropropagation helps to supply growers with more performant and healthy genotypes, and a better production schedule can be envisaged for cultivation and packaging.

Key words: *Ranunculus asiaticus* L., Micropropagation, Axillary bud stimulation, Commercial production

Abbreviations

BA N^6 -Benzyladenine
-IBA Indole-3-butyric acid
NAA 1 -Naphthalenic acetic acid

1. Introduction

The ornamental species *Ranunculus asiaticus* L., belonging to *Ranunculaceae*, is a widely appreciated cut flower and pot plant. It is native to the eastern Mediterranean basin, from the Greek Islands to Turkestan from where it spread to the western part of Europe, South Africa, California, Israel and Japan. Meynet [1] estimated that 30–50 ha are used in tuberous root production and 65 ha in cut flower production (France 31%, Italy 46% and

S.M. Jain and S.J. Ochatt (eds.), *Protocols for In Vitro Propagation of Ornamental Plants*, Methods in Molecular Biology, vol. 589
DOI 10.1007/978-1-60327-114-1_4, © Humana Press, a part of Springer Science+Business Media, LLC 2010

Israel 23%). During the last decade, this species gained economical importance: 13 million Ranunculus flowers were sold in the auctions of the Netherlands in 1994, 22 in 1999 and 34 in 2001 [2]. Data recorded at the Flower Market of San Remo, the most important market in Italy, shows that it is a leading product with a high percentage (about 84%) of dealers involved in its commercialization.

The former varieties were classified [3] into four groups according to their presumed origin and morphological characteristics: "Turban Ranunculus", "Persian Ranunculus", "French Ranunculus" and "Peony-flowered Ranunculus". Hybridizations of these have resulted in the present commercial lines, e.g. "Grazia", "Elegance", "Riviera Style", "Friandine", "Pauline", "Giga", "Reinette", "Tecolote" and "Victoria" for cut flowers and "Bloomingdale" and "Maché" for potted plants.

Propagation by sexual means is the predominant technique in *Ranunculus*. The breeding programs for the selected lines are quite complex, and homogeneity of the population is not guaranteed. Plantlets grow from seeds, flower during spring, and yield tuberous roots at the end of the first cultivation year; these can be used for winter production during the second year. Vegetative propagation by division of the tuberous roots is possible, but it does not ensure healthy and disease-free plants, involves dependence on season, and yields slow multiplication rates [1]. Micropropagation was proven to encourage a further diffusion of this production, because of its potential in efficient and rapid multiplication to propagate parental lines in breeding programs and for the prompt commercial production of new cultivars; growers can be supplied with more profitable genotypes and a better production schedule can be encouraged [4, 5].

Two different methods of propagation of *Ranunculus* via tissue culture have been considered in the different papers published on this subject: axillary bud stimulation and somatic embryogenesis. Different authors have documented that *Ranunculus* can be regenerated *in vitro* from different tissue sources [6–9]. In our laboratory, we have developed a system for somatic embryogenesis from thalamus-derived calli of *R. asiaticus* which, although allowed to reach a mass clonal propagation, needs to take some more evaluation [5, 10, 11]. The main constraints concern a genotype-dependent response, and a variable percentage (±20%) of abnormal embryos which did not convert to the normal ones; the true-to-type is not always guaranteed. The most common deviations of the regenerated somatic embryos concern the leaf morphology, plant habit and flowering delay; they can persist for different cultural cycles as permanent deviation, or they can be observed during the first cultural cycle as transitional (plant vigor and high percentage of aberrant leaves) [12].

Taking into account the need for further investigations to evaluate suitability of somatic embryogenesis as an alternative approach for mass propagation, we describe the micropropagation protocol through axillary bud stimulation in this chapter. This system, developed in our laboratory, has enabled us to bring clones in the market (*see* **Note 6**).

2. Materials

1. Tuberous roots collected at the end of the culture cycle and stored till the start of the experiment.
2. Refrigerator, laminar-flow cabinet, magnetic stirrer, glass-bead sterilizer, forceps, scalpels.
3. Ethanol, sodium hypochloride (bleach), distilled water, pipettes, culture tubes, culture vessels, plastic cap, (Bellco Kaputs ™).
4. Autoclave, Culture room.
5. Media (Table 4.1).
6. PPM (Plant Preservative Mixture, Plant Cell Technology, Inc™, USA.
7. Polyethylene film (Domopack ™, USA).

Table 4.1
Media for micropropagation through axillary bud stimulation of *Ranunculus asiaticus*

	Initiation culture R-1	Shoot multiplication R-2	Rooting R-3
Major elements (mM)	Murashige and Skoog [21]		Murashige and Skoog [21]; half strength
$CaCl_2$	2.99		1.50
KH_2PO_4	1.25		0.63
KNO_3	18.79		9.40
$MgSO_4$	1.50		0.75
NH_4NO_3	20.61		10.30
Iron and minor elements (μM)	Murashige and Skoog [21]		Murashige and Skoog [21]; half strength
$CoCl_2\ 6H_2O$	0.11		0.06
$CuSO_4\ 5H_2O$	0.10		0.05

(continued)

Table 4.1 continued

	Initiation culture R-1	Shoot multiplication R-2	Rooting R-3
FeNaEDTA	100.0		50.0
H_3BO_3	100.27		50.14
KI	5.00		2.5
$MnSO_4\ H_2O$	100.00		50.0
$Na_2MoO_4\ 2H_2O$	1.03		0.52
$ZnSO_4\ 7H_2O$	29.91		14.96
Vitamins (μM)	Murashige and Skoog [21]		
Glycine	26.64		
Myo-Inositol	554.94		
Nicotinic acid	4.06		
Pyridoxine HCl	2.43		
Thiamine HCl	0.30		
Organic addenda			
Sucrose (mM)	87.64		43.82
Fructose (mM)	–		44.40
Citric acid (mM)	–		0.39
Ascorbic acid (mM)	–		0.28
Yeast extract (mg/L)	500	–	–
Peptone (mg/L)	100	–	–
Growth regulators (μM)			
NAA	0.11		
BA	2.22		
Kinetin	9.29		
IBA	–		4.92

Preparation to make 1 L of medium: Weigh all the chemicals and dissolve in 300–400 mL of bi-distilled water (conductivity: $2 \pm 0.5\ \mu S/cm$). For microelements, vitamins and growth regulators, stock solutions will be used and an adequate amount of mL will be added to the bi-distilled water. Adjust the volume of the solution to 500 mL. In the meanwhile, add 8 g of agar–agar to 500 mL of bi-distilled water and heat for 10–12 min, taking care to avoid evaporation. Add the hot agar-solution to the previously prepared ingredient solution and mix for some seconds. The pH of all the media are adjusted to 5.85 ± 0.01 while the media are in sol-status ($T = 76$ C ± 1 C). The media are poured into tubes (10 mL; initiation phase) and vessels (100 mL; multiplication and rooting phase) and autoclaved at 120 C, 101 kPa, 15 min

3. Methods

The propagation method can be divided into five main steps: preparation of the tuberous roots, initiation phase, multiplication, rooting and acclimatization.

3.1 Preparation of the Tuberous Roots (Stage 0)

In order to ensure healthy propagation, the tuberous roots used to start the culture should be obtained from mother plants cultivated as candidate nuclear stock under insect-proof greenhouse conditions and submitted to rigorous controls regarding the phytosanitary status (*see* **Note 1**).

1. Drench the dried tuberous roots in water overnight.
2. Put the hydrated tuberous roots in reusable trays filled with moistened perlite taking care that the upper part stays out of the artificial soil.
3. Incubate at 10–12 C 8 h light for over about 15–20 days; when the buds sprout, the culture can be started (Fig. 4.1a).

3.2 Initiation of Culture (Stage 1)

1. Buds (about 0.5 cm) are dissected from the tuberous roots.
2. They are sterilized by a quick dip in 70% (v/v) ethanol followed by a 15 min treatment of NaOCl, 1% available chlorine.
3. The buds are rinsed three times with sterile water.
4. A second disinfection step is carried out by shaking the explants in PPM, 50%, v/v; 10 min.
5. The buds are then transferred to a gelled culture medium R-1 (Table 4.1; Test tubes 25 × 150 mm, closed by a plastic cap, 10 mL medium) and incubated in a climate room (16 h light/8 h dark, 50–60 μmol/m^2/s) at 19 ± 1 C.

3.3 Shoot Multiplication (Stage 2)

1. After 7–8 weeks of the *in vitro* initiation, the explants are transferred to a multiplication medium (R-2; Table 4.1). 10–15 shoots/container are grown in vessel cultures (7.5 cm; 320 mL) closed with polycarbonate lids and sealed by a polyethylene film (100 mL medium); the culture conditions are the same as described for the initiation phase (Fig. 4.1b).
2. At stage 2 the initial explants develop in shoot clusters after about 3–4 weeks of culture (Fig. 4.1c). The *vitro shoots* formed are isolated in a laminarflow cabinet by transverse section, and subsequently separated and subcultured on fresh R-2 medium (Table 4.1). Continued subculturing can be carried out over a period of 12–24 months. Each shoot produces 1.5–4.5 new shoots at each subculture, depending on the genotype. Moreover, different factors have been shown to

Fig. 4.1. Micropropagation of *Ranunculus asiaticus* through axillary bud stimulation. (**a**) Tuberous roots with sprouted buds ready for *in vitro* inoculation (Stage 0). (**b**) Mass propagation via tissue culture. (**c**) Shoot cluster at the end of the multiplication phase (Stage 2). (**d**) Ex vitro plantlets 2 weeks after the start of acclimatization. (**e**) Cultivation of micropropagated plants. (**f**) Flower packaging for *Ranunculus* clones.

greatly affect the *in vitro* performance of *Ranunculus* shoots (*see* **Note 2**): among others, the water status and availability of nutrients [13, 14], the agar brand [14–16], type of sugar, plant density, frequency of subculture, and ventilation of tissue culture vessel [17] (*see* **Notes 4** and **5**).

3.4 Rooting (Stage 3)

1. Shoots 1.5–2.0 cm long are cut and transferred to the rooting medium R-3 (Table 4.1; 10–15 shoots for each vessel culture (previously described at **stage 2**).
2. Three to four weeks after the *in vitro* rooting phase is undertaken, shoots will root more or less pronounced way, depending

on the genotype. In terms of further growth of plants under *in vivo* conditions, shoots which don't show root development can be successfully transferred under *in vivo* conditions. In any case, root induction on medium containing auxin (medium R-3, Table 4.1) is advisable.

3.5 Acclimatization (Stage 4)

1. Rooted and unrooted shoots from stage 3 are transplanted in a peat-perlite substrate (1:1) under unheated greenhouse conditions and protected by a white polypropylene film (17 g/m^2) and green shading net (70%) over a 2 week period. Every day, the plants are checked to ensure that they are kept at constant humidity (R.H. 90–100%).
2. Two weeks after the start of the acclimatization phase, the polypropylene film is gradually removed and for a month from transplanting, only the shading net is left. At this time, the fertilizer can be applied with a 0.1% salt concentration, in the ratio 1:0.7:1.1 (N:P:K) (Fig. 4.1d).
3. After **steps 1** and **2**, the plantlets are ready to be transferred for cultivation and production (Fig. 4.1e, f). The usual cultural practices for nutrition, irrigation and pest control can be applied ([18], *see* **Note 3**).

4. Notes

1. Production of disease-free plants is a major concern because of different viral infections that can affect *Ranunculus asiaticus* [19]. No valuable methods are yet available to eradicate virus by meristem culture of *Ranunculus* [1, 20] and consequently, an integrated production schedule should be performed in order to guarantee virus-free starting material for healthy large scale production (Table 4.2).
2. In the field, micropropagated plants show an appreciable vegetative growth. Three months after planting, their development is generally more advanced than that reached by seedlings of the same age, although the degree of development is genotype-dependent. Generally, *ex vitro* plantlets have homogeneous flowering and an increased number of shoots slightly earlier, and at the end of the culture cycle, they yield more than one tuberous root, although this production per plant is genotype-dependent [12].
3. The acclimatization period could influence both the tuberous root production and size; under Mediterranean conditions it is advisable to carry out the hardening phase from October till January in order to guarantee an efficient tuberous root production.

Table 4.2
Different steps of production and operations which should be carried out in the nursery to ensure that selected and healthy genotypes could be obtained through a propagation scheme via tissue culture

	Phase	Goal of each phase
1.	Candidate nuclear stock	To provide selected genotypes for which controls on the phytosanitary condition are carried out
2.	Nuclear stock (Stage 0)	To grow mother plants as "nuclear stock" in insect-proof greenhouses
		Controls on phytosanitary conditions and genetic characteristics are performed
3.	*In vitro* propagation (Stage 1–2 and 3)	To perform a mass propagation via tissue culture through axillary bud stimulation
4.	Acclimatization of ex-vitro plantlets (Stage 4)	To provide the grower with ex vitro plantlets or with tuberous roots gathered from ex vitro plant material at the end of the cultivation period. Plantlets are grown under controlled greenhouse conditions and "pool" controls are carried out to test the phytosanitary conditions
5.	Cultivation for commercial purposes	To provide the consumer with new and improved forms of *Ranunculus*

4. Generally, the propagation method through axillary bud stimulation provides a true-to-type production. Our research carried out over the past 15 years showed that no significant phenotypic differences could be scored for a wide number of genotypes [5]; however, appropriate selection remains advisable for each micropropagated line.
5. Continuous subculturing must be given attention because a lower flowering percentage and the appearance of some abnormalities (plant vigor, increased percentage of aberrant stems and flowers) can be noticed in some lines if *in vitro* culture is carried out for more than 12–22 months [12].
6. The results of the protocol outlined in this chapter are now used for commercial propagation, and about 1,600,000/2,000,000 micropropagated plantlets have been produced. The cost price of *in vitro* produced plants (±0.45–0.50) is higher than that of traditional propagation units [seedlings (0.14–0.17) and tuberous roots from seedlings (0.29–0.34)]; however, many benefits can be envisaged: micropropagation helps to supply growers with better performers and healthy genotypes, and a better production schedule can be envisaged for cultivation and packaging.

References

1. Meynet J (1993) Ranunculus. In: De Hertogh A, Le Nard M (eds) The physiology of flower bulbs. Elsevier Science Publishers, The Netherlands, pp 603–610
2. Statistisch Jaarboek (2001) Vereniging van Bloemenveilingen in Nederland (VBN). Leiden, Holland, p 117
3. Krabbendam P, Baardse AA (1967) Bijgoed Bloembollenteelt. W.E.J. Tjeenk Willink, Zwolle, The Netherlands, pp 349–360
4. Beruto M (2002) Micropropagation brings new varieties. FlowerTech 5(6):31–33
5. BerutoM,DeberghP(2004)Micropropagation of *Ranunculus asiaticus*: a review and perspectives. Plant Cell Tissue Organ Cult 77:221–230
6. Pugliesi C, Rabaglio M, Cecconi F, Baroncelli S (1992) Plant regeneration from tissue cultures of Persian buttercup (*Ranunculus asiaticus* L.). Plant Cell Tissue Organ Cult 28:125–128
7. Meynet J, Duclos A (1990) Culture *in vitro* de la renoncule des fleuristes (*Ranunculus asiaticus* L.). I. Néoformation et multiplication végétative *in vitro* de plantes à partir de tronçon de thalamus. Agronomie 2:157–162
8. Meynet J, Duclos A (1990) Culture *in vitro* de la renoncule des fleuristes (*Ranunculus asiaticus* L.). II. Production de plantes par culture d'anthères *in vitro*. Agronomie 10:213–218
9. Meynet J, Duclos A (1990) Culture *in vitro* de la renoncule des fleuristes (*Ranunculus asiaticus* L.). III. Etude des plantes produites par embryogenèse somatique à partir des tissus superficiels de l'anthère. Agronomie 10:285–290
10. Beruto M, Debergh P (1992) Somatic embryogenesis in *Ranunculus asiaticus* L. hybr. thalamus cultivated *in vitro*. Plant Cell Tissue Organ Cult 29:161–165
11. Beruto M, Curir P, Debergh P (1996) Callus growth and somatic embryogenesis in thalamus tissue of *Ranunculus asiaticus* L. cultivated *in vitro*: cytokinin effect and phenol metabolism. *In Vitro* Cell Dev Biol Plant 32:154–160
12. Beruto M, Cane G, Debergh P (1996) Field performance of tissue-cultured plants of *Ranunculus asiaticus* L. Sci Hortic 66:229–239
13. Beruto M (1997) Agar and gel characteristics with special reference to micropropagation systems of Ranunculus asiaticus L. *PhD Thesis*, Gent, Faculteit Landbouwkundige en Toegepaste Biologische Wetenschappen. Univeristy Gent, Belgium, pp 21–44 125–184
14. Beruto M, La Rosa C, Portogallo C (2001) Effects of agar impurities on *in vitro* propagation of *Ranunculus asiaticus* L. Acta Hortic 560:399–402
15. Beruto M, Curir P, Debergh P (1999) Influence of agar on *in vitro* cultures: II. Biological performance of Ranunculus on media solidified with three different agar brands. *In Vitro* Cell Dev Biol Plant 35:94–101
16. Beruto M, Curir P (2006) Effecrs of agar and gel characteristics on micropropagation: *Ranunculus asiaticus*, a case study. In: Teixeira da Silva JA (ed) Floriculture, ornamental and plant biotechnology – advances and topical issues, vol II. Global Science Books, UK, pp 277–284
17. Beruto M, Portogallo C (2000) Factors affecting the growth of *Ranunculus in vitro*. Acta Hortic 520:163–170
18. Sparago E, Beruto M, Bergamasco F, Mancini G, Restuccia P (2004) La coltivazione del ranuncolo: l'importanza della nutrizione. Colture protette 10:55–62
19. Turina M, Ciuffo M, Lenzi R, Rostagno L, Mela L, Derin E, Palmaro S (2006) Characterization of four viral species belonging to the family *Potyviridae* isolated from *Ranunculus asiaticus*. Phytopathology 96:560–566
20. Maia E, Bettachini B, Beck D, Marais A (1973) Régénération de renoncules par culture d'apex *in vitro*. Ann Phytopathol 5:125–129
21. Murashige T, Skoog F (1962) A revised medium for rapid growth and bioassays with tobacco tissue cultures. Physiol Plant 15:473–497

Chapter 5

Micropropagation of *Crataeva adansonii* D.C. Prodr: An Ornamental Avenue Tree

Purnima Tyagi, P.K. Sharma, and S.L. Kothari

Abstract

In this chapter, we describe multiplication of the superior and elite tree of *Crataeva adansonii* using plant tissue culture techniques. An ornamental and avenue tree, it is not available in abundance because of poor seed germination and seedling establishment. It reproduces in nature by root suckers, but that restricts its distribution to very limited areas. Efficient procedures are outlined for plant regeneration through direct shoot bud formation, indirect organogenesis, and somatic embryogenesis through callus formation. Different explants were utilized for separate pathways of regeneration. Murashige and Skoog's (MS) medium containing 3 mg/L BA and 0.05–0.1 mg/L NAA is most effective in direct induction of axillary buds from nodal explants and shoot tips. Adventitious shoots developed from leaves on MS medium containing 3 mg/L BA and 0.1 mg/L NAA. De novo shoots were obtained from the anthers on MS medium supplemented with 3 mg/L BA. Somatic embryos developed on half strength MS medium containing 0.1 mg/L 2, 4-D. Roots were induced at the cut ends of shoots on MS basal medium devoid of growth regulators. The plantlets were then transferred to pots.

Key words: *Crataeva adansonii*, Axillary shoots, Adventitious shoot buds, Somatic embryos

1. Introduction

Crataeva adansonii D.C. Prodr., family Capparaceae, is a medium sized deciduous tree, commonly called Barna or Narnohi. It is often planted on the roadside as an avenue tree and in gardens because of its lovely flowers.. Flowering takes place in late March (Fig. 5.1a). Besides ornamental value it has medicinal and edible value also (1, 2); nevertheless, it has limitations as it is a rare tree (3). The leaves of this plant are demulcent and are used in rheumatism, stomach ache and as a tonic. Its bark is used to stop vomiting and fever (4). Natural propagation takes place by root suckers and seeds. The ratio of flower to fruit formation is very low. Pulp coated seeds are attacked by insects when they fall on

S.M. Jain and S.J. Ochatt (eds.), *Protocols for In Vitro Propagation of Ornamental Plants*, Methods in Molecular Biology, vol. 589
DOI 10.1007/978-1-60327-114-1_5, © Humana Press, a part of Springer Science+Business Media, LLC 2010

Fig. 5.1. Shoot bud differentiation and plant regeneration from different explants of *Crataeva adansonii*. (**a**) Tree of *C. adansonii* in flowering. (**b**) Proliferation of axillary buds from nodal explants. (**c**) Proliferation of axillary buds from shoot tip. (**d**) Shoot organogenesis from petiole. (**e**) Adventitious shoot development from leaves. (**f**) Somatic embryos in half strength MS medium supplemented with 2, 4-D (0.1 mg/L). (**g**) Shoot elongation and multiplication. (**h**) Plantlet with roots on MS medium without plant growth hormones.

the ground after dehiscence of fruits, making seed propagation further difficult.

Explants from mature trees face problems in propagation by tissue culture methods. In general, juvenile tissues from forest trees are more responsive to in vitro manipulation than mature tissues. The longer life span of the trees adds to the problem of contamination in vitro by the symbiotic association of microorganisms. Besides the age of the tree, the response of the explants is primarily determined by genotype, the physiological state of the tissue, and the time of the year when the explants are collected and cultured. In general, three modes of in vitro plant regeneration have been in practice: organogenesis, embryogenesis and axillary proliferation. Micropropagation via axillary and adventitious shoot proliferation is an advantageous option as there is no intervening callus phase, and the genetic stability of the resultant plants is maintained. But for augmenting quantity, de novo shoot formation via callus culture is also desirable as callus induced plants are more in number per subculture, and the maintenance of callus is easier. The present study was carried out in *C. adansonii* with the objective of developing all possible morphogenic protocols that are effective and reproducible.

2. Materials

2.1. Plant Material

Cultures of *Crataeva adansonii* should be initiated from young shoots collected between May and July (when the plant shows fresh growth) and mature shoots collected in the remaining months of the year.

Explants are obtained from young shoots:

1. Nodal segments (1–1.5 cm long).
2. Shoot tips (1–2 cm long).
3. Leaves (1.5–2.0 cm) from III to V nodes and petioles.
4. Anthers from 8 to 10 mm floral buds.

Surface-sterilizations should be carried out in 250 mL sterile flasks; the shoot cuttings cultured in sterile, transparent Borosil flasks.

2.2. Media

Micropropagation medium is Murashige and Skoog medium (5)

1. *Macronutrients:* 1,650 mg/L ammonium nitrate, 1,900 mg/L potassium nitrate.
2. Use 5 mL in total volume of 1 L medium (Stock: Dissolve 370 mg/L Magnesium Sulfate 7 hydrated, 170 mg/L KH_2PO_4, 27.8 mg/L $FeSO_4 \cdot 7H_2O$, 37.3 mg/L EDTA. $Na_2 \cdot 2H_2O$.).

3. Minor salt stock: Use 1 mL in total volume of 1 L medium (Stock: Dissolve 248 mg boric acid, 676 mg $MnSO_4$, 344 mg $ZnSO_4 \cdot 7H_2O$, 50 mg $Na_2MoO_4 \cdot 2H_2O$, 5 mg copper sulfate, 5 mg cobalt chloride in 200 mL distilled water. Store at –200° C) (*see* Note 1).
4. Vitamin stock I: Use 1 mL in total volume of 1 L medium (Stock: Dissolve 50 mg Nicotinic acid, 10 mg Thiamine hydrochloride, 50 mg pyridoxine hydrochloride, 200 mg Glycine in 100 mL distilled water. Can be stored at –20° C for up to 3–4 months.).
5. Vitamin stock II: Use 10 mL in total volume of 1 L medium (Stock: Dissolve 1 g myoinositol in 100 mL distilled water. Can be stored at –20° C for up to 3–4 months.) (*see* Note 1).
6. Calcium chloride stock: Use 5 mL in total volume of 1 L of the medium (Stock: Dissolve 8.8 g Calcium chloride dihydrate in 100 mL distilled water. Can be stored at 4° C for up to 3–4 months).
7. Potassium iodide stock: Use 5 mL in total volume of 1 L medium (Stock: Dissolve 16.6 mg potassium iodide in 100 mL distilled water; store in a dark glass bottle at 4° C. Make fresh every 1–2 months).
8. Add sucrose (30 g/L). Adjust pH of the medium to pH 5.8 with 1 N NaOH and add 8 g/L agar (Merck, India); autoclave at 121° C for 20 min.

2.3. Other Supplies

1. Glassware and Borosilicate glass from Borosil, India.
   a. Erlenmeyer conical flasks: 100–250 mL capacity, 1,000–3,000 mL capacity.
   b. Round bottom flasks: 500–1,000 mL capacity.
   c. Test tubes: 25 × 150 mm.
   d. Reagent bottles: 100–500 mL capacity.
   e. Petriplates: 100 × 50 mm and 150 × 75 mm diameter.

3. Methods

3.1. Explant Preparation

1. Obtain shoot cuttings, 6–7 cm long with 5–6 nodes and unopened flower buds (8–9 mm long) of green color and with 3 mm long anther lobes from *C. adansonii* plants (*see* Note 2).
2. Surface sterilize the shoot cuttings and carry out the surface sterilization steps with 0.1% (w/v) mercuric chloride solution for 5 min followed by 5–6 rinses in autoclaved distilled water (*see* Note 3) with detergent 0.5 mL 20% Extran in 50 mL and 0.1% (v/v) Tween-80 in 250 mL sterile conical flasks by continuous shaking for 20 min.

3. In the laminar hood, rinse both shoot cuttings and buds with sterile distilled water at least four times. Take the shoot cuttings out in Petri plates and cut with the help of sterile scalpel and forceps to obtain nodal segments, shoot tips, leaves and Petiole. Remove the anther from the buds; care should be taken to remove the filament from the anther lobes.

3.2. Shoot and Root Initiation

1. Place the sterilized explants (five per flask) vertically in 40 mL micropropagation medium in sterile flasks.
2. Incubate flasks under a 16 h light regime (50 uE m^{-2} s^{-1}) at a constant temperature of 25° C+2° C.
3. When shoots (axillary, adventitious and de novo shoots) develop from different explants (in 3–4 weeks), Section the cultures obtained from nodal segments in half.
4. Subculture on the multiplication medium (MS+1 mg/L BAP). Incubate under the same growth conditions as above.
5. The developed shoots can be cut and rooted to obtain plantlets. The remaining stump can be subcultured for another 4 weeks (*see* Note 4).
6. Culture anthers on solid MS medium with callus induction medium for 3 weeks. Transfer the resultant callus on morphogenesis induction medium and keep for 3 weeks. Then, transfer this callus medium to de novo shoot induction medium to obtain shoots.
7. Culture anthers in Liquid embryogenesis induction medium for 5 weeks. Transfer the embryoids to gelled MS medium devoid of hormone and gelled with Phytagel for germination (*see* Note 5).
8. The Sixth to ninth nodal segments respond best on MS medium supplemented with 3 mg/L BAP, 0.05 mg/L NAA (Fig. 5.1b).
9. Shoot tips show axillary shoot multiplication effectively on MS medium containing 3 mg/L BAP and 0.1 mg/L NAA. The axillaries multiply quickly and form 6 to 8 shoot buds without any trace of callus (Fig. 5.1c).
10. Five to six shoot buds are formed on petiole when cultured on MS medium supplemented with 0.5 mg/L BAP and 0.1 mg/L TDZ (Fig. 5.1d).
11. Leaves respond the best from the second week of May to June. 3 mg/L BAP with 0.1 mg/L NAA induce shoot buds from the midvein part of the leaves (Fig. 5.1e). The shoots formed are adventitious in nature.
12. Anther inoculation on 1.0 and 2.0 mg/L BAP with 0.2 mg/L Picloram induces harder, pale green, nodular callus. Transfer

this callus to 0.2 mg/L BAP and 0.2 mg/L 2, 4-D supplemented medium to obtain pale green, hard, nodular callus with morphogenetic response. Transfer the callus on to 3 mg/L BAP to obtain de novo shoot formation (15–17) (6).

13. Culture the anthers in liquid half MS medium containing 0.1 mg/L 2, 4-D in order to obtain somatic embryos (Fig. 5.1f). These embryos reach the cotyledonary stage in the fifth week of culture. Select and transfer the cotyledonary stage embryos to MS medium devoid of hormone and gelled with Phytagel for germination. These rooted plantlets should be subjected to hardening treatment before transferring to outer conditions.
14. Elongated shoots (Fig. 5.1g) are detached and transferred to rooting medium, while the remaining basal portion can be further subcultured on the medium amended with 0.3 mg/L BAP 0.05 mg/L NAA, for continuous production of shoots. Shoot cuttings induce roots on MS medium with 1 mg/L NAA in 4–5 wk (Fig. 5.1h).
15. *Crataeva adansonii* shows adaptability to tissue culture methods of propagation. Shoot morphogenic responses vary according to the explants and the hormonal concentrations (Table 5.1) (Fig. 5.2).

Table 5.1
Regeneration responses from different explants of *Crataeva adansonii* on different PGRs

Explants	Regeneration medium	Type of response
Shoot buds	3 mg/L BAP and 0.1 mg/L NAA	Axillary multiplication
Nodal segments	3 mg/L BAP and 0.05 mg/L NAA	Axillary multiplication
Petioles	BAP 0.5 mg/L and TDZ 0.1 mg/L	Adventitious shoot formation
Leaves	3 mg/L BAP and 0.1 mg/L NAA	Adventitious shoot formation
Anthers	1.0–2.0 mg/L BAP and 0.2 mg/L Picloram 0.2 mg/L BAP and 0.2 mg/L 2, 4-D 3 mg/L BAP	Callus Morphogenic callus De novo shoot formation
Anthers	0.1 mg/L 2, 4-D	Somatic embryogenesis

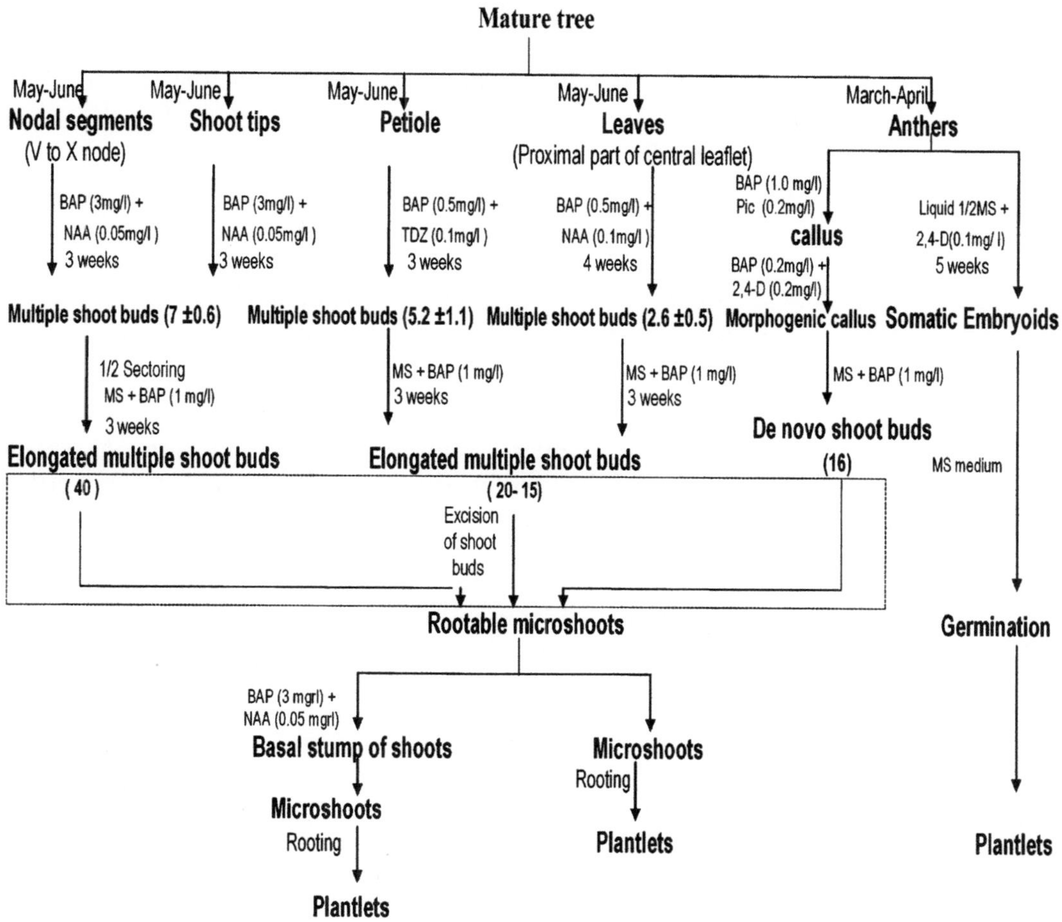

Fig. 5.2. Outline of in vitro micropropagation of *Crataeva adansonii*.

3.3. Acclimation and Transfer of Plantlets to Soil

1. Rinse the roots of the regenerated plantlets in tap water to remove tissue culture medium and transfer into pots containing the potting mix (*see* Note 6).
2. Place the regenerated shoot into the mix by making a hole, depth depending upon the root length, in the soil. Compact the soil around the shoot and water lightly.

4. Notes

1. MS vitamin stocks and minor salts are stored frozen, and all solutions should be completely thawed before taking any aliquots. All prepared stock solutions of vitamins can be stored at 4° C.

2. The explant should be collected in the month of May–June when the tree shows a fresh growth of branches. Floral buds should be collected in March–April when the tree is in full bloom.
3. A maximum of five explants can be used per flask.
4. 90–100 % of the regenerated shoots are able to root on the rooting media. NAA (1 mg/L) can be used for the rooting of the regenerated shoots. Root initiation usually occurs within two weeks on NAA containing medium.
5. Plantlets that are ready for the transfer to pots develop a minimum of 6–7 roots, 3–4 cm long, and with 4–5 nodes per plantlet.
6. Clean the plantlets of any trace of agar medium. Use well-cleaned, disinfected or new pots. Add fungicide to the potting mix to kill the fungi present in the soil. Transfer the plantlets individually to pots containing 1:1 garden soil: farmyard compost and harden within a chamber with a glass top, allowing penetration of light and maintenance of high humidity. The plantlets are gradually acclimatized through periodic exposure to the low relative humidity of ex vitro conditions.

References

1. Rajwar GS (1983) Low altitude medicinal plants of south Garhwal. Bull Med Ethnobot Res 4:14–28
2. Bhattacharjee SK (2001) Hand book of medicinal plants. Pointer Publisher, Jaipur
3. Tyagi P, Kothari SL (2005) Anther culture in *Crataeva adansonii* (DC.)Prodr. Phytomorphology 55:49–54
4. Shivarajan VV, Balachandran I (1994) Aurvedic drugs and their plant sources. Oxford and IBH Publisher, Calcutta, India
5. Murashige T, Skoog FA (1962) Revised medium for rapid growth and bioassays with tobacco tissue cultures. Physiol Plant 15: 473–497
6. Tyagi P, Kothari SL (2005) Anther culture in *Crataeva adansonii* (DC.)Prodr. Phytomorphology 55:49–54

Chapter 6

Strategies for the Micropropagation of Bromeliads

Miguel Pedro Guerra and Lirio Luiz Dal Vesco

Abstract

Bromeliads are tropical plants that are native to the Americas with a wide distribution in the rain forests, deserts and coastal areas. They are mostly epiphytes and terrestrial, diverse and important from the ecological point of view, they are found in microhabitats in strong interactions with fauna. Most of the ecosystems where they naturally occur are now endangered. Bromeliads are also one of the bases of the ornamental industry worldwide, being commercialized according to the features and colour of the foliage and flowers. This industry relies also on breeding programmes generating new hybrid bromeliads with improved bloom and foliage. Thus, advanced propagation systems based on micropropagation are valuable tools for both bromeliad germplasm conservation and for the mass clonal propagation of a selected genotype. In this chapter, a series of in vitro regenerative protocols for bromeliad mass propagation are described in detail. Emphasis is placed on the Brazilian native species, mainly those from the Atlantic Forest biome. The morphogenetic pathway based on nodule cluster culture is stressed, as (a) it seems to be recurrent in bromeliads, (b) it results in high regenerative rates, and (c) it is easy to handle. Protocols based on somatic embryogenesis and encapsulated units using hydrogel are presented. Finally, the use of temporary immersion systems (TIS) for mass propagation is described in detail for two bromeliads: *Aechmea fasciata*, and *Vriesea brusquensis*.

Key words: Bromeliaceae, Bromeliads, Ornamental industry, Micropropagation, Protocols, Nodule cluster cultures, Temporary immersion systems, Encapsulate units, Mass propagation, Tissue culture

1. Introduction

The *Bromeliaceae* family includes 56 genera and *ca.* 3,146 species and subspecies; more than half are epiphytes: that is, their roots can attach to tree trunks, rocks or other substrates, and they gather moisture from the air or dew rather than from the ground (1, 2). All of them are native to the Americas with the exception of *Pitcairnia feliciana,* which is native to Africa (3). The individual bromeliad typically consists of a short stem bearing a few to many, closely placed, alternate, usually spirally arranged, strap-shaped to filiform leaves (4).

S.M. Jain and S.J. Ochatt (eds.), *Protocols for In Vitro Propagation of Ornamental Plants*, Methods in Molecular Biology, vol. 589
DOI 10.1007/978-1-60327-114-1_6, © Humana Press, a part of Springer Science+Business Media, LLC 2010

The Brazilian Atlantic Forest biome contains high levels of genetic diversity. Bromeliads are important components of this biome and, among them, species of *Vriesea* are normally found in the South and Southwest regions of Brazil (3). *Vriesea* comprises 257 species distributed in the American continent (5). In Santa Catarina State, South Brazil, 31 native *Vriesea* species have been described, some of them endemic (3, 6, 7).

Although horticulturists use bromeliads extensively as ornamental plants because of their beautiful rosettes and flowers (which bloom purple, white, red, blue, yellow, orange, and even brown), surprisingly little is understood about their reproductive strategies (1). In the last decades, horticulturists have relied on the development of plant tissue culture based techniques in order to develop new strategies for bromeliad mass propagation.

The in vitro morphogenesis in bromeliads from different explant sources has been associated with organogenesis in *Vriesea fosteriana* (8), *Dyckia macedoi* (9), and *Aechmea fasciata* (10). In *Dyckia distachya*, in vitro morphogenesis occurred from flower stalks segments (11), and in *Cryptanthus sinuosos* from young leaves (12). The induction and development of monopolar structures may generate adventitious shoots with high frequencies in bromeliads. These structures, often isolated from a single explant, may be responsive and generate adventitious shoots from nodular cultures (13).

Tissue culture techniques comprise valuable tools for the mass propagation of bromeliads (14). It has been shown that the basal region of bromeliad leaves presents vascular elements whose cells may be competent for re-differentiation when activated by growth regulators (15). Such explants were successfully used in several bromeliad in vitro systems (14, 16–19). Specifically in the case of *Vriesea* species, several lines of evidence show that the in vitro regenerative route follows a specific pattern associated with nodule cluster cultures. We have shown this regenerative pattern to be recurrent in several *Vriesea* species from the Atlantic Forest biome (7, 14, 20).

Nodule clusters, normally originating from the basal region of explants in bromeliads, were induced and described in *Vriesea reitzii* (20) as competent morphogenic structures that are globular in form, translucent to yellowish in colour, and compact in texture. In *V. reitzii*, their development was observed starting from the parenchimatic tissues of explants after 3 weeks in culture. The high regenerative potential of leaf explants was ascribed to the presence of intercalary axillary meristems in the basal region of bromeliad leaves (15). Under adequate conditions, such nodule clusters develop radially into a number of small shoots (0.5 cm long) called microshoots, which elongate to give shoots. This terminology will be employed here regarding the in vitro regenerative systems in bromeliads.

2. Materials

2.1. General Procedures for Mother Plants

1. An important step in the establishment of in vitro regenerative protocols is the choice and maintenance of the mother plant (*see* stage Zero, Fig. 6.1). Specifically for bromeliads, nutrition and sanity are the main concerns in order to guarantee the establishment of viable and responsive cultures. Improving nutritional conditions includes the weekly application of a commercial chelated fertilizer 30-10-10 (NPK, %), and Nitrofoska®. The application of 10 mL α-naphthaleneacetic acid (NAA) (5 μM), and Vitamin B1 (Thiamine) directly in the rosette enhances the formation of offshoots to be used as a source of explants.
2. The establishment of aseptic and vigorous cultures is facilitated when the mother plants are sprayed with benomyl and mineral oil, and maintained for at least one month in a phytotron at 25 ± 2° C under 16 h light of an intensity of 300 μmol/m^2/s, supplied by cool white fluorescent (Sylvania®) and high pressure sodium vapour lamps (Empalux® – VST).
3. Bacterial contamination during culture establishment may be avoided by supplementing 20 mL/L Kanamycin and/or 100 mg/L Agrimicin® to the medium. Fungal contaminations are efficiently controlled with 20 mg/L miconazol nitrate. These substances should be filter sterilized before being added to the culture medium in order to keep their efficiency.

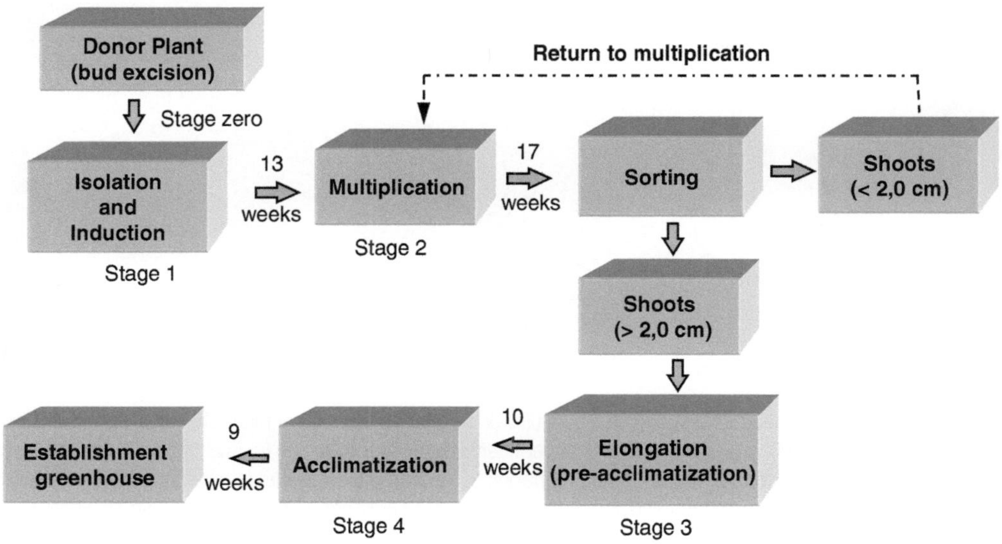

Fig. 6.1. Process for ornamental bromeliad mass micropropagation (modified from (28)).

2.2. General Procedures for Scale Up Micropropagation of Bromeliads

1. *Isolation and induction* (*see* stage 1, Fig. 6.1): The axillary buds are isolated from the young offshoots, 1.0–2.5 cm long, of the mother plant (Fig. 6.2a).
2. The young offshoots are reduced to size 0.5 cm by carefully removing the external leaves without damaging the axillary buds. The final reduced size offshoot contains the apical meristem and 3–5 lateral buds. This offshoot is washed thoroughly in tap water; then, in an aseptic chamber, it is submitted to disinfection with 70% ethanol for 2 min, 1.5% sodium

Fig. 6.2. (**a**) Isolated axillary bud from young offshoots (1.0–2.5 cm) from the mother plant; (**b**) Induction of buds in culture medium with NAA (2 μM), and BAP (4 μM) over paper filter bridges after 9 weeks; (**c**) Shoots develop after 13 weeks in culture; (**d**) Subculture clusters of shoots to 300 ml flasks; (**e**) Growth after 9 weeks in medium with NAA (1 μM) and BAP (2 μM); (**f**) After 10 weeks, >3.0 cm shoots are obtained; (**g**, **h**) Shoots are transferred for acclimatization, pruning roots (if present) to 1 cm long; and (**i**) Transferring them to trays of 128 cells (60 cm^3 each) filled with a mix commercial substrate. Bar = 1 cm.

hypochlorite for 20 min, and 2–3 drops of Tween 20; finally, it is rinsed three times in sterile water.

3. The buds are inoculated on paper filter bridges in test tubes (25 mm × 150 mm) containing 15 mL liquid culture medium consisting of MS (21) salts (Table 6.1), Morel vitamins (22),

Table 6.1
Culture media for bromeliad micropropagation based on Murashige and Skoog (MS; (21)) and Morel vitamins (22)

Component	Concentration (mg/L)
Macronutrients	
NH_4NO_3	1,650
KNO_3	1,900
$CaCl_2 \cdot 2H_2O$	440
KH_2PO_4	170
$MgSO_4 \cdot 7H_2O$	370
FeEDTA solution	
$Na_2EDTA \cdot 2H_2O$	37.3
$FeSO_4 \cdot 7H_2O$	27.8
Micronutrients	
$MnSO_4 \cdot H_2O$	16.9
$ZnSO_4 \cdot 7H_2O$	8.6
H_3BO_3	6.2
KI	0.83
$CuSO_4 \cdot 5H_2O$	0.025
$Na_2MoO_4 \cdot 2H_2O$	0.25
$CoCl_2 \cdot 6H_2O$	0.025
Morel vitamins	
Myo-inositol	100
Thiamine–HCl	1.0
Pyridoxine–HCl	1.0
Nicotinic acid	1.0
Ca pantothenate	1.0
Glycine	0.01

Add sucrose (30 g/L). Adjust the pH of the medium to 5.5–5.8 with 0.5 N NaOH, autoclave at 121° C for 16 min sucrose (30 g/L), 2 μM NAA and 4 μM 6-benzylaminopurine (BAP). Alternatively, the culture medium may be free of plant growth regulators. The medium is autoclaved at 121° C for 15 min. The pH is adjusted to 5.5–5.8 according to the bromeliad to be micropropagated.

4. The cultures are kept at 25 ± 2° C, RH 60 ± 5%, under 50–60 μmol/m^2/s photosynthetic photon flux density with a 16-h light period.
5. The induction of new buds over the paper filter bridges occurs after 9 weeks (Fig. 6.2b) and a shoot develops after 13 weeks of culture (Fig. 6.2c).
6. *Multiplication* (*see* stage 2, Fig. 6.1): The shoots are subcultured in 300 mL flasks containing 15 mL of the same culture medium (Fig. 6.2d). After 9 weeks, 5–8 clusters are subcultured in 300 mL flasks containing 25 mL of the same culture medium described in **item 3**, with NAA and BAP lowered to 1 μM and 2 μM, respectively (Fig. 6.2e).
7. *Elongation*: The shoots are sorted out (*see* stage 3, Fig. 6.1). Shoots <2 cm long return to the multiplication medium. Shoots >2 cm long are subcultured in 300 mL flasks containing 25 mL liquid medium, either hormone-free or supplemented with 5 μM GA_3 , as defined for each protocol. After 10 weeks, shoots longer than 3 cm are obtained (Fig. 6.2f) and are transferred to the acclimatization stage. See summary in Table 6.2.

2.3. General Procedures for the Acclimatization of Bromeliads

Our results on the micropropagation of several bromeliad species show that the main factor for the success of acclimatization is the size of the shoots. The minimum size is 3.0 cm (*see* stage 4, Fig. 6.2f), below which shoots return to the elongation phase according to the culture medium composition defined for each protocol until they reach this size (*see* stage 3). The presence of adventitious roots is not necessary (Fig. 6.2g, h). Thus, the rooting stage may be performed ex vitro in a phytotron or nebulization tunnel. The eventual roots present at the base of shoots should be reduced to 1 cm at the time of transfer to trays of 128 cells (60 cm^3 each) with a commercial mix substrate of 2:1:1 (v:v:v) of Plantmax® HA: pine bark: carbonized rice coat (stage 4, Fig. 6.2i). An alternative substrate is a 2:2:1 (v:v:v) mixture of peat, vermiculite and sand. New roots will appear after 6 weeks depending on the species. During acclimatization, the plantlets should be sprayed every week with a solution of ¼MS salts. The trays should be kept in a greenhouse with controlled mist under 50% light intensity reduction for 9 weeks (stage 5, Fig. 6.1).

Table 6.2
Summary of bromeliad micropropagation protocols

Species	Culture medium composition
Aechmea fasciata and Vriesea brusquensis	
Temporary immersion systems	Induction: MS + NAA (2 μM) + BAP (4 μM) Multiplication: MS + PBZ (6 μM) for *A. fasciata* and MS + PBZ (2 μM) for *V.brusquensis* Elongation: MS + GA_3 (10 μM)
Dyckia distachya	
Somatic embryogenesis (19)	Induction: MS + Picloram (5 μM) Establishment/multiplication: MS + 2-iP (5 μM) and NAA (0.5 μM) Conversion: MS free of PGR
Direct organogenesis (14)	Induction: MS + BA (2 μM) and Kin (2 μM) Multiplication: NAA (2 μM), BAP (4 μM) and PBZ (6 μM) Elongation: MS + GA_3 (5 μM)
Vriesea fosteriana, V. hieroglyfica and Alcantarea imperialis	
Direct organogenesis	Induction: MS + NAA (2 μM) + BAP (4 μM) Multiplication: MS + Thidiazuron (0.1 μM) Elongation: MS free of PGR
Vriesea fosteriana	
Encapsulated units	Induction of encapsulated units: MS + NAA (2 μM) + BAP (4 μM) and PBZ (4.0 μM) Encapsulation: Shoots 0.5–1.0 cm long in sodium alginate (1%) + ½ MS salt + GA_3 (5 μM) and complexed in a $CaCl_2$ (50 mM) Sowing: Decomplexation with KNO_3 (100 mM) Plantlet conversion: Vermiculite substrate and ¼ MS solution, keep the trays in a phytotron
Vriesea friburgensis Mez var. *paludosa*	
Direct organogenesis (18)	Induction: MS + 2,4-D (5 μM) + Kin 1(μM) Multiplication: MS + 2-iP (12 μM) Elongation: MS + GA_3 (5 μM)
Vriesea reitzii	
Adventitious shoots from nodule cluster (7)	Induction: MS + NAA (2 μM) + BAP (4 μM) Multiplication: MS + NAA (1 μM) and BAP (2 μM) Elongation: MS + GA_3 (10 μM)
Regeneration from nodule cluster cultures (20)	Induction: MS + 2,4-D (20 μM) and Kin (1 μM) Establishment: MS + 2-iP (2.5 μM) and NAA (0.5 μM) Multiplication: MS free of PGR Elongation: MS + GA_3 (10 μM)
Vriesea splendens	
Adventitious shoots from nodule cluster	Induction: MS + NAA (1 μM) + BAP (2 μM) Establishment: MS free of PGR Multiplication: MS + Thidiazuron (0.1 μM) Elongation: MS + GA_3 (10 μM)

3. Methods

3.1. Regeneration of *Vriesea friburgenesis* from Nodule Cluster Cultures

3.1.1. Methods

See details of this species in Note 1.

1. Microshoots in proliferation (*see* Subheading 2.2) are the explant source (*see* Fig. 6.3).
2. *Induction* – Inoculate microshoots in Petri dishes containing 25 mL culture medium consisting of MS salts, Morel vitamins, 30 g/L sucrose, 7 g/L Agar Sigma®, 5 μM 2,4-dichlorophenoxyacetic acid (2,4-D), and 1 μM Kinetin (Kin). Keep the cultures as described in Subheading 2.2.
3. After 13 weeks in culture, yellow coloured nodular cluster cultures arise.
4. *Establishment and development* – Subculture 0.05 g nodular cluster cultures over paper filter bridges in test tubes (25 mm × 15 mm) containing 15 mL liquid culture medium described in step 2, free of Plant Growth Regulators (PGR).
5. *Multiplication* – Subculture 0.5 g nodular cultures in 300 mL glass flasks containing 15 mL of the same liquid culture medium described in step 2, replacing the PGR by 12 μM (2-isopentyl) adenine (2-iP) (stage 2).
6. Clusters of microshoots develop after 9 weeks.
7. *Elongation* – Subculture twice, at 4 week intervals, in 300 mL glass flasks containing 15 mL of the same liquid culture medium described in step 2, replacing 2-iP by 5 μM Gibberellic acid (GA_3) (stage 3).
8. *Acclimatization* – Shoots at least 3 cm long from stage 3 are transferred to trays with 128 cells (60 cm^3 each) containing a mix substrate of 1:1 (v:v) of carbonized rice coat and Turfa Fértil® mineral supplement (N:4%, P_2O_5:14% K_2O:8%) (stage 4).

3.2. Adventitious Shoots from Nodule Cluster of Vriesea reitzii

See details of this species in Note 2.

1. Shoots in proliferation (*see* Subheading 2.2) are the source of explants.
2. After 10 weeks, subculture microshoots in the same culture medium described in Subheading 2.2 with the levels of NAA and BAP lowered to 1 μM and 2 μM, respectively.
3. *Induction* – Inoculate 5–8 microshoot clusters in 300 mL flasks with 20 mL liquid medium composed of MS salts, Morel vitamins, 30 g/L sucrose, 2 μM NAA and 4 μM BAP.
4. Yellow and green nodular clumps develop at the basal region of the microshoots within 10 weeks (Fig 6.4a). Subculture 0.05 g nodular clumps twice in 300 mL flasks with 15 mL of

Fig. 6.3. General aspect of the mother plants used as source of explants.

the same liquid medium at 9-week intervals. They will evolve to nodular cluster cultures bearing microshoots (Fig. 6.4b).

5. Long term maintenance of these nodular clumps is achieved by keeping them in 300 mL flasks with 15 mL liquid culture medium described in step 3, now free of PGR (Fig. 6.4c1).

Fig. 6.4. (**a–c**) Adventitious shoots from nodule cluster of *Vriesea reitzii*: (**a**) In vitro protuberances developed in presence of 2 μM NAA and 4 μM BA (*arrow*); (**b**) In vitro shoot multiplication; (**c**) Multiplication (c1), GA_3 results in the uniform elongation of shoots (c2); (**d–g**) Regeneration of *V. reitzii* from nodule cluster cultures; (**d**) Yellow nodule clusters induced from the basal region of leaves in MS medium with 2,4-D (5 μM) and Kin (0.5 μM); (**e**) Nodule cultures in medium with 2-iP (2.5 μM) and NAA (0.5 μM); (**f**) Clusters of shoot-buds after 6 weeks in culture; (**g**) Clusters of microshoots in medium with GA_3 (10 μM); (**h–k**) Adventitious shoots from nodule cluster of *Vriesea splendens:* (**h**) After 12 weeks, nodule cluster cultures arise from the explants; (**i**) Cluster cultures over filter paper bridges in liquid culture free of PGR; (**j**) Subcultured nodular cluster cultures in liquid medium with TDZ (0.1 μM) (**k**) Elongation of shoots after 12 weeks in medium with GA_3 (10 μM). Bar = 1.0 cm.

6. *Multiplication 1* – Subculture 0.05 g nodule cluster cultures in 300 mL flasks containing 10 mL of the same liquid culture medium described in step 3.
7. *Multiplication 2* – Inoculate shoots 1.0–2.5 cm long that have been previously elongated (item 3.1-7) in liquid culture

medium, in 300 mL flasks containing 20 mL liquid culture medium described in step 3, with NAA and BAP lowered to 1 µM and 2 µM, respectively. The multiplication rate may reach 20:1 after 12 weeks.

8. *Elongation* – Subculture ten microshoot clusters in each 300 mL flask with 15 mL liquid medium replacing the PGR by 10 µM GA_3 during 8 weeks, and subsequently, for a further 8 weeks, in flasks with the same culture medium, now free of PGR (Fig. 6.4c2).
9. *Acclimatization* – Transfer shoots >3 cm to trays of 72 cells (120 cm^3 each) with a mix substrate 1:1 (v:v) of carbonized rice coat and Turfa Fértil® mineral supplement (N:4%, P_2O_5:14% K_2O:8%). Keep the trays in a greenhouse under controlled mist for 9 weeks until the minimum size is 4–5 cm. General view of acclimatization in tunnel (Fig. 6.5a).

3.3. Regeneration of Vriesea reitzii *from Nodule Cluster Cultures*

1. Inoculate buds as described (*see* Subheading 2.2) over filter paper bridges in 25 mm × 150 mm test tubes containing 15 mL liquid culture medium composed of MS salts, Morel vitamins, 30 g/L sucrose, 2 µM NAA, and 4 µM BAP. After 8 weeks, nodule cluster cultures arise from the explant; subculture them for 8 more weeks. They evolve to clusters bearing microshoots.
2. *Induction* – Remove 0.5 mm size segments from the basal region of the young leaves of microshoots that originated in step 1. Inoculate 5–8 of them in Petri dishes containing 25 mL culture medium described in step 1, replacing PGR by 20 µM 2,4-D, and 1 µM Kin. Keep the cultures as described in Subheading 2.2.
3. After 7 weeks, nodular cluster cultures arise from the basal region of the leaves (Fig. 6.4d).
4. *Establishment and development* – Subculture 0.25 g nodular cluster cultures in the same culture medium as in step 1, replacing the PGR by 2.5 µM 2-iP and 0.5 µM NAA for 8 weeks (Fig. 6.4e).
5. *Multiplication* – Subculture 0.5 g nodular clusters for 6 more weeks onto the same culture medium, now free of PGR. Clusters of microshoots evolve from the nodular cultures (stage 2, Fig. 6.4f).
6. *Elongation* – Subculture 8–10 clusters of microshoots in 300 mL flasks with 15 mL liquid medium as described in step 1, replacing the PGR by 10 µM GA_3 (stage 3, Fig. 6.4g).
7. *Acclimatization* – Shoots longer than 3 cm are transferred to trays of 128 cells (60 cm^3 each) containing substrate 1:1 (v:v)

of carbonized rice coat and Turfa Fertil® mineral supplement (N:4%, P_2O_5:14% K_2O:8%) in a nebulization tunnel. The survival rate of acclimatization is 95 % after 9 weeks (Fig. 6.5b). Transfer the plantlets to pots in a greenhouse with controlled mist for 9 more weeks (Fig. 6.5c).

Fig. 6.5. Acclimatization of bromeliads: (**a**) General view of acclimatization in tunnel; (**b**) Plantlets of *Vriesea reitzii* in trays; (**c**) In pots growing in greenhouse; (**d–f**) Acclimatization of plantlets in mix substrate with pine bark: (**d**) *V. splendens*, (**e**) *V. fosteriana*; (**f**) *V. hieroglyfica*; (**g**) Plantlets of *Alcantharea imperialis var. vinicolor* acclimatized in substrate 2:1:1 (v:v:v) of Plantmax® HA: pine bark: carbonized rice coat; (**h**) Plantlets of *V. fosteriana* growing in pots in greenhouse; (**i**) General view of plants in different stages.

3.4. Adventitious Shoots from Nodule Cluster of Vriesea splendens

See details of this species in Note 3.

1. *Induction* – Inoculate buds as described (*see* Subheading 2.2) over filter paper bridges in 25 mm × 150 mm test tubes with 15 mL of liquid culture medium containing MS salts, Morel vitamins, sucrose (30 g/L), 1 μM NAA and 2 μM BAP. After 12 weeks, adventitious shoots from nodule cluster cultures arise from shoots (Fig. 6.4h).
2. *Establishment and development* – Subculture 0.10 g nodular cluster cultures twice (8 weeks each) over filter paper bridges in 25 mm × 150 mm test tubes containing 15 mL liquid culture described in step 1, but devoid of PGR (Fig. 6.4i).
3. *Multiplication* – Subculture 0.5 g nodular cluster cultures in 300 mL flasks with 25 mL liquid medium as in step 1, replacing PGR by 0.1 μM Thidiazuron (TDZ) (*see* Fig. 6.4j).
4. *Elongation* – After 12 weeks, subculture 8–10 clusters bearing microshoots in each 300 mL flask with 15 mL liquid medium as in step 1, replacing PGR with 10 μM GA_3 (Fig. 6.4k).
5. *Acclimatization* – Shoots >3 cm are transferred to trays of 128 cells (60 cm^3 each) containing substrate composed of 2:1:1 (v:v:v) of Plantmax® HA: pine bark: carbonized rice coat. The survival rate is 95% after 8 weeks (Fig. 6.5d). Keep the trays in a greenhouse with controlled mist for 9 weeks.

3.5. Regeneration of Vriesea fosteriana, V. hieroglyfica *and* Alcantarea imperialis

See details of these species in Note 4.

1. *Induction* – Inoculate buds as described (*see* Subheading 2.2) over filter paper bridges in 25 mm × 150 mm test tubes with 15 mL of liquid culture medium containing MS salts, Morel vitamins, sucrose (30 g/L), NAA (2 μM), and BAP (4 μM). After 10 weeks, nodule cluster cultures arise from the explants. Subculture 3 times, every 6 weeks, in the same medium.
2. *Multiplication* – Subculture 5–8 microshoot clusters in each 300 mL flask with 15 mL of liquid medium as in step 1, replacing PGR by TDZ (0.1 μM). The multiplication rate may reach 14:1 in *Vriesea fosteriana,* and 6:1 in *Alcantarea imperialis var vinicolor* and *V. hieroglyfica.* When considering TDZ levels compared to those employed for BAP and NAA as well as their concentrations, the final cost is considerably lowered with TDZ.
3. *Elongation* – Subculture 8–10 microshoots clusters during 10 weeks in each 300 mL flask containing 15 mL of liquid culture medium described in step 1, but free of PGR.
4. *Acclimatization* – Shoots >3.0 cm are transferred to trays of 128 cells (60 cm^3 each) with a substrate composed of 2:1:1 (v:v:v) Plantmax® HA: pine bark: carbonized rice coat.

Keep the trays in a nebulization tunnel during 9 weeks. The survival rate is 98 % for *Vriesea fosteriana* (Fig. 6.5e), 95 % for *V. hieroglyfica* (Fig. 6.5f), and 98% for *Alcantarea imperialis* var *vinicolor* (Fig. 6.5g).

3.6. Micropropagation for Vriesea fosteriana *Using Encapsulated Units*

1. *Induction* – Inoculate buds as described in Subheading 2.2 over filter paper bridges in 25 mm × 150 mm test tubes with 15 mL of liquid medium containing MS salts, Morel vitamins, sucrose (30 g/L), NAA (1 μM), and BAP (2 μM). After 8 weeks, new buds and shoots develop from the explants.
2. *Multiplication* – After 3–4 subcultures every 5–7 weeks, 3–5 of the clusters in each 300 mL flask containing 25 mL result in multiple shoot proliferation.
3. *Induction of encapsulated units* – Subculture 5–8 clusters in each 300 mL flask containing 20 mL of the same culture medium described in step 1, replacing PGR by Paclobutrazol (PBZ) (4.0 μM). After 8 weeks, the multiplication rate may reach 13:1.
4. *Encapsulate units* – Encapsulate shoots 0.5–1.0 cm long in a matrix (Fig. 6.6a) of sodium alginate (1%) diluted in a half strength MS salt solution supplemented with GA_3 (5 μM). The capsules are complexed in a $CaCl_2$ (50 mM) solution during 10 min. Wash in tap water and store the capsules for 4 weeks at 5° C (Fig. 6.6b).
5. *Sowing* – Before sowing the capsules in trays, submit them to a decomplexation treatment with KNO_3 (100 mM) during 20 min. Then, transfer the synthetic seeds to trays of 220 cells (13 cm^3 each) containing vermiculite (Fig. 6.6c). Keep the trays in a phytotron adjusted to 25 ± 2° C temperature, 16 h light period and 400 μmol/m^2/s light intensity. The trays are placed inside a plastic box covered with glass to allow light entry and to reduce water exchange (Fig. 6.6d). Wet the plantlet germinating capsules periodically with a 25% solution of MS salts (Fig. 6.6e). After 5 weeks, remove the trays to a greenhouse, transferring the plantlets to trays of 128 cells (60 cm^3 each) with a substrate composed of 1:1 (v:v) of carbonized rice coat and Turfa Fertil® mineral supplement (N:4%, P_2O_5:14% K_2O:8%) in a nebulization tunnel, with intermittent mist. After 8 weeks in this condition, 85% survival rate is observed (Fig. 6.6f). Keep the pots in the greenhouse with controlled mist for ten more weeks (Fig. 6.5h).

3.7. Somatic Embryogenesis in Dyckia distachya

See details of this species in Note 5.

1. *Induction* – Collect the green capsules from the stalks of mother plants. Disinfect them with 70% Ethanol (2 min), sodium hypochlorite 1.5% (20 min), and Tween 20 (2–3 drops); then, rinse three times in sterile water (*see* Subheading 2.2).

2. Open the capsules and inoculate the ten seeds in each Petri dish with 20 mL of medium with MS salts, Morel vitamins, sucrose (30 g/L), Agar (6 g/L), Picloram (5 μM). Keep the cultures in the dark at 25 ± 1° C, 60 ± 5 % relative humidity.
3. *Development* – After 8 weeks in culture, proembryogenic cellular masses are observed, and after six more weeks, these masses evolve in to embryogenic cultures.
4. Subculture 0.5 g of embryogenic cultures in Petri dishes with 20 mL of the culture medium described in step 2, replacing Picloram by 5 μM 2-iP, and 0.5 μM NAA. Keep the cultures as described in Subheading 2.2.
5. After 30–45 days, somatic embryos develop to mature stage and convert to plantlets.
6. Subculture plantlets in Petri dishes containing 20 mL fresh culture medium described in step 2, but free of PGR for the initial development of plantlets.
7. *Acclimatization* – Shoots >3 cm long are transferred to 72-celled trays (120 cm^3 each) with a mix substrate 1:1 (v:v) carbonized rice coat and Turfa Fértil® mineral supplement (N:4%, P_2O_5:14% K_2O:8%). An alternative substrate is a 2:2:1 (v:v:v) mixture of peat, vermiculite and sand. Keep the trays in a greenhouse with controlled mist for 6 weeks.

3.8. Direct Organogenesis Induction from Seeds

1. Collect seed-containing capsules from the mother plants. Seal them in polyethylene bags.
2. Disinfect the capsules, and keep cultures as described in Subheading 2.2.
3. *Induction* – In an aseptic chamber, open the capsules and inoculate ten seeds in each Petri dish with 20 mL of culture medium containing MS salts, Morel vitamins, sucrose (30 g/L), Agar Sigma® (6 g/L), BAP (2 μM) and Kin (2 μM).
4. *Multiplication* – After 6 weeks, subculture 5–8 shoot clusters in each 300 mL flask containing 15 mL of liquid culture as described in step 3, replacing PGR with NAA (2 μM), BAP (4 μM), and PBZ (6 μM).
5. *Elongation* – Subculture 5–8 clusters during six more weeks in 300 mL flasks with 20 mL of liquid culture described in step 2, replacing PGR by GA_3 (5 μM).
6. *Acclimatization* – Shoots >3.0 cm are transferred to trays of 128 cells (60 cm^3 each) containing substrate composed of 2:2:1 (v:v:v) mixture of peat, vermiculite and sand. The survival rate is 92% after 18 weeks.

3.9. Temporary Immersion Systems for Mass Propagation of Vriesea brusquensis and Aechmea fasciata

At least three different TIS (*see* Note 6) have been assayed for bromeliad mass propagation (Fig. 6.6g), including *Ananas comosus* (23). The protocol for *Vriesea brusquensis* (Fig. 6.3h) and *Aechmea fasciata* (Fig. 6.3i) by TIS (Fig. 6.6g–i) developed in the Laboratory of Plant Developmental Physiology and Genetics (LFDGV/CCA/UFSC) is described below.

1. Inoculate the buds and keep the cultures as described in Subheading 2.2.
2. *Induction* – After 8 weeks, clusters of cultures are obtained, each containing 5–8 shoots of average 0.5 cm.

Fig. 6.6. Encapsulation of microshoots: (**a**) Encapsulated units of *Vriesea fosteriana*; (**b**) Encapsulated microshoots in simple layer system (**c**) Ex vitro conversion of encapsulated microshoots; (**d**) Trays are placed inside a plastic box covered with glass to allow light entry and reduce water exchange; (**e**) Regrowth to plantlets (**f**) Plantlets in a nebulization tunnel. After 8 weeks, 85% survival rate is observed; (**g–j**) Temporary immersion system for mass propagation: (**g**) Standard TIS apparatus used in the bromeliad micropropagation; (**h**) Rita® apparatus; (**i**) TIS apparatus developed at LFDGV/CCA/UFSC and; (**j**) Detail of proliferation in TIS shown in (**g**).

3. *Multiplication* – Transfer 8–10 clusters to each TIS unit, which contains 300–500 mL according to the container (Fig. 6.6g–i), of MS salts, Morel vitamins, sucrose (3%) and PBZ (6 μM) for *Aechmea fasciata* and PBZ (2 μM) for *Vriesea brusquensis.*
4. Adjust the timer system for a cycle of 3 h stationary: 3 min immersion.
5. After 6 weeks, replace the container with culture medium, which normally acquires a dark-brown colour.
6. *Elongation* – After 6 more weeks, replace the culture medium is replaced, substituting PBZ by GA_3 (10 μM) in order to synchronously elongate the microshoots. By this time, the regenerative rate may reach 30:1 (Fig. 6.6j).
7. *Acclimatization* – Shoots longer than 3 cm are transferred to trays of 128 cells (60 cm^3 each) containing substrate composed of 2:1:1 (v:v:v) of Plantmax® HA: pine bark: carbonized rice coat. Keep the trays in a greenhouse with controlled mist for 6 weeks.

4. Notes

1. *Vriesea friburgensis* Mez var. *paludosa* (L. B. Smith) L. B. Smith (Fig. 6.3a) is found in epiphytic or terrestrial habitats in the southern and southeastern Brazilian regions of the Atlantic Forest domain. They grow preferentially as terrestrials, in sandy and dry soils found on the seacoast, in which the general vegetation is typically that found on sand dunes. In this habitat, they usually form dense groups. They are typically helophytic but capable of surviving under diffuse light, showing few requirements in terms of air humidity (3). In its natural habitat, this bromeliad is threatened by extinction (24). For this reason, it is necessary to develop strategies for its conservation and large-scale propagation (18).
2. *Vriesea reitzii* is an endemic bromeliad in South Brazil revealing excellent potential as an ornamental plant. This species presents red and yellow coloured inflorescences (Fig. 6.3b). It occurs in regions with altitudes from 750 to 1,200 m in the states of Santa Catarina, Paraná and Rio Grande do Sul, its name being a tribute to the so called ′father of bromeliads′, the Brazilian priest Raulino Reitz (2, 25). The ecosystem in which this bromeliad is distributed, the Araucarian Mixed Forest, was severely devastated during the last century (25), and now *V. reitzii* is threatened with extinction.
3. *Vriesea splendens* (Fig. 6.3c), commonly known as "flaming sword" is a bromeliad showing lance-shaped or linear, green

or purplish, foliage 30–45 cm long and 2.5–4 cm wide. The leaves have smooth margins and may have colourful bracts at the leaf bases. The foliage is often arching, forming a funnel-shaped rosette. From the centre of this rosette, emerges an upright inflorescence, consisting of a flattened stem of brilliant red or orange overlapping bracts, and yellow flowers which are short-stalked, generally in flattened spike-like racemes or panicles in two rows (2). The petals are often tubular and short-lived, while the bracts are brightly coloured and last for months. In natural populations, it grows as an epiphyte, and it is terrestrial in the lower levels of the dense forests of tropical Trinidad, Venezuela, Guyana and Surinam.

4. *Vriesea fosteriana* is a large bromeliad species spreading from Brazil to Mexico (Fig. 6.3d). It shows broad, mid-green leaves with a reddish-brown band on both sides. This foliage can grow up to 1-m long forming a dense rosette from the centre of which a conspicuous flower stalk reaches upwards to 1.5 m , developing a yellow flower from each bract (2). *Vriesea hieroglyphica* is a bromeliad native to southern Brazil (Fig. 6.3e). Its ornamental importance relies mainly on the extremely handsome foliage, which shows glossy, bright green leaves with wide and irregular horizontal bands of dark green. The leaves are 90 cm long and 7.5 cm wide. The inflorescence with long green bracts and yellow flowers is trivial as compared to the foliage (2). The imperial bromeliad (*Alcantarea imperialis* syn. *Vriesea imperialis*) is one of the largest bromeliads (Fig. 6.3f) of the world occurring naturally in the rocky south-eastern regions of the Mata Atlantica (1). This impressive, magnificent giant bromeliad is a full sun reaching up to 90–120 cm high, but can grow up to 2 m. The spineless leaves with a dark green superior side and a purple inferior side form a rosette from the centre of which emerges the flower stalk. The huge inflorescence blooms from mid summer to early fall bearing white flowers protected by large bracts which range from purple to dark green in color. In its natural habitat, it may take this bromeliad 50 years to bloom, but in cultivation, this time is reduced to 5–7 years. Nowadays, *A. imperialis* is one of the most cultivated bromeliads in gardens worldwide.

5. *Dyckia* genus (*Pitcairnioideae*) comprises 121 species, occurring exclusively in the southwest of South America (26). Among these species, *Dyckia distachya* (Fig. 6.3g) is an endemic and endangered bromeliad native to the western part of the Santa Catarina State, southern Brazil. This bromeliad shows several ornamental features. It occurs in the Brazilian Atlantic Forest, which is considered one of the most important plant diversity centres in the world and is a hotspot of biodiversity. Bromeliads are frequent is this biome, most of

them being endemic. Several studies in the Laboratory of Plant Developmental Physiology and Genetics of the Graduate Group in Plant Genetic Resources of the Federal University of Santa Catarina, Santa Catarina State, South Brazil, culminated with the development of some regenerative protocols for this species.

6. Temporary immersion systems (TIS) have been used for the mass propagation of ornamental plants. In organogenic regenerative systems, their benefits are associated with an increase in the multiplication rates (shoot proliferation), and improved responses in the acclimatization stage. Furthermore, these systems decrease the production costs by reducing the manipulation, work, area and number of containers (27).

Acknowledgments

The authors gratefully acknowledge the financial support of CAPES, CNPq, FAPESC, and FAPESP for the grants and fellowships.

References

1. Martinelli G (2000) The Bromeliads of the Atlantic forest. Sci Am 282:86–93
2. Baensch U, Baensch U (1994) Blooming Bromeliads. Rasch Druck Verlag, Bramsche, GE
3. Reitz R (1983) Bromeliáceas e a malária – bromélia endêmica. Flora ilustrada Catarinense, Itajaí
4. Benzing DH (2000) Bromeliaceae: profile of an adaptive radiation. Cambridge University Press, Cambridge
5. Smith LB, Downs RJ (1977)Tillandsioideae (Bromeliaceae). Flora Neotrop. Hafner Press, New York
6. Alves GM, Rech Filho A, Puchalski A, Reis MS, Nodari RO, Guerra MP (2004) Allozymic markers and genetic characterization of a natural population of *Vriesea friburgensis* var. *paludosa*, a bromeliad from the Atlantic Forest. *P.* Genet Res 2:23–28
7. Rech Filho A, Dal Vesco LL, Nodari RO, Lischka RW, Muller CV, Guerra MP (2005) Tissue culture for the conservation and mass propagation of *Vriesea reitzii* Leme and Costa, a bromeliad threatened of extinction from the Brazilian Atlantic Forest. Biodivers Conserv 14:1799–1808
8. Mercier H, Kerbauy GB (1992) In vitro multiplication of *Vriesea fosteriana*. Plant Cell Tissue Organ Cult 30:247–249
9. Mercier H, Kerbauy GB (1993) Micropropagation of Dychia macedoi – an endangered endemic Brazilian bromeliad. Bot Gard Microprop News 1:70–72
10. Vinterhalter B, Vinterhalter D (1994) True-to-the type in vitro propagation of *Aechmea fasciata* Baker. Sci Hortic 57:253–263
11. Daquinta M, Almeida AP, Guerra MP (1998) In vitro morphogenesis of immature flower and buds of flower stalk in *Dyckia distachya*. J Bromeliad Soc 49:72–76
12. Carneiro LA, Cândido MSD, Araujo RFG, Fonseca MHPB, Crocomo OJ, Mansur E (1998) Clonal propagation of *Cryptanthus sinuosos* L.B. Smith, an endemic stoloniferous Bromeliaceae species from Rio de Janeiro, *Brazil.* Plant Tissue Cult Biotechnol 4:153–158
13. George EF (1993) Plant propagation by tissue culture: the technology. Exegetics, Edington
14. Pompelli MF, Guerra MP (2005) Micropropagation enables the mass propagation and conservation of *Dyckia distachya* Hassler.Crop Breeding Apl Biotechnol 5:117–126

15. Hosoki T, Asahira T (1980) In vitro propagation of bromeliads in liquid culture. HortScience 15:603–604
16. Mercier H, Kerbauy GB (1997) Micropropagation of ornamental bromeliads (*Bromeliaceae*). In: Bajaj YPS (ed) Biotechnology in agriculture and forestry. Springer, Berlin, pp 43–57
17. Carneiro LA, Araujo RFG, Brito GJM, Fonseca MHPB, Costa A, Crocomo OJ, Mansur E (1999) In vitro regeneration from leaf explants of *Neoregelia cruenta* (R. Graham) L.B. Smith, an endemic bromeliad from Eastern Brazil. Plant Cell Tissue Organ Cult 55:79–83
18. Alves GM, Guerra MP (2001) Micropropagation for mass propagation and conservation of *Vriesea friburgensis* var. *paludosa* from microbuds. J Bromeliad Soc 51:202–212
19. Pompelli MF, Fernandes D, Guerra MP (2005) Somatic embryogenesis in *Dyckia distachia* Hassler (*Bromeliaceae*) - An endangered bromeliad from South Brazil. Propagation Ornament Plants 5:192–198
20. Alves GM, Dal Vesco LL, Guerra MP (2006) Micropropagation of the Brazilian endemic bromeliad *Vriesea reitzii* trough nodule clusters culture. Sci Hortic 110:204–207
21. Murashige T, Skoog F (1962) A revised medium for rapid growth and biossays with tobacco tissue cultures. Physiol Plant 15:473–497
22. Morel GM, Wetmore RH (1951) Tissue culture of monocotyledons. Am J Bot 38:138–140
23. Feuser S, Nodari RO, Guerra MP (2001) Comparative efficiency of stationary culture and temporary immersion systems for pineapple micropropagation. R Bras Frutic 23:6–10
24. Klein RM (1990) Espécies raras ou ameaçadas de extinção – Estado de Santa Catarina: mirtáceas e bromeliáceas. IBGE, Rio de Janeiro
25. Leme EMV, Costa A (1991) A new species from Southern Brazil, a tribute to Father Raulino Reitz. J. Bromeliad Soc 41:195–198
26. Smith LB, Downs RJ (1974) *Bromeliaceae* (*Pitcairnioideae*). Flora Neotr. New York, Hafner Press
27. Etienne H, Berthouly M (2002) Temporary immersion systems in plant micropropagation. Plant Cell Tissue Organ Cult 69:215–231
28. Dal Vesco LL, Pinto AA, Zaffari GR, Nodari RO, Reis MS, Guerra MP (2001) Improving pineapple micropropagation protocol through explant size and medium composition manipulation. Fruits 56:143–154

Chapter 7

Micropropagation of Poinsettia by Organogenesis

Marcos Castellanos, J. Brian Power, and Michael R. Davey

Abstract

Poinsettia (*Euphorbia pulcherrima*) is one of the most popular ornamental pot plants. Conventional propagation is by cuttings, generally focused on a period prior to the most intensive time of sales. Rapid multiplication of elite clones, the production of pathogen-free plants and more rapid introduction of novel cultivars (cvs.) with desirable traits, represent important driving forces in the poinsettia industry. In recent years, different strategies have been adopted to micropropagate poinsettia, which could assist breeders to meet consumer demands. The development of reliable in vitro regeneration procedures is likely to play a crucial role in future production systems. Stem nodal explants cultured on semi-solid MS-based medium supplemented with benzylaminopurine (BAP) and naphthalene acetic acid (NAA) develop shoots from adventitious/axillary buds after 7 weeks of culture. Rooting of *in vitro* regenerated shoots is achieved in semi-solid MS-based medium containing the auxin indole-3-acetic acid (IAA). Four to six weeks after transfer to root-inducing medium, regenerated plants can be transferred to compost and acclimatized in the glasshouse. Direct shoot regeneration from cultured explants is important to minimize somaclonal variation in regenerated plants.

Key words: *Euphorbia pulcherrima*, Ornamental plants, Shoot regeneration, Stem nodal explants, Tissue culture

1. Introduction

Poinsettia has become the most popular Christmas ornamental plant in the USA, Canada, Australia and in several European countries, including France, Norway and the UK (1, 2). In 2006, poinsettias contributed 170 million dollars in sales to the U.S.A. economy and, in total, many times that amount to the economies of other countries (3). Conventionally, it is mass propagated by stem cuttings, but generally this method is restricted to a short pre-Christmas period, with the rooting of cuttings taking up to 8 weeks (4).

S.M. Jain and S.J. Ochatt (eds.), *Protocols for In Vitro Propagation of Ornamental Plants*, Methods in Molecular Biology, vol. 589
DOI 10.1007/978-1-60327-114-1_7, © Humana Press, a part of Springer Science+Business Media, LLC 2010

Micropropagation through tissue culture permits the regeneration of large numbers of plants from small pieces (explants) of stock plants in a relatively short period and, crucially, without seasonal restrictions (5). In general, the number of publications on different aspects of the culture of poinsettia is limited, with emphasis on micropropagation through indirect shoot regeneration from callus (6–8), meristem-tip culture (9–11) and somatic embryogenesis (12–15), the latter with variable success. Direct shoot regeneration from cultured explants provides an efficient and reproducible in vitro regeneration baseline that can underpin the biotechnological options for plant improvement, such as *Agrobacterium*- and particle bombardment-mediated gene delivery. This gives poinsettia producers the opportunity to develop new varieties in order to increase and sustain the commercial value of this ornamental plant. In this paper, the protocol is described for shoot induction from cultured explants, with subsequent rooting and acclimatization of regenerated plants.

2. Materials

2.1. Surface Sterilization of Source Material

1. Tap water.
2. Ethanol 70% (v:v).
3. Commercial bleach solution (e.g. "Domestos" bleach; Diversey Lever, Northampton, UK; 5% (v/v) NaClO), diluted 1:10 (v:v) with tap water.
4. Autoclaved reverse-osmosis water; 150 mL aliquots in 250 mL screw capped bottles.
5. Autoclaved Whatman No. 1 filter paper.
6. Magnetic stirrer, magnetic bar, 500 mL beaker (autoclaved).
7. Tissue culture facilities – Instruments (scalpel, forceps, spirit burner to flame sterilize instruments), laminar flow bench, culture room.
8. Potted plants as a source of explants.

2.2. Culture Media

1. Media based on the formulation of Murashige and Skoog (MS; (16)) for (a) shoot regeneration from stem nodal explants and (b) root induction on regenerated shoots. Media formulations are given in Table 7.1.
2. Petri dishes (9 cm diameter; Bibby-Sterilin, Stone, UK).
3. Nescofilm (Bando Chemical Ind., Kobe, Japan; distributed by Alfresa Pharma Corp., Osaka, Japan).
4. Screw-capped Duran glass bottles, 500 mL capacity (Schott AG, Mainz, Germany); screw-capped Powder Round glass

Table 7.1
Culture media based on the formulation of Murashige and Skoog (MS; (16))

Component	Concentration (mg/L)
$CaCl_2 \cdot 2H_2O$	440
NH_4NO_3	1,650
KNO_3	1,900
KI	0.835
$CoCl_2 \cdot 6H_2O$	0.025
KH_2PO_4	170
H_3BO_3	6.2
Na_2MoO_4	0.25
$MgSO_4 \cdot 7H_2O$	370
$MnSO_4 \cdot 4H_2O$	22.3
$CuSO_4 \cdot 5H_2O$	0.025
$ZnSO_4 \cdot 7H_2O$	8.6
$FeSO_4 \cdot 7H_2O$	27.8
Na_2EDTA	37.3
Glycine	2
Myo-inositol	100
Nicotinic acid	0.5
Pyridoxine HCl	0.5
Thiamine HCl	0.1
Sucrose	30,000
Agar	8,000
pH	5.8

Add BAP at 1.0 mg/L and NAA at 0.1 mg/L to induce shoots on cultured explants
Add IAA at 4.0 mg/L to induce roots on regenerated shoots

bottles, 175 mL capacity (Beatson Clarke and Co. Ltd., Rotherham, UK).

2.3. Acclimation of Regenerated Plants to Ex Vitro Conditions

1. Tap water.
2. Plastic pots (9 cm diameter).

3. Potting medium consisting of Levington M3 compost (The Scotts Company Ltd., Ipswich, UK), vermiculite (Vermiperl; Silverperl Ltd., Lincoln, UK) and perlite (Silvaperl, Gainsborough, UK) (2: 1: 1 by volume).
4. Transparent plastic bags.

3. Methods

The micropropagation of poinsettia by direct organogenesis involving shoot formation from cultured explants, has the advantage that a callus phase is omitted, significantly reducing the number of steps in the culture procedure and variation in the regenerated plants. This protocol involves three stages. First, the selection of source material and the establishment of axenic cultures of stem nodal explants. In poinsettia, the surface sterilization of nodal explants can be time-consuming because of the presence of milky latex within the stem tissues (*see* Note 1). Second, the culture and maintenance of explants on a shoot regeneration medium, with a high cytokinin to auxin concentration ratio, to promote bud development. Third, the induction of roots on regenerated shoots using a culture medium with a high concentration of auxin. Regenerated plants of poinsettia require a period of acclimatization to ex vitro conditions with a gradual reduction of the humidity to permit the development of stomata and surface waxes, with an associated reduction in transpiration (13, 17, 18).

3.1. Preparation and Sterilization of Culture Media

1. Prepare culture media for (a) shoot regeneration from stem nodal explants and (b) root induction on regenerated shoots according to the formulations given in Table 7.1. The medium for shoot regeneration contains 1.0 mg/L BAP and 0.1 mg/L NAA as growth regulators. The medium to induce roots on regenerated shoots has IAA at 4.0 mg/L.
2. For convenience, use powdered medium sourced from a commercial supplier (e.g. Sigma-Aldrich Co. Ltd., Poole, UK). Check the formulation of the commercial preparation.
3. Add other components, such as sucrose and growth regulators, as required. Add the correct volume of growth regulators from 1.0 mg/mL stock solutions (*see* Note 2) to give the required final concentrations.
4. Adjust the pH of the media to 5.8 using 1 M HCl or 1 M KOH.
5. Add 0.8% (w:v) agar (Type IV; Sigma-Aldrich).

6. Dispense the media into suitable containers, e.g. 200 mL aliquots into 500 mL capacity Duran glass bottles, or 40 mL aliquots into 175 mL Powder Round glass bottles. Use both Duran and Powder Round bottles for the shoot regeneration medium. Medium in Duran bottles will be dispensed into 9 cm diameter Petri dishes for the initial stages of culture of stem nodal explants. Dispense medium for root induction on regenerated shoots into Powder Round bottles.
7. Sterilize the media by autoclaving at 121° C for 20 min (118 Kpa steam pressure).
8. Store the autoclaved media at room temperature in the dark for a maximum of 1 month.

3.2. Plant Source Material and Surface Sterilization

1. Take cuttings from parental stock plants and root the cuttings in the potting mixture described in Subheading 2.3, step 3. Maintain the rooted, potted cuttings under glasshouse conditions and use the rooted cuttings as a source of stem nodal explants. The latter should be taken about 28 days after potting the cuttings (*see* Note 3).
2. Excise stem nodal sections (each 15–20 cm in length) from the rooted potted cuttings and wash the sections for 30 min under cold, running tap water.
3. Surface sterilize the stem sections by immersion in 70% (v:v) ethanol for 1 min, followed by 10% (v:v) "Domestos" bleach solution for 30 min in a sterile beaker on a magnetic stirrer (*see* Note 4). Sterilize the magnetic bar before use by immersion in 70% (v:v) ethanol; allow to dry in the laminar flow bench.
4. Wash the explants three times with sterile reverse-osmosis water for 5, 10 and 30 min respectively, before blotting the explants dry on sterile Whatman No. 1 filter paper.
5. Excise nodal sections (each approx. 4 mm in diameter, 6 mm in length) from surface sterilized material using a sterile scalpel. Any white latex produced by excision can be removed using sterile filter paper (*see* Note 5).

3.3. Culture and Maintenance of Explants

1. Liquify, by heating in a microwave oven, the shoot regeneration medium contained in Duran bottles and dispense 25 mL aliquots into 9 cm diameter Petri dishes (*see* Note 6).
2. Place 5–8 nodal explants on the surface of the medium in each dish. Prepare a minimum of 120 explants per genotype. Seal the dishes with Nescofilm.
3. Incubate the explants at 22 ± 2° C with a 16 h photoperiod (70 ± 3 μmol/m^2/s, White fluorescent light; 18 W/33, Philips TLD, Philips Electronics Ltd., Guildford, UK).

Fig. 7.1. Shoot regeneration from cultured stem nodal explants. (**a**, **b**) Adventitious shoot formation on explants after 4 weeks on MS-based medium containing 1.0 mg/L BAP and 0.1 mg/L NAA. Bars = 1.3 cm (**a**), 0.24 cm (**b**), (**c**) Regenerated shoots, each with 3–4 leaves, after 7 weeks of culture. Bar = 1.3 cm, (**d**) Regenerated shoots transferred to MS-based medium containing 4.0 mg/L IAA to induce root formation. Bar = 2.4 cm, (**e**) A rooted, regenerated shoot after 6 weeks on MS-based medium with 4.0 mg/L IAA. Bar = 2.1 cm, (**f**) Acclimated plant 7 weeks after transfer to compost. Bar = 5.1 cm, (**g**) Mature poinsettia plants regenerated from stem nodal explants. Bar = 17 cm.

4. Transfer the explants to a new medium of the same composition every 14 days to maintain active growth.
5. After 4 weeks, small shoots (each 0.5–1.0 cm in height) should be formed on at least 70% of the cultured explants, with a mean of about five shoots per explant (Fig. 7.1a, b) (*see* Note 7).
6. After about 7 weeks of culture, regenerated shoots should each be approximately 2 cm in height with 3–4 leaves (Fig. 7.1c). During later transfers, subculture the explants to 40 mL aliquots of shoot regeneration medium in 175 mL capacity screw-capped glass jars (*see* Note 8).
7. Evaluate the shoot regeneration capacity of cultured explants (frequency of shoot induction and number of regenerated shoots per explant for each genotype).
8. Excise well developed shoots when 5–6 cm in height and transfer to MS-based medium containing IAA for root induction.

3.4. Root Development on Regenerated Shoots In Vitro

1. Insert the bases of regenerated shoots into 40 mL aliquots of MS-based medium containing 4.0 mg/L IAA in 175 mL screw-capped jars (Fig. 7.1d) (*see* Note 9).

2. Incubate the cultures at 22 ± 2° C with a 16 h photoperiod (70 ± 3 μmol/m²/s; White fluorescent light).
3. Subculture the shoots to new root inducing medium every 2 weeks.
4. After 6 weeks, regenerated shoots will develop root systems consisting of some long roots with smaller lateral roots (Fig. 7.1e). Some callus may also develop at the bases of the shoots prior to root formation.

3.5. Acclimation of Regenerated Shoots to Ex Vitro Conditions

1. Remove the rooted shoots from the culture medium and wash the roots carefully with warm (approx. 35° C) tap water to remove excess agar (*see* Note 10).
2. Transfer plants individually to 9 cm diameter plastic pots each containing a mixture of compost, vermiculite and perlite (2: 1: 1 by volume; *see* Subheading 2.3, step 3) (*see* Note 11).
3. Water the plants thoroughly.
4. Enclose the plants within their pots in transparent plastic bags. Maintain the regenerated plants in a growth room at 20 ± 2° C with a 16 h photoperiod and 300 μmol/m²/s White fluorescent light.
5. After 7 days, remove a corner from the top of each bag to reduce the relative humidity and to facilitate acclimation of the plants to ex vitro conditions.
6. Remove the remaining top corner of the bags after a further 2 weeks; discard the bags after an additional 2 weeks (Fig. 7.1f).
7. Transfer the plants to the glasshouse.
8. After 4–6 weeks in the glasshouse, induce the plants to develop their characteristic pigmented bracts (Fig. 7.1g) (*see* Note 12).

4. Notes

1. Most species of the family *Euphorbiaceae* are characterized by the presence of a milky latex that, in many cases, can be toxic and cause skin irritation. The use of disposable gloves at all times is recommended when handling plant material of poinsettia, in order to avoid skin irritation.
2. Stock solutions of auxins and cytokinins should be prepared separately at a concentration of 1.0 mg/mL. Dissolve the compounds in small volumes (approx. 500 μL/100 mg) of 1.0 M KOH or 1.0 M HCl, respectively. Adjust the volume with reverse-osmosis water. Store the solutions of growth regulators in the dark at 4° C. Stock solutions should be replaced every 2 weeks.

3. Poinsettia stock plants are used as a source of stem cuttings. Take the cuttings, each 5–8 cm in length with 2–3 mature leaves, and insert into compost (Subheading 2.3, step 3) in 9 cm diameter plastic pots. Water every 3–4 days and maintain under glasshouse conditions until required for explants.
4. The establishment of axenic cultures of poinsettia may be difficult because source material is often contaminated by micro-organisms. Surface sterilization by immersion in 70% (v:v) ethanol followed by 10% (v:v) "Domestos" bleach solution, is normally effective in removing contaminating microorganisms.
5. The production of excessive white latex is more common when explants are excised from donor potted plants that are more than 2 months old. This latex affects shoot regeneration from cultured explants significantly. Optimum results are obtained from explants excised from young donor plants approximately 1 month after the cuttings from stock plants have been transferred to compost.
6. Take extreme care when heating bottles containing culture medium in a microwave oven. Loosen the lids on the bottles. Use only thick walled glass bottles of the Duran type in a microwave oven.
7. Shoot regeneration from stem nodal explants of poinsettia is genotype-specific. In general, explants from red-bract cultivars regenerate more shoots than explants from donor plants of other colours, such as cultivars with white bracts. However, the protocol described in this chapter has been effective for inducing shoot regeneration from at least 70% of the explants from all poinsettia cultivars evaluated. Some regenerated buds may fail to develop and these cultures should be discarded.
8. The headspace above the culture medium is greater than in Petri dishes, facilitating growth of regenerated shoots.
9. Indole-3-acetic acid (IAA) is the preferred auxin for inducing roots on regenerated shoots of poinsettia.
10. Prior to the transfer of regenerated plants to compost, it is recommended that the longer roots are reduced to 2–3 cm in length and any dead or broken lateral roots are removed.
11. Generally, regenerated plants of poinsettia can be maintained ex vitro in several soil substitutes, including perlite, vermiculite and sand with added organic material, such as peat, humus and bark (13, 17, 19).
12. Poinsettias are classified as short-day plants, which develop coloured bracts after 8–9 weeks under optimum glasshouse conditions, with a photoperiod of 12–12.25 h (4, 20).

References

1. Jerardo A (2002) Holiday sales look bright for Christmas trees and poinsettias. Agric Outlook 297:14–15
2. Clarke JL, Klemsdal SS, Floistad E, Hvoslef-Eide AK, Haugslien S, Moe R, Blystad DR (2006) Genetic engineering of poinsettia with the aim of enhancing its resistance to poinsettia mosaic virus. Acta Hortic 722:321–325
3. Jerardo A (2007) Floriculture and nursery crops yearbook. Online. Economic Research Service. United States Department of Agriculture Publications. Retrieved from http://www.ers.usda.gov/Publications/Flo/2007/09Sep/Flo2007.pdf. Accessed 4 Feb 2008
4. Hartley DE (1992) Poinsettias. In: Larson RA (ed) Introduction to floriculture. Academic Press, New York, USA, pp 306–331
5. Preil W, Florak P, Wix U, Back A (1988) Towards mass propagation by use of bioreactors. Acta Hortic 226:99–107
6. Pickens KA, Cheng ZM, Trigiano RN (2005) Axillary bud proliferation and organogenesis of *Euphorbia pulcherrima* Winter Rose™. In Vitro Cell Dev Bio-Plant 41:770–774
7. Lin M (2006) Induction and differentiation of callus in *Euphorbia pulcherrima*. J Southwest Univ Sci Technol 21:109–112
8. Pickens KA, Cheng ZM, Kania SA (2006) Effects of colchicines and oryzalin on callus and adventitious shoot formation of *Euphorbia pulcherrima* Winter Rose™. HortScience 41: 1651–1655
9. Paludan N, Begtrup J (1986) Inactivation of poinsettia mosaic virus and poinsettia cryptic virus in *Euphorbia pulcherrima* using heat treated mini-cuttings and meristem-tip culture. Tidsskrift für Planteavl 90:283–290
10. Preil W (1994) In vitro culture of Poinsettia. In: Stromme E (ed) The Scientific Basis of Poinsettia Production. Advisory Service, Agricultural University of Norway, Norway, pp 49–55
11. Beck K, Rasmussen K (1996) *Euphorbia pulcherrima*. Methods to eliminate poinsettia mosaic virus (PnMV) and reinfection by different methods to reveal the "nature" of the branching factor. Acta Hortic 432:176–186
12. Osternack N, Saare-Surminski K, Preil W, Lieberei R (1999) Induction of somatic embryos, adventitious shoots and roots in hypocotyl tissue of *Euphorbia pulcherrima* Willd. ex. Klotzsch: comparative studies on embryogenic and organogenic competence. J Appl Bot 73:197–201
13. Jasrai YT, Thaker KN, D'Souza C (2003) In vitro propagation of *Euphorbia pulcherrima* Willd. through somatic embryogenesis. Plant Tissue Cult 13:31–36
14. Castellanos M, Power JB, Davey MR (2006) Somatic embryogenesis in red- and white-bract cultivars of poinsettia. Prop Ornamental Plants 6:61–66
15. Ying X, Gang W (2006) Studies on high quality somatic embryogenesis applied to artificial seed of *Euphorbia pulcherrima* Willd. Acta Hortic Sin 33:175–178
16. Murashige T, Skoog F (1962) A revised medium for the rapid growth and bioassays with tobacco tissue cultures. Physiol Plant 15:473–496
17. Geier T, Beck A, Preil W (1992) High uniformity of plants regenerated from cytogenetically variable embryogenic suspension cultures of poinsettia (*Euphorbia pulcherrima* Willd. ex. Klotzsh). Plant Cell Rep 11:150–154
18. Wilkerson D, Barnes L, Drees B, Hall C (1999) The Texas poinsettia producers guide. Online. Texas Agricultural Extension Service, Texas A&M. Retrieved from http://aggie-horticulture.tamu.edu/greenhouse/nursery/guides/poinsettia/. Accessed 30 Jan 2008
19. Roy SK, Jinnah M (2001) In vitro micropropagation of poinsettia (*Euphorbia pulcherrima* Willd.). Plant Tissue Cult 11:133–140
20. Kristoffersen T (1994) Early Norwegian studies of growth and development in poinsettia. In: Stromme E (ed) The Scientific Basis of Poinsettia Production. Advisory Service, Agricultural University of Norway, Norway, pp 15–28

Chapter 8

Micropropagation of *Phalaenopsis* Blume

Pinaki Sinha, M. Firoz Alam, and M. Lokman Hakim

Abstract

For high frequency regeneration of *Phalaenopsis*, young leaf segments are cultured on gelrite-gelled ½ MS medium supplemented with 2% sucrose, 2.0 mg/L BA, 0.5 mg/L NAA, 10% coconut water (CW), 2 g/L peptone and 1 g/L activated charcoal. Cultures are incubated at 24 ± 2° C under fluorescence light 50 μmol/m^2/s for 16 h photoperiod per day. The PLBs (protocorm like bodies) are induced within 12 weeks of culture and are subcultured to proliferate on the fresh nutrient culture medium for 8 weeks. The PLBs clumps are dissected and cultured on ½ MS medium containing 2% sucrose, 10% coconut water (CW), 2 g/L peptone, 150 mg/L l-glutamine and 1 g/L activated charcoal. The PLB sections elongate to form shoots and new PLBs are induced from the base within 8 weeks of culture. For plantlets formation the shoots are cultured on ½ MS medium supplemented with 2% sucrose, 10% CW, 2 g/L peptone, 1 g/L activated charcoal and 50 g/L banana pulp. The regenerated plants are acclimatized and cultivated in the nursery, where they bloom within 2 years.

Key words: Phalaenopsis, Leaf segment, In vitro culture, Clonal propagation, Acclimatization, Blossom

1. Introduction

Phalaenopsis is known as the "Moth Orchid" and is very popular to the hobbyists and commercial growers. *Phalaenopsis* orchids are found throughout the tropical rainforests of South and South-East Asia, Australia and New Guinea. They are monopodial epiphytic plants, and consist of only a few leathery leaves. It is certainly the most important commercial genus of orchids grown. Millions of plants are annually sold worldwide and has created a huge trade for the pot flowering plant. The demand for *Phalaenopsis* species and hybrids is gradually increasing. It has been estimated that in 2004 more than 31 million plants of *Phalaenopsis* species or hybrids were produced in German in vitro laboratories alone (1).

S.M. Jain and S.J. Ochatt (eds.), *Protocols for In Vitro Propagation of Ornamental Plants*, Methods in Molecular Biology, vol. 589
DOI 10.1007/978-1-60327-114-1_8, © Humana Press, a part of Springer Science+Business Media, LLC 2010

Phalaenopsis is a monopodial genus and as such does not normally form offshoots. Sometimes under unfavorable culture conditions, they produce vegetative offsets, called keikis (from the Hawaiian word for "little child"), from the stem or inflorescence axis. So, large scale natural clonal propagation is not possible in this genus. Therefore in vitro clone propagation is the only method for high frequency regeneration of this plant. Several methods for direct micropropagation of *Phalaenopsis* were developed (**2–5**). However, all of these methods can not be used for commercial micropropagation because of differences in survival rate, PLB induction and plantlet formation. Though the plant has a high market value as cut flower and pot plant throughout the world, its high frequency regeneration protocol is yet to be determined (**6**). Ishii et al. (**7**) and Tokuhara and Mii (**6, 8**) established a highly efficient system of micropropagation of *Phalaenopsis* through embryogenic callus induction and cell suspension culture, but callus mediated plant regeneration does not ensure plant homogeneity. Park et al. (**9**) reported in vitro propagation method for *Phalaenopsis* by using leaf explants derived in vitro from flower stalk node but the method was highly tedious since the method was accomplished in two steps: in the first step, leaves were induced from flower stalk node explant and in the second step, PLBs were induced from this leaf explant and the individual PLBs were cut for further regeneration of PLBs. Kuo et al. (**10**) described a protocol for regenerating a *Phalaenopsis* cultivar through direct somatic embryogenesis. However, the method is not so efficient for high frequency regeneration of different cvs of *Phalaenopsis* hybrid. Moreover, the method is not feasible for commercial propagation, which warrants development and establishment of efficient protocol for mass clonal propagation and consequently commercial exploitation of these lucrative ornamental plants.

This chapter describes an efficient and superior protocol for high frequency plant regeneration of *Phalaenopsis* and their establishment in the garden up to flowering.

2. Materials

2.1. Plant Materials

1. Potted plants as a source of explants.

2.2. Surface Sterilization of Plant Materials

1. Tap water.
2. Ethanol 70% (v/v).
3. Tween 80 5% (v/v).
4. Mercuric chloride 2% (w/v).

5. Autoclaved water 150 mL aliquots in 250 mL cotton plugged conical flask.
6. Magnetic stirrer, magnetic bar, 500 mL beaker (autoclaved).
7. Tissue culture facilities – laminar air flow bench, instruments, culture room.
8. White fluorescent light, (Philips Electronics Ltd., Dhaka, Bangladesh)

2.3. Culture Media

Half strength macro- and micro salts of MS (**11**) medium with different vitamins and hormones as detailed below:

1. *Macronutrients:* 220 mg/L $CaCl_2{\cdot}2H_2O$, 13.9 mg/L $FeSO_4{\cdot}H_2O$, 950 mg/L KNO_3, 85 mg/L KH_2PO_4, 185 mg/L $MgSO_4{\cdot}7H_2O$, 825 mg/L NH_4NO_3, 18.65 mg/L Na_2.EDTA. (To prepare stock solution, tenfolds of each compound are dissolved in 1 L distilled water and stored at 4° C. One hundred mL stock solution is used to prepare a total volume of 1 L medium (*see* Note 1).
2. *Micronutrients:* 0.0125 mg/L $CoCl_2{\cdot}6H_2O$, 0.0125 mg/L $CuSO_4{\cdot}5H_2O$, 3.1 mg/L H_3BO_3, 0.415 mg/L KI, 22.15 mg/L $MnSO_4$, 0.125 mg/L $Na_2MoO_4{\cdot}2H_2O$, 8.3 mg/L Zn SO_4. (To prepare stock solution, 1,000-folds of each compound are dissolved in 1 L distilled water and stored at 4° C. Use 1 mL stock solution to prepare total volume of 1 L medium.).
3. *Vitamin:* 0.5 mg/L nicotinic acid, 0.5 mg/L pyridoxine HCl, 0.1 mg/L thiamine HCl, 2.0 mg/L glycine, and 100 mg/L myoinositol (To prepare stock solution 100-folds of each compound are dissolved in 100 mL distilled water and stored at –20° C up to 4 months (*see* Note 2). One milliliter stock solution is used to prepare 1 L medium.).
4. *Organic supplements*: Coconut water – 10%, it is taken from green fruit and stored at –20° C (*see* Note 2). Banana pulp – 50 g/L, ripe banana is peeled off, smashed in a mortar with pestle and used in medium preparation. 2 g/L peptone and 1 g/L activated charcoal are directly added in the medium during preparation.
5. Sucrose (20 g/L) is added directly to the medium. The pH of the medium is adjusted to 5.6 with 1 N NaOH and 2.2 g/L gelrite is added, and the medium is autoclaved at 121° C for 20 min.

2.4. Acclimation of Regenerated Plants to Ex Vitro Conditions

1. Tap water.
2. Plastic basket (25 cm diameter).
3. Plastic pots (10 cm diameter).

4. Charcoal chips – Wood-charcoal is prepared by burning tree-wood and 10 mm^3 chips are prepared.
5. Coconut husk – Husk of the ripe coconut fruit is extracted and 20 mm^3 sized husks are prepared by cutting.
6. Fertilizer – 30N-10P-10K and 20N-20P-20K mixtures are collected from PROSHIKA, Bangladesh. 3 g mixture is dissolved in 1 L water.

3. Methods

3.1. Preparation and Sterilization of Culture Media

1. Prepare culture media for (a) PLBs induction from leaf explants, (b) PLBs proliferation and development of shoots and (c) root induction in regenerated shoots according to the formulation given in Table 8.1. The medium for shoot regeneration contains 2.0 mg/L BA and 0.5 mg/L NAA as growth regulators and 10% coconut water and 1 g/L activated charcoal.
2. Adjust the pH of the media to 5.6 using 1 N HCl or 1 N NaOH.
3. Add 2.2 g/L gelrite (Dachifa) and boil the medium for melting the gelrite.
4. Dispense the medium into suitable glass vessels, e.g. 60 mL aliquots into 250 mL capacity conical flasks or 250 mL round glass bottles.
5. Sterilize the media by autoclaving at 121° C for 20 min.
6. Store the autoclaved media at room temperature in the dark for a maximum of 2 weeks.

3.2. Surface Sterilization and Preparation of Explant

1. Take young leaf segments from nursery grown mature plants.
2. Wash them under running tap water followed by 5% (v/v) Tween 80 for 5 min.
3. Take the explants then in a laminar air flow cabinet and transfer to a sterile conical flask.
4. Wash the explants in double-distilled water, rinse in 70% ethanol for 30 s. and surface sterilize them by rinsing in 0.2% (w/v) mercuric chloride solution for 5 min.
5. Wash the explants thoroughly with sterile double-distilled water 3–4 times.
6. Prepare the surface sterilized leaf explants for inoculation. Cut the explants into small pieces (1.5 cm^2).

Table 8.1
Composition of media used for the different stages of plantlets formation from leaf segment explants of *Phalaenopsis*

Compounds[a] (mg/L)	Medium 1 PLBs induction (mg/L)	Medium 2 PLBs proliferation and development of shoots (mg/L)	Medium 3 Formation of plantlets (mg/L)
$CaCl_2{\bullet}2H_2O$	220	220	220
$FeSO_4{\bullet}2H_2O$	13.9	13.9	13.9
KNO_3	950	950	950
KH_2PO4	85	85	85
$MgSO_4{\bullet}7H_2O$	185	185	185
NH_4NO_3	825	825	825
Na_2 EDTA	18.65	18.65	18.65
$CoCl_2{\bullet}6H_2O$	0.0125	0.0125	0.0125
$CuSO_4{\bullet}5H_2O$	0.0125	0.0125	0.0125
H_2BO_3	3.1	3.1	3.1
KI	0.415	0.415	0.415
$MnSO_4$	11.15	11.15	11.15
$Na_2MoO_4{\bullet}2H_2O$	0.125	0.125	0.125
$ZnSO_4$	4.3	4.3	4.3
Myo-inisitol	100	100	100
Nicotinic acid	0.5	0.5	0.5
Pyridoxine HCl	0.5	0.5	0.5
Thiamine HCl	0.1	0.1	0.1
Glycine	2.0	2.0	2.0
Sucrose	20,000	20,000	20,000
Peptone	2,000	2,000	2,000
BA	2.0	–	
NAA	0.5	–	–
Coconut water	10%	10%	10%
L-glutamine	–	150	–
Activated charcoal	1,000	1,000	1,000
Banana pulp	–	–	50,000
pH	5.6	5.6	5.6

[a]All macro- and micro elements are the ingredients of ½ MS medium

3.3. Explants Inoculation and PLBs Induction

1. Inoculate the prepared explants (leaf segments) aseptically on the surface of the prepared culture medium 1 (Table 8.1) containing 2% sucrose, 2.0 mg/L BA, 0.5 mg/L NAA, 10% coconut water (CW), 2 g/L peptone, 1 g/L activated charcoal.
2. Incubate cultures at 24 ± 2° C under white fluorescent light, 50 μmol/m²/s, 16 h photoperiod per day.
3. Subculture the explants to freshly prepared medium of the same composition every 4-week to maintain active growth.
4. At the end of second subculture, 12-week after inoculation, small PLBs should be formed on at least 80% of the cultured explants, with 150–200 PLBs per culture. The PLBs are globular in shape.

3.4. Proliferation of PLBs and Their Development Into Shoots

1. Subculture the initially induced PLB clumps in the same nutrient medium, as mentioned in **Subheading 3.3**, step 1, for further 8 weeks. During this subculture, the number of PLBs should be increased up to an average of 600 (Fig. 8.1a).
2. Dissect the PLB clumps longitudinally so that each section contains an average of 150 PLBs.
3. Transfer each section of PLB clump on gelrite-gelled ½ MS medium supplemented with 2% sucrose + 10% CW + 2 g/L peptone + 1.0 g/L AC + 150 mg/L l-glutamine (Table 8.1, medium 2). After 8 weeks of culture in this medium, the PLBs should be developed into shoots and elongated properly. Moreover, a large number of secondary PLBs should be induced from the base of the developing shoots (Fig. 8.1b, c).

3.5. Formation of Plantlets

1. Subculture the well developed shoots (4–6 cm long) on gelrite-gelled ½ MS medium supplemented with 2% sucrose + 10% CW + 2 g/L peptone + 1 g/L AC + 50 g/L banana pulp (Table 8.1, medium 3).
2. Incubate the cultures at 24 ± 2° C with a 16 h photoperiod, 50 μmol/m²/s; white fluorescent light. In 8 weeks, 4–6 roots should be induced from the base of the shoots (Fig. 8.1d).
3. Within the first 40 weeks after initiation of culture, an average of 960 plantlets as well as huge amount of PLBs are produced from a single explant of leaf segment.
4. Through repeated subculture of shoots on rooting medium and PLBs on proliferation medium regeneration of plantlets may be continued, and on an average 2,000,000 plantlets could be produced every 32 weeks.

3.6. Acclimatization and Potting of the Regenerated Plants for Establishment in the Nursery

1. Transfer the culture vessels with rooted plantlets from the growth room to the open room and keep there for seven days. When the culture vessels with plantlets are kept in normal room temperature (30 ± 2° C) and indirect day-light, the plantlets should be acclimatized to some extent.

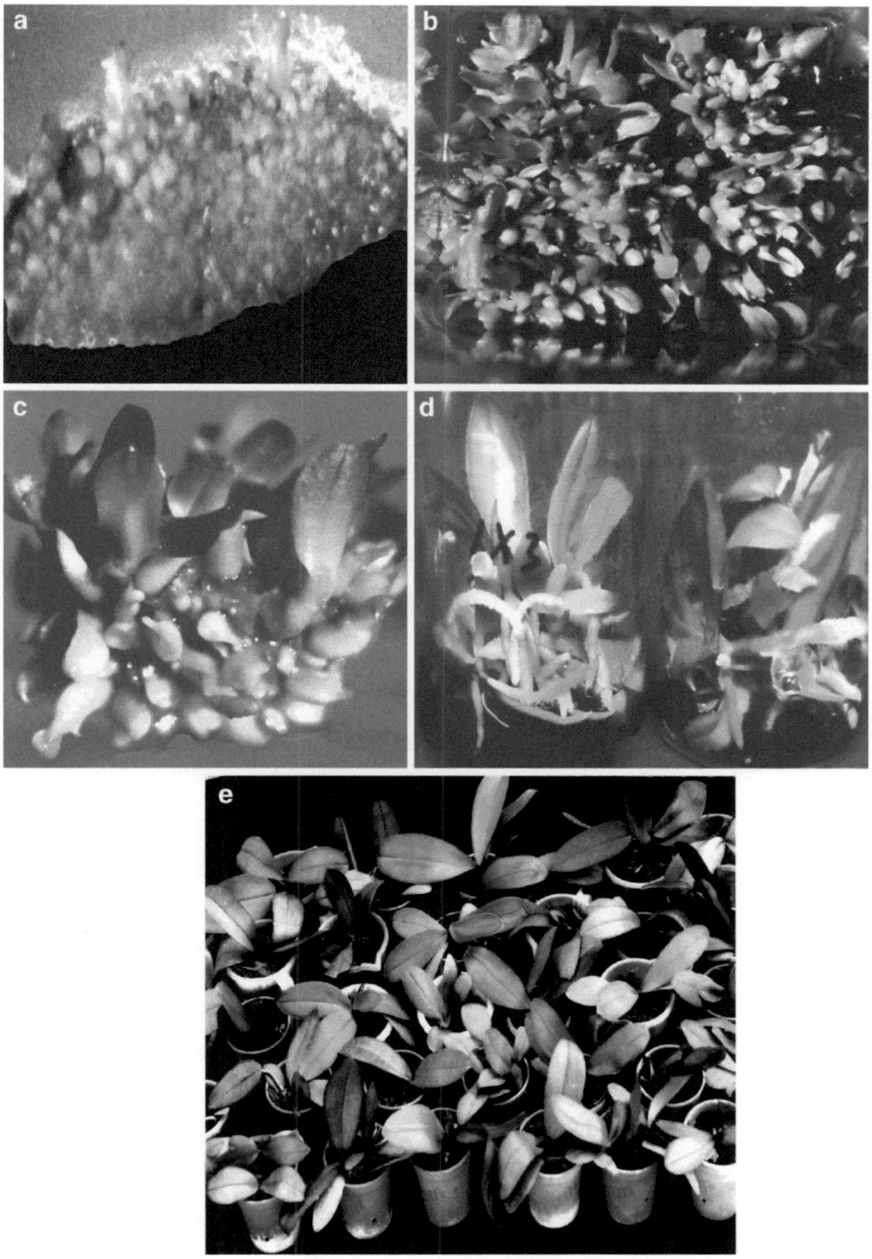

Fig. 8.1. (**a–e**) Regeneration of *Phalaenopsis* through in vitro culture of leaf segment. (**a**) Induction of PLBs from explant cultured on ½ MS + 2% sucrose + 2.0 mg/L BA + 0.5 mg/L NAA + 10% CW + 2 g/L peptone + 1 g/L AC. (**b**) Development of shoots from the old PLBs and proliferation of new PLBs from the base of the old ones, cultured on ½ MS + 2% sucrose + 150 mg/L l-glutamine + 10% CW + 2 g/L peptone + 1 g/L AC. (**c**) A segment of (**b**) showing proliferation of new PLBs. (**d**) Eight-week-culture of regenerated shoots with induction of roots as well as shoot growth on ½ MS + 2% sucrose + 10% CW + 2 g/L peptone + 1 g/L AC + 50 g/L banana pulp. (**e**) One-year-old regenerated plants.

2. Remove the rooted shoots from the culture medium and wash the roots carefully with warm tap (approx. 35° C) water to remove the gel adhered to the roots.
3. Transfer the plantlets to 25 cm diameter plastic basket (50 plantlets in a basket) containing coconut husk (20 mm^3) and charcoal chips (10 mm^3) (2:1) (*see* Note 3).
4. Keep the baskets containing the plantlets in indirect day light and mist the plantlets of 6-h intervals for 30 days, in which 82.3–89.6% plants survive (*see* Note 4).
5. Transfer the plants individually to 10 cm diameter plastic pots containing coconut husk (20 mm^3) and charcoal chips (10 mm^3) (2:1).
6. Rear the plantlets up to 12 months in plastic pots (**Fig. 8.1e**) by spraying of 3 g/L 30N-10P-10K solution at 10-day intervals.
7. After 12 months, fertilize the plants by spraying with 3 g/L 20N-20P-20K solution at 10-day intervals with watering every alternate day. The regenerated 2-year-old plants should start flowering.

4. Notes

1. Half strength MS macro- and micro salts stock solutions are stored at 10°C up to 4 months.
2. Auxins, cytokinins and vitamins stock solution are stored at –20°C and thus ensures the solution is completely thawed before taking any aliquots.
3. The potting mix (coconut husk and charcoal chips) should be autoclaved, to kill bacteria and fungi, before use.
4. Plants should be kept in well aerated orchid house in not too much sunlight. The most important factor to remember is that the leaves of these plants will burn easily if exposed to too much sunlight.

References

1. Winkelmann T, Thomas G, Priel W (2006) Commercoal in vitro plant production in Germany in 1985–2004. Plant Cell Tissue Organ Cult 85:319–327
2. Tanaka M, Sakanishi Y (1977) Clonal propagation of *Phalaenopsis* by leaf culture. Am Orchid Soc Bull 46:733–737
3. Tanaka M (1992) Micropropagation of *Phalaenopsis* spp. In: Bajaj YPS (ed) Biotechnology in agriculture and forestry, vol 20, High tech and micropropagation IV. Springer, Berlin, Heidelberg, New York, pp 246–268
4. Tokuhara K, Mii M (1993) Micropropagation of *Phalaenopsis* and *Doritaenopsis* by culturing shoot tips of flower stalk buds. Plant Cell Rep 13:7–11
5. Arditti J, Ernst R (1993) Micropropagation of orchids. Wiley, New York 649 pp

6. Tokuhara K, Mii M (2001) Induction of embryogenic callus and cell suspension culture from shoot tips excised from flower stalk buds of *Phalaenopsis* (Orchidaceae). In Vitro Cell Dev Biol Plant 37:457–461
7. Ishii Y, Takamura T, Goi M, Tanaka M (1998) Callus induction and somatic embryogenesis of *Phalaenopsis*. Plant Cell Rep 17:446–450
8. Tokuhara K, Mii M (2003) High-frequency somatic embryogenesis from cell suspension cultures of *Phalaenopsis* orchids by adjusting carbohydrate source. In Vitro Cell Dev Biol Plant 39:635–639
9. Park SY, Murthy HN, Paek KY (2002) Rapid propagation of *Phalaenopsis* from floral stalk-derived leaves. In Vitro Cell Dev Biol Plant 38:168–172
10. Kuo HL, Chen JT, Chang WC (2005) Efficient plant regeneration through direct somatic embryogenesis from leaf explants of *Phalaenopsis* 'Little steve'. In Vitro Cell Dev Biol Plant 41:453–456
11. Murashige T, Skoog F (1962) A revised medium for rapid growth and bioassaya with tobacco tissue cultures. Physiol Plant 15: 473–497

Chapter 9

Genetic Transformation of Carnation (*Dianthus caryophylus* L.)

Chalermsri Nontaswatsri and Seiichi Fukai

Abstract

This chapter describes a rapid and efficient protocol for explant preparation and genetic transformation of carnation. Node explants from greenhouse-grown plants and leaf explants from in vitro plants are infected with *Agrobacterium tumefaciens* AGL0 harboring pKT3 plasmid, consisting of GUS and NPTII genes. Explant preparation is an important factor to obtain the transformed plants. The GUS-staining area was located only on the cut end of explants and only explants with a cut end close to the connecting area between node and leaf, produced transformed shoots. The cocultivation medium is also an important factor for the successful genetic transformation of carnation node and leaf explants. High genetic transformation efficiency of node and leaf explants cocultured with *Agrobacterium tumefaciens* was achieved when the explants were cocultivated on a filter paper soaked with water or water and acetosyringone mixture (AS).

Key words: Carnation (*Dianthus caryophylus L.*), *Agrobacterium*, Genetic transformation, Adventitious shoot regeneration, Node explants, Leaf explant, Gus (β-glucuronidase), nptII (neomycin phosphotransferase), AS (acetosyringone)

1. Introduction

Carnation is highly valuable for the cut flower industry and the demand for new and improved varieties is growing. The development of an effective *Agrobacterium*-mediated transformation protocol will allow the introduced novel characteristic to carnation. Several protocols for genetic transformation of carnation have been reported. However, genetic transformation of carnations is currently limited because of a lack of efficient routine technique. One of the most serious problems with genetic transformation of carnation is low-transformation efficiency. Many factors affect successful genetic transformation, such as the strain of *Agrobacterium* used, plant condition, inoculation methods, and

S.M. Jain and S.J. Ochatt (eds.), *Protocols for In Vitro Propagation of Ornamental Plants*, Methods in Molecular Biology, vol. 589
DOI 10.1007/978-1-60327-114-1_9, © Humana Press, a part of Springer Science+Business Media, LLC 2010

cocultivation methods. Carnation explants derived from greenhouse plants were more susceptible to *Agrobacterium* than in vitro explants (1). Many reports indicate the importance of explant preculture by phytohormones for successful genetic transformation (2–4). Cocultivation is one of the most important steps for the genetic transformation of plants. Some researchers successfully transfer genes to explants cocultivated with a medium, rich with hormones, such as *Cyclamen persicum* Mill. (5) and Tomato (6). Some have reported successful genetic transformation by reducing the concentration of the cocultivation medium to half that of the regeneration medium (e.g., for *Drosera rotundifolia* L. (7)). Successful transformation of Eustoma has been achieved by using a low-nutrient cocultivation medium (8). Studies have also indicated that acetosyringone (AS) is an important substance, which affects the success of genetic transformation (9). AS activates the virulence genes of *Agrobacterium* and enhances the transfer of foreign genes into plant genomes (9). Pretreatment of explants with AS in the S phase of the cell cycle could stimulate bacterial attachment to cell walls (10). Many researchers are interested in the attachment behavior of *Agrobacterium* to plant cells and how this affects the success of genetic transformation. Polar orientation attachments of single bacteria were found at the beginning of attachment (11, 12). Bacteria then aggregate to form massive clusters with a cellulose fibrillar net (11, 13). These fibrils cause the bacteria to bind tightly to the surface of the plant cells (13). Nevertheless, it is still not clear which type of attachment is essential for T-DNA transfer. Sometimes, the polar orientation attachment of bacteria influences the genetic transformation ability of *Agrobacterium*. The attachment of bacteria to plant cells is essential for the transfer of T-DNA and the investigation show that bacteria attach to plant cells in a polar manner (14).

This chapter describes an easy and effective protocol for adventitious shoot regeneration and gene transfer to carnation node and leaf explants. Both explants preparation and cocultivation medium are important to increase transformation efficiency of carnation. We have successfully obtained transgenic carnation plants from five genotypes (Killer, Laurella, Master, Otome, and Tanga). The transformation efficiencies vary from 4 to 95% (β-glucoronidase (GUS) positive shoot/100 node explants infected) depending on genotypes (15, 16).

2. Materials

2.1. Node Explants

5- to 8-cm long carnation shoots, taken from the greenhouse mother carnation plants, maintained at a minimum temperature of 15°C under natural day length.

2.2. Node Explants Culture Media and Stock Solutions

1. *Thidiazuron (TDZ) stock (100 mg/L).* Dissolve 10 mg TDZ in a few drops of 1 N NaOH and make up the volume to 100 mL with double distilled water (ddH$_2$O) and store at –20°C.
2. α-*napthalene acetic acid (NAA) stock (100 mg/L).* Dissolve 10 mg NAA in a few drops of 1 N NaOH and make up the volume to 100 mL with ddH$_2$O and store at –20°C.
3. *N6-benzyladenine (BA) stock (100 mg/L).* Dissolve 10 mg BA in a few drops of 1N NaOH and make up the volume to 100 mL with ddH$_2$O and store at –20°C.
4. *Medium I.* Murashige and Skoog (MS) medium (NH_4NO_3, 1,650 mg/L, KNO_3, 1,900 mg/L, $MgSO_4{\cdot}7H_2O$, 370 mg/L, $CaCl_2{\cdot}2H_2O$, 440 mg/L, KH_2PO_4, 170 mg/L, H_3BO_3, 6.2 mg/L, $MnSO_4{\cdot}4H_2O$, 22.3 mg/L, $ZnSO_4{\cdot}7H_2O$, 8.6 mg/L, $CoCl_2{\cdot}6H_2O$, 0.025 mg/L, $CuSO_4{\cdot}5H_2O$, 0.025 mg/L, KI, 0.83 mg/L, $Na_2MoO4{\cdot}2H_2O$, 0.25 mg/L, $FeSO_4{\cdot}7H_2O$, 27.8 mg/L, $Na_2EDTA{\cdot}2H_2O$, 37.3 mg/L, *myo*-inositol, 100 mg/L, thiamine HCl, 0.1 mg/L, pyridoxine HCl, 0.5 mg/L, nicotinic acid, 0.5 mg/L, and glycine, 2.0 mg/L) containing 1 mg/L TDZ , 0.1 mg/L NAA, and 20 g/L sucrose. Adjust pH to 5.75 using NaOH and add 2 g/L Gellan gum (Wako (cat.no. 075-03075)), and autoclave at 121°C for 15 min.
5. *Medium II.* MS medium containing 1 mg/L BA, 0.1 mg/L NAA, and 20 g/L sucrose. Adjust pH to 5.75 using NaOH and add 2 g/L Gellan gum, and autoclave at 121°C for 15 min.
6. *Hyponex medium.* 3 g/L Hyponex fertilizer (6.5:6:19) (Hyponex Co.Ltd.), 20 g/L sucrose, and 2 g/L Gellan gum. Autoclave at 121°C for 15 min.

2.3. Leaf Explants

First to sixth leaf pairs taken from in vitro mother plants, incubated at 25°C, 16 h photoperiod, 36 μmol/m^2/s provided by cool white fluorescent lamps for 4 weeks.

2.4. In Vitro Mother Plants Culture Media and Stock Solutions

1. *N6-benzyladenine (BA) stock (100 mg/L).* Dissolve 10 mg BA in a few drops of 1N NaOH and make up the volume to 100 mL with ddH$_2$O and store at –20°C.
2. α-*napthalene acetic acid (NAA) stock (100 mg/L).* Dissolve 10 mg NAA in a few drops of 1N NaOH and make up the volume to 100 mL with ddH$_2$O and store at –20°C.
3. *Mother Plants Culture Medium.* MS medium containing 1 mg/L BA, 0.1 mg/L NAA, and 20 g/L sucrose. Adjust pH to 5.75 using NaOH and add 2 g/L Gellan gum, and autoclave at 121°C for 15 min.

2.5. Leaf Explants Culture Media and Stock Solutions

1. *Indole-3-butyric acid (IBA) stock (100 mg/L).* Dissolve 10 mg BA in a few drops of 1N NaOH and make up the volume to 100 mL with ddH_2O and store at –20°C.
2. *Thidiazuron (TDZ) stock (100 mg/L).* Dissolve 10 mg TDZ in a few drops of 1N NaOH and make up the volume to 100 mL with double distilled water (ddH_2O) and store at –20°C.
3. *Leaf Explants Culture Medium.* MS medium containing 0.5 mg/L IBA, 0.22 mg/L TDZ, and 20 g/L sucrose. Adjust pH to 5.75 using NaOH and add 2 g/L Gellan gum, and autoclave at 121°C for 15 min.

2.6. Agrobacterium Strain and Plasmid Vectors

Agrobacterium tumefaciens, strain AGL0, having a binary plasmid pKT3 (10). This construct contains a screenable marker gene cassette with the nopaline synthase promoter (Pnos) driving the β-glucuronidase (GUS) gene. The gus gene has an intron derived from caster bean catalase. The selectable marker gene cassette is the CaMV 35S promoter driving the neomycin phosphotransferase (nptII) gene.

2.7. Agrobacterium Culture Medium and Cocultivation Medium

1. *Solid Luria Bertani (LB) medium.* 10 g/L tryptone, 5 g/L yeast eXtract, 10 g/L sodium chloride, and 8 g/L agar, pH 7.0, Autoclave and cool to between 50 and 60 °C, then supplemented with 50 mg/L kanamycin and 50 mg/L rifampicin.
2. *Liquid LB medium.* 10 g/L tryptone, 5 g/L yeast extract, and 10 g/L sodium chloride, pH 7.0. Autoclave and store at room temperature. Before using, supplement with 50 mg/L filter-sterilized kanamycin and 50 mg/L filter-sterilized rifampicin.
3. *Cocultivation medium.* 50 μM AS in sterilized water.

2.8. Antibiotics and Other Stock

All antibiotics are dissolved either in ddH_2O or dimethyl sulfoxide (DMSO), filter-sterilized, and stored in aliquots in –20°C. They should be added to various media after being autoclaved and cooled down to around 55°C.

1. *Kanamycin stock (50 mg/mL).* Dissolve 500 mg kanamycin (Wako) in 10 mL ddH_2O.
2. *G418 (geneticin) stock (50 mg/mL).* Dissolve 500 mg G418 (Wako) in 10 mL ddH_2O.
3. *Vancomycin stock (200 mg/mL).* Dissolve 1 g vancomycin (Wako) in 5 mL ddH2O.
4. *Cefotaxin stock (250 mg/mL).* Dissolve 1 g cefotaxin in 4 mL ddH_2O.
5. *Rifampicin stock (50 mg/mL).* Dissolve in DMSO.
6. *AS stock (19.62 mg/mL).* Dissolve 196.2 mg AS in 1 mL ethanol and make up the volume to 10 mL with ddH_2O

2.9. Selection Medium

2.9.1. Node Explant

1. *MS medium for selection.* Medium II plus 20 mg/L G418, 250 mg/L cefotaxin, and 400 mg/L vancomycin.
2. *Hyponex medium for selection.* Hyponex medium plus 60 mg/L G418, 250 mg/L cefotaxin, and 400 mg/L vancomycin.

2.9.2. Leaf Explant

1. *Leaf explants culture medium for selection.* Leaf explants culture medium plus 20 mg/L G418, 250 mg/L cefotaxin, and 400 mg/L vancomycin.
2. *Hyponex medium for selection.* Hyponex medium plus 60 mg/L G418, 250 mg/L cefotaxin, and 400 mg/L vancomycin.

3. Methods

3.1. Node Explants Culture (See Fig. 9.1)

All in vitro cultures are incubated at 25°C under a 16-h photoperiod, 36 µmol/m^2/s provided by cool white fluorescent lamps.

1. Shoots, 5–8 cm long, are harvested from mother plant growing in green house.
2. Shoots are surface sterilized for 15 min in sodium hypochlorite (1% available chloride) and one drop of Tween-20 and then rinsed twice for 10 min each in sterilized water.
3. Leaves are removed carefully from the nodes (*see* Note 1).
4. The first three nodes below the shoot meristem are cut into 1-mm thick explants, each containing a node and leaf scar (*see* Fig. 9.1).
5. Explants are cultured on Medium I for 10 days.
6. Node explants are cut into eight radial segments and subcultured onto Medium II for 4 weeks.
7. The regenerated shoots are subcultured onto Hyponex medium.

3.2. Leaf Explants Culture (See Fig. 9.2)

1. Six pairs of expanded leaves are removed carefully from the top of the mother plant stem .
2. The leaf explants are cut into 3 mm long, in each explant containing a connecting area between the leaf base and the stem (*see* Note 2).
3. Explants are cultured on leaf explants culture medium.
4. The regenerated shoots are subcultured onto Hyponex medium.

3.3. Explant and Agrobacterium Culture Preparation

1. For genetic transformation, prepare carnation node explants as in Subheading 3.1, steps 1–4 but preculture them on Medium I for 3–5 days. In leaf explants, prepare leaf explants as in subheading 3.2, steps 1 and 2 (preculturing not required).

Fig. 9.1. (**a**) Longitudinal section of carnation shoot (**b**) Shoot preparation and position of nodes suitable for culture. (**c**) 1-mm thick node explant (**d**) GUS staining area in connecting area between leaf and node (**e**) GUS positive shoot primordia regenerated from connecting area between leaf and node.

2. A frozen *Agrobacterium* culture is inoculated into liquid LB medium and incubated overnight at 27°C on a rotary shaker (120 rpm).
3. Solid LB medium is streaked with bacteria suspension and incubated at 27°C for 1–2 days. These bacteria can be kept for 1 month at 4°C.

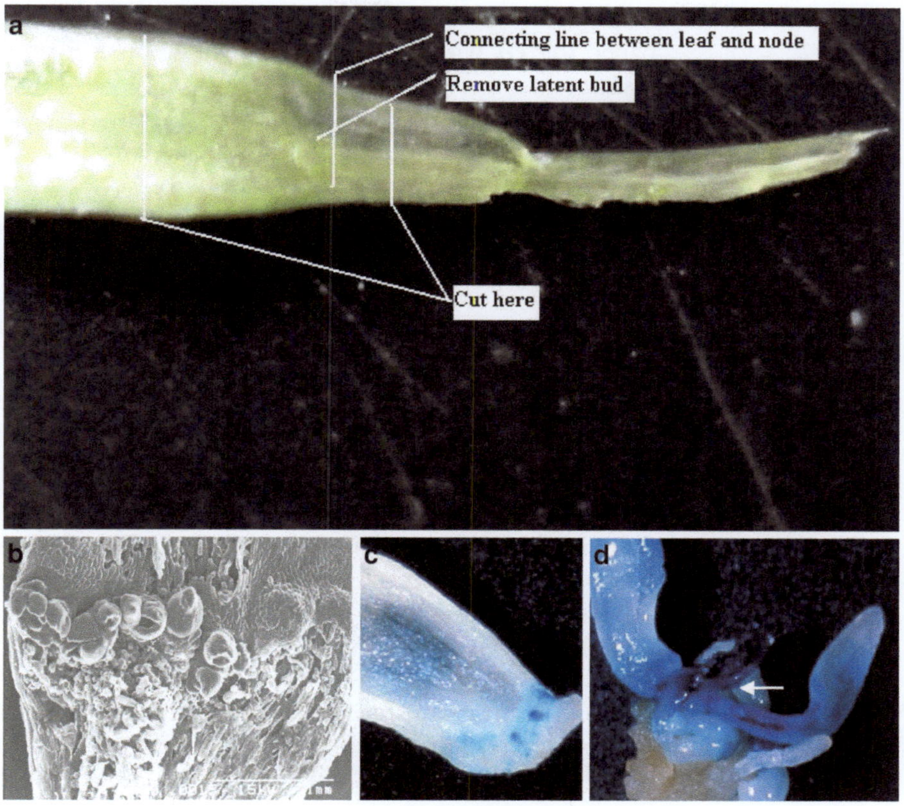

Fig. 9.2. (**a**) Leaf explant preparation (**b**) Scanning electron microscope observation show shoot regenerated from connecting line between leaf and node. (**c**) GUS staining area was located on connecting area between leaf and node. (**d**) GUS positive shoot regenerated from connecting area between leaf and node (*arrow*).

4. *Agrobacterium* culture from LB solid medium are cultured overnight in 20 mL LB medium containing 50 mg/L kanamycin and 50 mg/L rifampicin at 27°C on a rotary shaker (120 rpm).
5. One milliliter of *Agrobacterium* is transferred to 1.5-mL Eppendorf tube and centrifuged for 1 min at 2,000*g*. The supernatant is removed and the pellet resuspended thoroughly in 10 mM glucose and 100 μM AS and centrifuged twice.
6. The density of bacteria is adjusted to 0.5 (OD600) with 10 mM glucose and 100 μM AS.

3.4. **Agrobacterium** *Infection and Cocultivation*

1. Node or leaf explants are submerged into the *Agrobacterium* suspension for 10 min (*see* Note 3).
2. Node or leaf explants are blotted briefly on sterilized filter paper.

3. Transfer the infected node explants onto three layers of filter paper soaked with water and 50 µM AS in a 90-mm Petri dish.
4. Node or leaf explants are cocultivated with *Agrobacterium* for 5 days in dark conditions at 25°C.

3.5. Selection

1. After cocultivation, the node explants are cut into eight radial segments and subcultured onto the MS selection medium containing 20 mg/L G418. For the selection, subculture leaf explants onto Leaf explants culture medium containing 20 mg/L G418.
2. The explants are subcultured every month onto a fresh MS selection medium or leaf explants culture medium for selection with step-wise increased G418 concentrations from 20 to 30 mg/L and 40 mg/L.
3. Antibiotic resistant shoots are harvested from explants and cultured on Hyponex selection medium, until transformed shoots are big enough (about 4–6 cm) for acclimation and greenhouse growth (*see* Notes 4–9).

3.6. Plant Acclimation and Greenhouse Care

1. Carnation shoots are acclimatized by cutting 3- to 4-cm long shoots, placing them into cutting media, and maintaining under shading with high humidity condition for 2 weeks.
2. The rooted cuttings are transplanted into 20-cm diameter pot and grown in special greenhouse for genetic transformation plants (closed glasshouse at 25°C under natural light intensity and photoperiod). Water moderately with water and fertilize with by 14:14:14 control release fertilizer (Nutricote) every 2 months. The plants flower after 4–5 months.

4. Notes

1. During preparation of node explants, it is important to remove leaves from the shoot by tearing not cutting, because the adventitious shoots will regenerate only from the tearing scar surface, not the cutting surface.
2. For preparing leaf explants tearing leaves from shoots carefully, ensure that the explants containing connecting area between leaf and shoot because adventitious shoots regenerate from connecting line between leaf and shoot.
3. Before incubating with *Agrobacterium* suspension, blot explants briefly on sterile filter paper if the surface of the node explants is too wet.

4. For leaf explants, preparation of explants was an important factor to obtain the transformed plants. The GUS-staining area was located only on the cut end of the leaf explants. Only those explants with a cut end close to the leaf base (connecting area between node and leaf) produce GUS-positive shoots.
5. A comparison of the genetic transformation efficiency between leaf explants from in vitro plants and node explants from greenhouse plants showed that node explants produced higher numbers of transformed shoots than leaf explants. The difference in transformation efficiency between the node and leaf explants may be due to the size of explants and their regeneration ability. Node explants from green house plants produce ten times more adventitious shoots than in vitro leaf explants. Therefore, the node explants have a higher chance of producing transformed shoots.
6. Even though they have a rather low transformation efficiency, the in vitro materials have some advantage, since transformants from in vitro starting materials are expected to be clean. Some genetic transformation studies, such as, an attempt to introduce virus-resistant genes into plants, required clean materials.
7. If the vitrification problem occurs, reduced sugar concentration and increased gellan gum concentration can reduce this problem.
8. Whenever performing the GUS assays, nontransformed explants should be included as negative controls, to rule out the background effect during the staining process.
9. This protocol is reproducible and applicable to other genotypes of carnation. This protocol was still effective when used to transfer genes into regenerable calli of *Dianthus hybrida* (Telstar scarlet).

References

1. Van Altvorst AC, Riksen T, Koehorst H, Dons HJM (1995) Transgenic carnations obtained by *Agrobacterium tumefaciens* mediated transformation of leaf explants. Trangenic Res 4:105–113
2. Chateau S, Sangwan RS, Sangwan-Norreel BS (2000) Competence of *Arabidopsis thaliana* genotypes and mutants for *Agrobacterium* tumefaciens mediated gene transfer: role of phytohormones. J Exp Bot 51:1961–1968
3. Sangwan RS, Bougeois Y, Brown S, Vasseur G, Sanewan-Norreel B (1992) Characterization of competent cells and early events of *Agrobacterium* mediated genetic transformation in *Arabidopsis thaliana*. Planta 188:439–456
4. Egnin M, Mora A, Prakash CS (1998) Factor enhancing *Agrobacterium tumefaciens* mediated gene transfer in peanut (Arachis hypogaea L.). In vitro Cell Dev Biol Plant 34:310–318
5. Aida R, Hirose Y, Kishimoto S, Shibata M (1999) *Agrobacterium tumefaciens* mediated transformation of *Cyclamen persicum* Mill. Plant Sci 148:1–7
6. Park SH, Morris JL, Park JE, Hirschi KD, Smith RH (2003) Efficient and genotype-independent *Agrobacterium*-mediated tomato transformation. J. Plant Physiol 160: 1253–1257

7. Hirsikorpi M, Kämäräinen T, Teeri T, Hohtola A (2002) *Agrobacterium* mediated transformation of round leaved sundew (*Drosera rotundifolia* L.). Plant Sci 162:537–542
8. Takanashi M, Irifune K, Hasegawa M, Kuroda H, Nishisawa S, Mrikawa H (1996) Starvation treatment increase transient expression frequency of a foreign gene introduced into Eustoma grandiflorum root tissues by particle bombardment. Plant Tissue Cult Lett 13:311–315
9. Stachel SE, Messens E, Montagu MV, Zambryski P (1985) Identification of the signal molecules produced by wounded plant cells that activate T-DNA transfer in *Agrobacterium tumefaciens.* Nature 318:624–629
10. Guivarc'h A, Caissard JC, Brown S, Marie D, Dewitte W, Van-Onckelen H, Chriqui D (1993) Localization of target cells and improvement of *Agrobacterium* mediated transformation efficiency by direct acetosyringone pretreatment of carrot root disc. Protoplasma 174:10–18
11. Pérez Hernándaz JB, Remy S, Galán Saúco V, Swennen R, Sági L (1999) Chemotactic movement and attachment of *Agrobacterium tumefaciens* of banana cells and tissues. J Plant Physiol 155:245–250
12. Douglas CJ, Halperin W, Nester EW (1982) *Agrobacterium tumefaciens* mutants affected in attachment to plant cells. J Bacteriol 152:1265–1275
13. Matthysse AG, Holmes KV, Gurlitz RHG (1981) Elaboration of cellulose fibrils by *Agrobacterium tumefaciens* during attachment of carrot cells. J Bacteriol 145:583–595
14. Matthysse AG (1987) Characterization of nonattaching mutants of *Agrobacterium tumefaciens.* J Bacteriol 169:313–323
15. Nontaswatsri C, Fukai S, Goi M (2004) Revised cocultivation condition produce effective *Agrobacterium*-mediated genetic transformation of carnation (*Dianthus caryophyllus* L.). Plant Sci 166:59–68
16. Nontaswatsri C, Fukai S (2005) *Agrobacterium*-mediated genetic transformation of carnation (*Dianthus caryophyllus* L.) using leaf explants derived from in vitro plants. Tech Bull Fac Agr Kagawa Univ 57:15–19

Chapter 10

Matrix Supported Liquid Culture and Machine Vision Analysis of Regenerated Shoots of Gladiolus

S. Dutta Gupta and V. S. S. Prasad

Abstract

Enhancement of efficiency and efficacy of plant regeneration are primary goals of micropropagation. An efficient method with rapid and improved shoot regeneration of gladiolus based upon matrix supported liquid culture technique has been demonstrated. This technique in addition ensured the quality of the regenerated shoots by effectively alleviating the phenomenon of hyperhydricity. Efficacy in plant regeneration can be brought about by the ability to sort and objectively select plants or group of plants with a desired feature. A reliable, noninvasive, machine vision-neural network based sorting method for clustering photometric variants of regenerated shoots of gladiolus has also been described.

Key words: Membrane raft, Shoot regeneration, Cluster analysis, Artificial neural network, Photometric feature, Adaptive resonance theory

1. Introduction

Liquid culture technique has immense potential to improve shoot organogenesis (1) particularly due to two main fundamental reasons. One is, the liquid medium increases the oxygen diffusion coefficient and secondly, the level of water potential is maintained at higher levels. These two salient features maintain the aeration levels and uninterrupted nutrient and growth regulator supply to the tissues. However, direct and prolonged contact of the cells and tissues with the liquid medium poses the risk of apoplastic accumulation of water inside the cells. The tissues affected in this manner are termed hyperhydric (2, 3). The hyperhydric plants are recalcitrant and susceptible to withering upon *ex vitro* transfer. The occurrence of hyperhydricity can be abbreviated by interfacing an inert absorbent or adsorbent support material between the

S.M. Jain and S.J. Ochatt (eds.), *Protocols for In Vitro Propagation of Ornamental Plants*, Methods in Molecular Biology, vol. 589
DOI 10.1007/978-1-60327-114-1_10,

tissue and liquid medium such that pooling of liquid medium around the tissue is avoided, yet maintaining the continuous medium supply while retaining all the advantages of liquid culturing. Out of many suggested support matrices, polypropylene membrane raft was particularly advantageous in increasing the shoot proliferation (4, 5) and reducing hyperhydricity in various monocotyledonous ornamental species.

Variations in culture conditions are inevitable in spite of controlled conditions. Due to limited air movement, CO_2, and humidity gradients occur in closed vessels. Light intensity also varies particularly in slender vessels (6, 7). These special inconsistencies affect the growth and qualitative uniformity of regenerated plants. The intrinsic and nonlinear variations in plant cell and tissue cultures are difficult for hands-on detection. A neural network based machine vision system, in its noninvasive capacity, offer pertinent solution for sorting such groups. Potential to cluster the plants into homogeneous groups enable us to objectively select plants or group of plants with salient biological property of interest (8, 9).

2. Materials

2.1. Matrix Supported Liquid Culture of Gladiolus

1. *Plant material.* Healthy and viable descaled corms of Gladiolus hybridus Hort. cv. Wedding Bouquet of approximately 3–4 cm diameter.
2. Systemic fungicide Bavistin® (Carbendezium; BASF, India Ltd, Mumbai) for preparation of 0.1% (w/v) solution.
3. Soil-rite mixture (Keltech Energies Pvt. Ltd., Bangalore, India).
4. 0.1% Mercuric chloride ($HgCl_2$); 0.1% (v/v) liquid detergent (Tween-20).
5. Dehydrated mixture of Murashige and Skoog's (10) basal salts (Himedia Laboratories Pvt. Ltd., Mumbai, India) and $CaCl_2$ stock solution.
6. α-Naphthalene acetic acid (NAA) and 6-Benzylaminopurine (BAP) (Sigma Aldrich Chemicals Pvt. Ltd.)
7. 0.1 N NaOH and/or 0.1 N HCl.
8. Magenta GA-7 plant tissue culture vessels (Magenta Corp., Chicago IL, USA)
9. Plant tissue culture grade agar (Himedia Laboratories Pvt. Ltd., Mumbai, India).
10. Pre-sterilized polypropylene membrane raft (OsmotekTM, Rehovat, Israel) with 0.3 μm pore size (MR).

11. Duroplast foam of 0.3 cm thickness (DF) (Polyurethane foam, Sheela Foam Pvt. Ltd., Ghaziabad, India).
12. NCSS (Number crunch statistical system, Utah) statistical software on Microsoft Windows® XP platform.

2.2. Photometric Clustering of Regenerated Plants

1. Hewlett Packard flatbed scanner series 11.0 containing charge coupled device (CCD) scanning element
2. Standard PC with input/output devices, 40-GB hard disk, 1.6-GHz clock speed and 256-MB RAM.
3. Adobe Photoshop® Ver.7.0 image processing software on Microsoft Windows® XP platform.
4. Executable "C" program of Adaptive resonance theory 2 algorithm (8).

3. Methods

3.1. Matrix Supported Liquid Culture of Gladiolus

3.1.1. Preparation of Medium and Culture Setup

1. Dehydrated basal salt of mixture of Murashige and Skoog (10) are used for the media preparation. Calcium chloride is added separately from the stock solution.
2. Growth regulators, α-naphthalene acetic acid (NAA) and 6-benzylaminopurine (BAP) are added prior to pH adjustment. The pH of the medium is adjusted to 5.7.
3. For semi-solid medium, 0.8 % (w/v) agar is added to the medium, and dispensed in the Magenta GA-7 culture vessels prior to autoclaving.
4. All media are autoclaved at 121°C under 103.42 kPa for 20min.
5. The media are allowed to cool (for liquid) and solidify (for semi-solid agar) overnight.
6. For liquid systems, the support systems are assembled in the laminar air flow hood as depicted in Fig. 10.1.

3.1.2. Preparation of Explants

1. The corms are treated with 0.1 % (w/v) Bavistin prepared in single distilled water for 1 h after which they are shallow planted in earthen pots containing soil-rite mixture and allowed to sprout till two-leaf stage in 16 h photoperiod, irradiance of 60 μmol/m^2/s PPFD.
2. 7- to 10-day-old primary leaves of the sprouted corms are excised from the base and washed under running tap water to remove soil particles and then washed with single glass distilled water. Thereafter, they are surface disinfected with 0.1% (w/v) Mercuric chloride ($HgCl_2$) and 0.1% (v/v) Tween-20 for 5 min and washed thoroughly 3–4 times with sterile double distilled water (DDW) in a laminar flow hood.

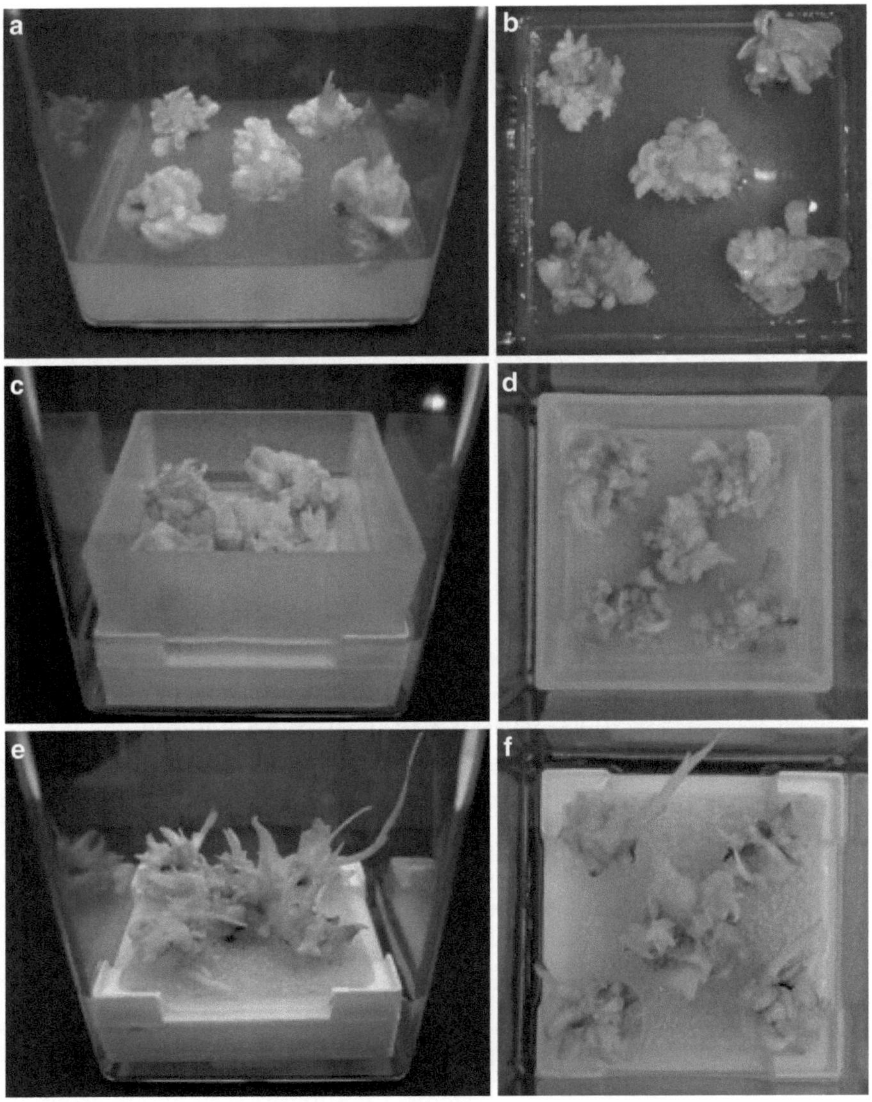

Fig. 10.1. Multiple shoot clusters in AS, MR, and DF culture systems at 35th day. (**a**) Regenerated shoot clusters upon semisolid agar support. (**b**) *Top view* of regenerated shoot clusters in AS system. (**c**) Setup of MR system with a buoyant support. (**d**) Regenerated shoot clusters in MR system. (**e**) Setup of DF system with a buoyant support. (**f**) Regenerated shoot clusters in DF system. With kind permission from Springer Science + Business Media.

3. The basal portion of the innermost leaf is dissected out and blot dried and used as initial source of explants.

3.1.3. Establishment of In Vitro Shoot Cultures

1. The explants (5 × 5 mm) are inoculated on semi-solid medium containing 2.0 mg/L NAA and incubated for 3 weeks for callus induction.

2. The calli are transferred onto MS medium supplemented with 0.2 mg/L. NAA and 2.0 mg/L BAP and incubated for three weeks for the induction of meristematic bud clusters (11).
3. The regenerated multiple shoot primordia are allowed to proliferate in AS, MR, and DF culture systems.
4. The culture media are supplemented with growth regulators NAA (0, 0.2, 0.5, and 1.0 mg/L) and BAP (0, 1, 2 and 4.0 mg/L) either alone or in factorial combination for optimization.
5. For AS culture, 0.8 % agar is used to solidify the media.
6. For liquid cultures, pre-sterilized polypropylene membrane raft with 0.3 μm pore size and Duroplast foam of 0.3 cm thickness are used as an interface between tissue and liquid media.
7. All the cultures are incubated at 25 ± 2°C for 35 days under a 16 h photoperiod provided by cool white fluorescent lamps, 50 μmol/m/s^2. Setup of culture systems and regeneration of multiple shoot clusters at 35 days of incubation is shown in Fig. 10.1.
8. Optimal medium composition for AS is MS + 0.2 mg/L NAA + 2.0 mg/L BAP whereas, that for MR and DF cultures are MS101 + 1.0 mg/L NAA + 2.0 mg/L BAP and MS + 0.5 mg/ L NAA + 4.0 mg/L BAP respectively. Highest regeneration of 33.15 shoots /cluster is obtained with MR system.

3.1.4. Poisson Regression Modeling of Shoot Count Data

In order to evaluate the in vitro shoot regeneration kinetics, Poisson regression analysis is carried out, the sequential steps of which are as follows:

1. The total number of shoots in each culture system is recorded at 7-days interval from the time of culture initiation throughout the incubation period.
2. Coefficient of variation (C.V. %) is calculated to estimate the variability of shoot numbers between culture systems through the incubation period. C.V. % = (Standard deviation / Mean) × 100.
3. The shoot multiplication data is subjected to log-link Poisson regression model considering incubation time as the independent variable and shoot number as the dependent. Both linear and quadratic effects are evaluated.
4. The probability of y_i shoot buds is given by:

$$\Pr(y_i) = \frac{\exp(-\mu_i) \times \mu^{y_i}}{\Upsilon_i!},$$

where, Υ_i, $i = 1, 2, \ldots, n$ represents shoot count with means (μ_i).

5. The coefficient of determination (R^2) is calculated for each equation. $R^2 = 1 - \text{Sum (Observed value} - \text{Expected value})^2 / \text{Sum (Observed value} - \text{Mean of observed values})^2$.
6. The kinetics (speed and acceleration) of the shoot multiplication rate during the incubation period is evaluated from the first and second derivatives, calculated from the regression models, respectively.

$$\text{Speed} = (b - 2 \times c \times X) \times Y$$

$$\text{Acceleration} = -2 \times c \times Y + (b - 2 \times c \times X)^2 \times Y$$

7. $Y = \exp\ (0.6419 + 1.2635 \times X - 0.1032 \times X \times X$ is the representative Poisson regression model for the AS system. The speed derivative (the rate of differentiation of new buds) of the fitted Poisson regression models suggest higher multiplication rate in MR system than the AS and DF.

3.1.5. In Vitro Corm Induction

1. The differentiated multiple shoot clusters are transferred onto basal MS medium devoid of growth regulators in five GA-7 Magenta vessels (five clusters per vessel each with 2–5 shoot buds) and incubated for 2 weeks for proliferation of shoots.
2. Developed shoots (1–2 cm long) with expanded lamina are sampled and individually inoculated onto 20 mL corm induction medium, MS + 0.5 mg/L NAA + 6% Sucrose, contained in test tubes (25 × 150 mm).
3. The cultures are incubated for 90 days for in vitro corm induction and development (Fig. 10.2).

3.2. Photometric Sorting of the Regenerated Shoots and Biological Validation

The general scheme of machine vision aided photometric clustering of regenerated shoots is presented in Fig. 10.3.

3.2.1. In Vitro Shoot Regeneration

The meristematic bud clusters are obtained as described in Subheadings 10.3.1.2 and 10.3.1.3. Five equally sized (~5 mm^3) meristematic bud clusters each with 2–5 shoot buds are placed on 50 mL MS medium supplemented with 0.5 mg/L NAA, 3% sucrose and 0.8% agar, contained in a GA-7 vessel and incubated for 3 weeks for the shoot proliferation and complete plantlet development.

3.2.2. Image Acquisition and Feature Extraction

The digitized versions of the outermost expanded leaves are secured from which the color histogram is extracted. The histogram describes the color distribution pattern of an image, that is,

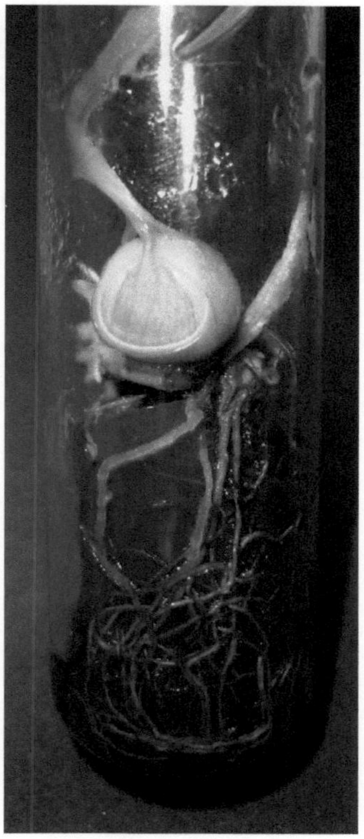

Fig. 10.2. In vitro corm developed from regenerated shoot of gladiolus at 90th day.

it provides the counts of the number of occurrences of each color in an image over an intensity scale ranging from 0 to 256.

1. The outermost expanded leaves of the regenerated plants are excised randomly from each cluster, washed, and blot dried.
2. Three sets of leaves representing training, test, and validation data sets are then scanned using Hewlett Packard flatbed scanner series 11.0 containing charge coupled device (CCD) scanning element at 300 pixels per inch (ppi) resolution, 24 bits color depth, and a constant luminosity to acquire the digitized images. The images are numbered for later identification.
3. A fixed number of pixels (4 × 4) are selected from the median portion of digital leaf images to obtain the color distribution histograms. Digital leaf images are saved in Adobe Photoshop; Adobe Systems Incorporated, USA (*.psd) format with 8 bits per pixel having 256 grey-scale levels. The pixel properties of the images are evaluated using Adobe Photoshop 7.0 software.
4. From the histograms of luminosity and trichromatic (Red, Green, and Blue) channels of the image, the mean brightness level, grey-scale level for the maximum pixel count are recorded.

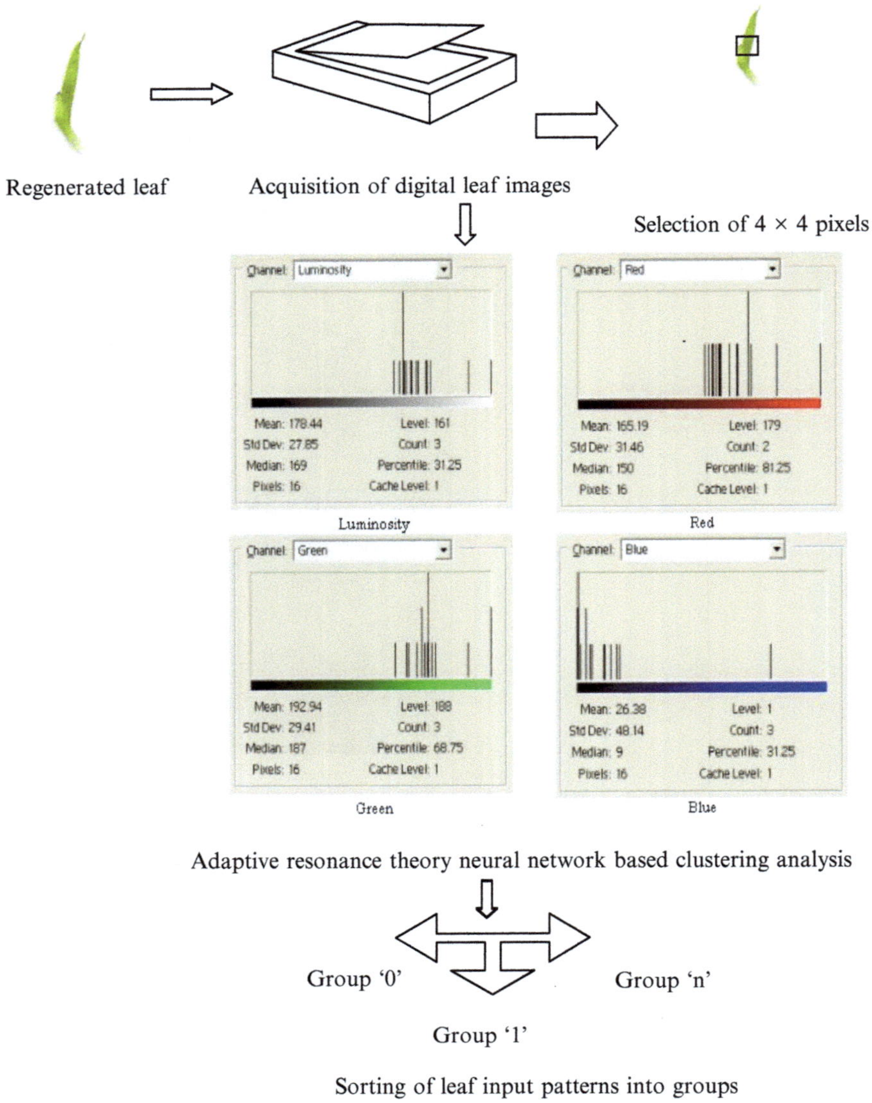

Fig. 10.3. Component steps of machine vision-neural network aided sorting of regenerated plants of gladiolus (Reproduced from Ref. 8).

3.2.3. Cluster Analysis by ART2 Algorithm

1. Twelve dimensional leaf input patterns comprising the training data set are serially subjected to ART 2 algorithm for learning.
2. The training data set (8) and validation data set (9) represented by 25 and 55 leaf input patterns, respectively are subjected to ART 2 neural network aided clustering analysis.
3. ART2 algorithm (12) coded in C language (8) is used for the analysis.

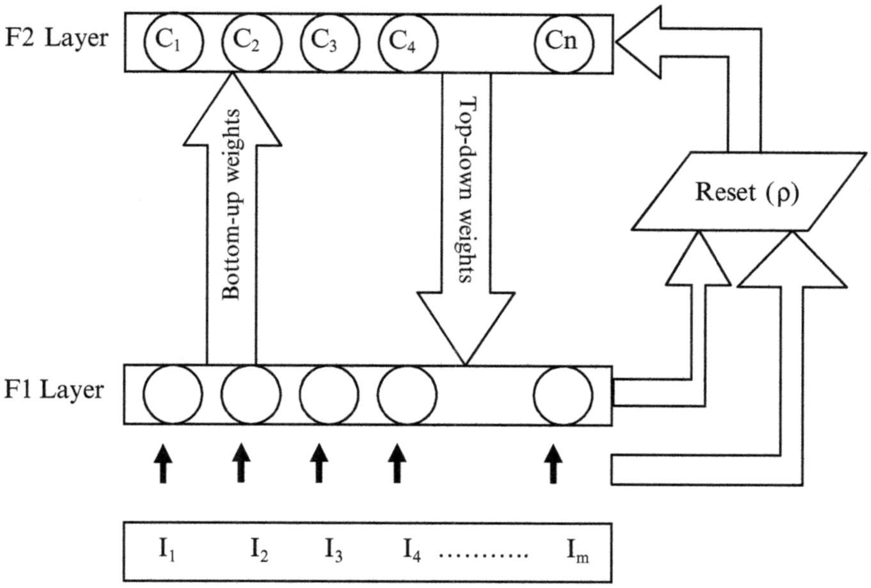

Fig. 10.4. Adaptive resonance theory 2 module.

ART 2 paradigm consists of two layers of nodes namely F1 and F2 (Fig. 10.4). F1 layer enhances the contrast and suppresses the noise of the input data patterns (I). The activation patterns of the F1 layer constitute short term memory. The top-down and bottom-up weights together constitute long-term memory. An F2 neuron that resembles the input pattern the most is termed winner category (C).

4. Various constant values are chosen as $a = 10$; $b = 10$; $c = 0.1$; $d = 0.8$; $q = 0.0001$. The parameter e is then typically set to a positive number considerably less than 1, $e = 0.000001$.
5. Clustering analysis is performed using VP value ranging from of 0.1 to 1.0 at an interval of 0.1. The VP value is then optimized, for efficient and logical sorting.
6. The network performance is verified by test data.
7. The validation input patterns, the inherent natural groups of which need to be identified, are presented to ART2 neural network with preset optimal VP value for autonomous clustering/sorting.
8. The training and validation data set are grouped into two clusters (group "0" and group "1"). Of 55 plants from validation set, 19 plants are sorted into group "0" and the rest 36 plants are sorted into group "1" at a VP of 0.099.
9. For gladiolus, a VP value of 0.999 is optimum for efficient and discrete clustering of training set input patterns.

Vigilance parameter (ρ) is a fixed scalar parameter that controls the resolution of the categories. Initially, all the weights are set to unity. If the calculated matches function value of winning category is greater than VP value, the weights are adjusted (learning) and the category gets established. On the contrary, if the match function of the winning category is less than VP, a new cluster is generated till Cn.

10. The sorted groups of validation set patterns are compared with the in vitro corm induction potential and corm developmental features such as diameter, perimeter, fresh weight, and relative dry weights of the corresponding shoots. A significantly high percentage of corm development with improved corm features is noted with group "1" plants. It indicates the biological potential of the regenerated shoots categorized in that group as well as the significance of photometric clustering of regenerated shoots. Early prediction of potential of the regenerated shoots for corm induction in a noninvasive manner appear to be feasible with ART 2 neural network.

4. Notes

1. During corm germination soil-rite mixture is kept moist and never water logged.
2. All the stock solutions meant for media preparation are stored at 4°C.
3. Duroplast foam is soaked in warm solution of 0.1 % (v/v) liquid detergent for at least 2 h, rinsed thoroughly in double distilled water and dried in oven prior to use.
4. Training set should comprise a significantly higher number of leaf input patterns than the test set. Number of patterns in validation set need to be of the same magnitude of that of training set.

References

1. Adelberg JW (2006) Agitated thin-films of liquid media for efficient Micropropagation. In: Dutta Gupta S, Ibaraki Y (eds) Plant tissue culture engineering. Springer, The Netherlands, pp 103–117
2. Chakrabarty D, Park SY, Ali BM, Shin KS, Paek KY (2005) Hyperhydricity in apple: ultrastuctural and physiological aspects. Tree Physiol 26:377–388
3. Hazarika BN (2006) Morpho-physiological disorders in in vitro culture of plants. Sci Hort 108:105–120
4. Dutta Gupta S, Prasad VSS (2006) Matrix-supported liquid culture systems for efficient micropropagation of floricultural plants. In: Teixeira da Silva JA (ed) Floriculture, ornamental and plant biotechnology: advances and topical issues, vol 2. Global Science Books, UK, pp 487–495

5. Prasad VSS, Dutta Gupta S (2006) In vitro shoot regeneration of gladiolus in semi-solid agar versus liquid cultures with support systems. Plant Cell Tiss Org Cult 87:263–271
6. Chen C (2004) Humidity in plant tissue culture vessels. Biosyst Eng 88(2):231–241
7. Huan C, Chen C (2005) Physical properties of culture vessels for plant tissue culture. Biosyst Eng 91(4):501–511
8. Mahendra, Prasad VSS, Dutta Gupta S (2004) Trichromatic sorting of in vitro regenerated plants of gladiolus using adaptive resonance theory. Curr Sci 87:348–353
9. Prasad VSS, Dutta Gupta S (2008) Photometric clustering of regenerated plants of gladiolus by neural networks and its biological validation. Comp Elect Agril 60:8–17
10. Murashige T, Skoog F (1962) A revised medium for rapid growth and bioassays with tobacco tissue culture. Physiol Plant 15: 473–497
11. Dutta Gupta S, Datta S (2004) Antioxidant enzyme activities during in vitro morphogenesis of gladiolus and the effect of application of antioxidants on plant regeneration. Biol Plant 47:179–183
12. Carpenter GA, Grossberg S (1987) ART 2: Stable self-organization of pattern recognition codes for analog input patterns. Appl Optics 26:4919–4930

Chapter 11

In Vitro Propagation of Carnation (*Dianthus caryophyllus* L.)

Jose L. Casas, Enrique Olmos, and Abel Piqueras

Abstract

Carnation (*Dianthus caryophyllus* L.) is one of the most popular ornamental plants worldwide and also among the most studied ones, mainly in cut flower postharvest physiology. Several protocols for the in vitro propagation of this species including nodal segment culture, somatic embryogenesis, and adventitious shoot induction are described in this chapter. The presence of hyperhydricity as an abnormality during micropropagation of carnation plants has also been the object of research for many years and different strategies to overcome this problem are also included in this study.

Key words: Carnation, Micropropagation, In vitro culture, Adventitious shoots, Hyperhydricity, Bottom cooling, Somatic embryogenesis, Organogenesis

1. Introduction

Carnation (*Dianthus caryophyllus* L., family Cariophyllaceae) is a top-selling ornamental crop worldwide and an important target for breeding new varieties with improved agronomic traits (1). They have high esthetic value in the floriculture industry . This genus has great phytogeographic importance because of the high rate of endemic species; many of them are included in National and European Red Lists. Also *D. caryophyllus* is widely used in physiological studies, especially with respect to flower senescence (2–4). The availability of a feasible and easy way of propagating carnation plants by means of in vitro techniques may be, therefore, of great interest for many purposes. In vitro carnation culture is frequently hampered by hyperhydricity. This is an abnormality in which leaves or shoots become water-soaked, translucent, or glassy. Due to this "like-glass" appearance, the phenomenon was formerly called vitrification or glassiness and recently redefined as hyperhydricity. Carnation is one of the most

S.M. Jain and S.J. Ochatt (eds.), *Protocols for In Vitro Propagation of Ornamental Plants*, Methods in Molecular Biology, vol. 589
DOI 10.1007/978-1-60327-114-1_11, © Humana Press, a part of Springer Science+Business Media, LLC 2010

Fig. 11.1. The appearance of hyperhydric (*left*) and normal (*right*) in vitro-derived plantlets from different carnation varieties: Alister (**a**), Oslo (**b**), and Killer (**c**).

affected plants by this abnormality (Fig. 11.1), as are all the species of the Caryophyllaceae. After the numerous efforts made by several research groups to overcome this syndrome, several solutions that can be applied for a successful carnation micropropagation are available.

2. Materials

1. *Basal medium.* Murashige and Skoog (MS) (macronutrients and micronutrients) powder (Duchefa); MS vitamin complement (Duchefa).
2. Plant agar (Duchefa).
3. 70% ethanol, 2% sodium hypochlorite, and Tween-20 (Sigma).
4. *Plant growth regulators.* Thidiazuron (TDZ); α-Naphtalene acetic acid (NAA); 2, 4-Dichlorophenoxyacetic acid (2, 4-D); Benzyladenine (BA); Giberellic acid (GA_3).
5. Sucrose.
6. Sand and compost.

3. Methods

3.1. Micropropagation from Nodal Segments (1, 5–7)

3.1.1. Explant Sterilization and In vitro Establishment

1. Rinse nodal segments three times in 70% (v/v) ethanol, then surface-sterilize for 8–10 min in 2% (w/v) sodium hypochlorite containing 0.01% (v/v) Tween-20, and rinse three times in sterile distilled water.
2. Trim segments to obtain a section of around 1-cm long with the node and axillary buds in the middle (*see* Note 1).
3. Transfer explants to 2.3 × 15-cm test tubes containing 15-mL sterile, full strength basal medium supplemented with 3% sucrose and solidified with 0.8% agar (*see* Note 2). The pH is adjusted to 5.8 prior to autoclaving.
4. The cultures are grown at 25 ± 1°C under an incident radiant flux of 30 μmol/m^2/s (*see* Note 3) provided by fluorescent "daylight" tubes, with 16-h photoperiod.

3.1.2. Multiplication, Elongation, and Rooting

1. Those explants that successfully establish in vitro after 15 days are transferred to multiplication medium (Table 11.1) consisting of full strength basal medium with 3% sucrose, 0.8% agar (*see* Note 4), and supplemented with 1 mg/L TDZ and 1 mg/L NAA. Culture conditions for multiplication and elongation are as above.

Table 11.1
Culture media used in the in vitro propagation of carnation

	Stage on in vitro regeneration		
Regeneration strategy	**Multiplication**	**Elongation**	**Rooting**
From nodal segments	Full strength basal medium	Full strength basal medium	Half strength basal medium
	3% sucrose	3% sucrose	No sucrose
	0.8% agar	0.8% agar	0.7% agar
	1 mg/L TDZ		5.4 μM NAA
	1 mg/L NAA		
Adventitious shoot formation from leaf explants	Full strength basal medium	Full strength basal medium	Half strength basal medium
	3% sucrose	3% sucrose	No sucrose
	0.7% agar	0.8% agar	0.7% agar
	4 μM BA		5.4 μM NAA
	1.6 μM NAA		

2. After 1 month in multiplication medium, explants are transferred to fresh hormone-free medium (Table 11.1) for a further 3–4 weeks for shoot elongation.
3. Proliferated shoots are transferred to rooting medium (Table 11.1) consisting in half strength basal medium with 0.7% agar + 5.4 μM NAA for 15 days.

3.1.3. Acclimatization

1. Regenerated shoots with well-developed roots are transferred to a soil mixture (1:1, w/w sand and compost) in plastic trays and maintained under controlled humidity (90%) in a greenhouse or growth chamber.
2. Relative humidity is progressively reduced at a rate of 5% per week and after 4 weeks plants can be successfully transferred to open field conditions.

3.2. Adventitious Shoot Formation from Carnation Leaf Explants (1, 8)

1. Nodal cuttings from greenhouse-grown plants are surface-disinfected as above.
2. Explants are taken from the three pairs of leaves just below the apex. The two leaves of each node are carefully removed making sure that the axillary buds remain attached to the stem.
3. Explants are cultured in shoot regeneration medium (Table 11.1) composed of full strength basal medium with 3% sucrose, 4 μM BA, 1.6 μM NAA, and 0.7% agar. The pH is adjusted to 5.7 before autoclaving.
4. After 8 days, active cell division begins in the basal region of the leaf explant. At that stage new meristems are formed along the complete explant base (Fig. 11.2a).
5. Fully developed shoots are developed after 4 weeks in culture (Fig. 11.2b). The regenerated shoots elongate and root as described previously (Table 11.1).

3.3. Somatic Embryogenesis (9, 10)

3.3.1. Explant Sterilization, Callusing, and Embryo Induction

1. Immature flower buds (1–1.5-cm long) (*see* Note 5) are harvested from greenhouse-grown plants and surface sterilized by 70% ethanol for 30 s and 2% sodium hypochlorite for 20 min followed by three rinses with sterilized distilled water.
2. Sepal and receptacles are removed from the buds and pieces of about 4 mm in length excised from the basal part of petals and placed on embryogenic medium.
3. Embryogenic medium (Table 11.2) consists of full strength basal medium supplemented with 9 μM 2, 4-D, 0.8 μM BA, and 9% sucrose.
4. All cultures are grown at 24 ± 2°C under an incident radiant flux of 30 μmol/m^2/s provided by cool white fluorescent lamps, with 16-h photoperiod.

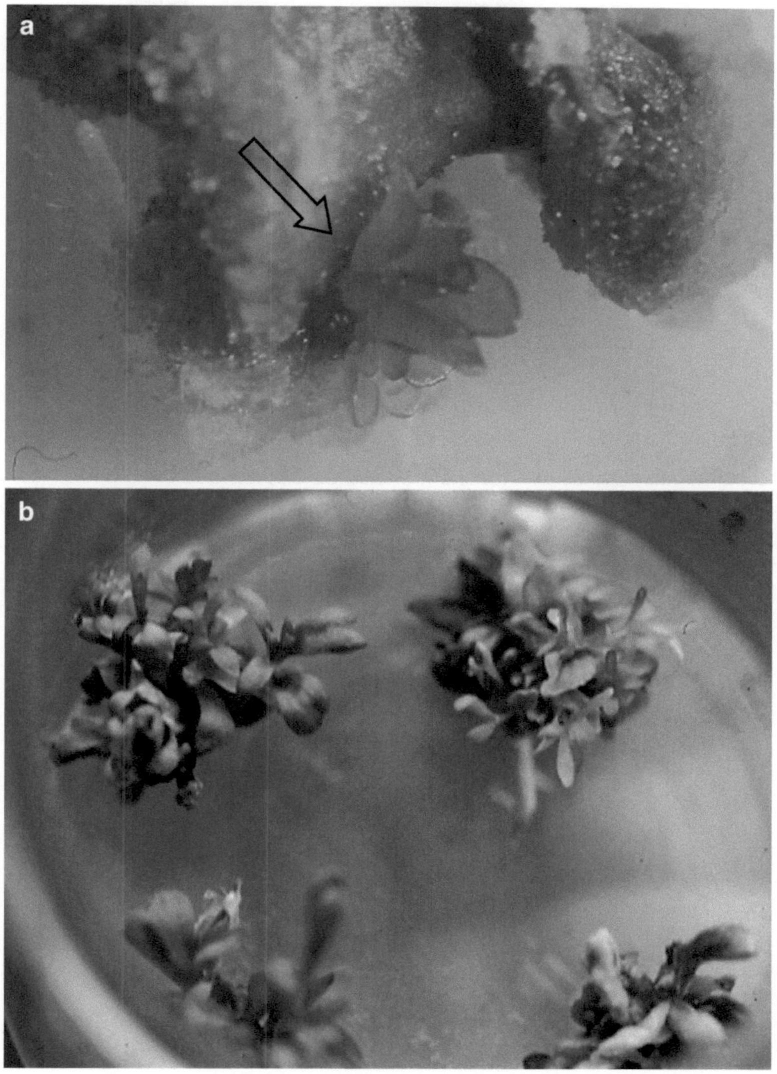

Fig. 11.2. (**a**) New shoot-forming meristems (*arrow*) arise from the basal portion of leaf explants. (**b**) Clusters of shoots are formed after 4 weeks of culture. Each shoot can now be isolated for elongation and rooting.

5. After 10 weeks of culture, calli are transferred to full strength hormone-free basal medium (Table 11.2) with 9% sucrose and maintained for a further 6 weeks for embryo development.

3.3.2. Embryo Growth and Plant Regeneration

1. Somatic embryos are transferred to half-strength hormone-free basal medium (Table 11.2) with 3% sucrose for plant regeneration.
2. Plantlets of at least 25 mm in length are cultivated in plastic pots containing a mixture of sand and compost (2:1 ratio) and

Table 11.2
Culture media used in carnation somatic embryogenesis

Embryogenic callus induction medium	Embryo induction	Plant regeneration
Full strength basal medium	Full strength basal medium	Half-strength basal medium
9% sucrose	3% sucrose	3% sucrose
9 μM 2,4-D		
0.8 μM BA		

maintained in a growth chamber at 18±2°C under 25 μmol/m^2/s of incident radiant flux with 16-h photoperiod.

3. After 4 weeks, well-established plants are placed into pots containing garden soil and left to stand in a greenhouse for a further 2 weeks to be acclimatized.

3.4. The Problem of Hyperhydricity (See Note 6)

3.4.1. Avoiding or Reducing Hyperhydricity

1. Improve the ventilation of the culture box (11) by using, for instance, filter-vented culture box.
2. Alternatively, hyperhydricity can be reduced by increasing iron and magnesium availability in the culture medium (*see* Note 7).

3.4.2. Reverting Hyperhydricity

1. Lowering the RH inside the culture tubes during the multiplication stage by placing the tubes on a bottom-cooling device inside the culture room (11) (*see* Note 8).
2. Changing the multiplication medium (c.f. Subheading 3.1.2) by MS medium supplemented with 0.5 mg/L GA_3 and 5 g/L bactopeptone (12).

4. Notes

1. When using apical buds, select 2-cm long shoot tips from the apex. Explants are prepared by excising stems around 0.6 mm below the apical dome (13).
2. A mixture of 0.4% BDH and 0.4% Roth agar can be used instead of Plant agar.
3. An incident radiant flux up to 50 μmol/m^2/s can also be used (6).
4. In some cases, an increase in agar strength up to 1.2% is used to enhance shoot multiplication (14).

5. Other explant types like internodal callus tissue may be used (13).
6. Hyperhydricity is a complex response and a serious problem in many commercial micropropagation strategies in which losses of up to 60% have been reported (15). The physiological basis of this alteration have not been clearly delimited, although many results have suggested the existence of an oxidative stress (16), deficiencies in cuticular waxes development (17), or changes in the methylation pattern of pectins in cell walls (18).
7. Different carnation varieties show different requirements of metal ions. It is necessary to optimize metal supply for each case (7).
8. By bottom cooling system we refer to any device capable of cooling the base of vessels, thus causing water in the vessel inner atmosphere to condense into the medium. This can be easily achieved by making a mixture of antigel fluid and water previously cooled by a cryostat to circulate within metal pipes. Vessels are then placed on metal shelves over the pipes. With the temperature of the circulator adjusted to 20°C, and for an average room temperature of 25°C, the relative humidity inside the culture flask is 75%.

References

1. Casanova E, Valdés AE, Zuker A, Fernández B, Vainstein A, Trillas MI, Moysset L (2004) *rolC*-transgenic carnation plants: adventitious organogenesis and levels of endogenous auxin and cytokinins. Plant Sci 167:551–560
2. Cook EL, Van Staden J (1988) The carnation as a model for hormonal studies in flower senescence. Plant Physiol Biochem 26:793–807
3. Serrano M, Romojaro F, Casas JL, Acosta M (1991) Ethylene and polyamine metabolism in climacteric and nonclimacteric carnation flowers. HortScience 26:894–896
4. Van Altvorst AC, Bovy AG (1995) The role of ethylene in the senescence of carnation flowers, a review. Plant Growth Regul 16:43–53
5. Piqueras A, Cortina M, Serna MD, Casas JL (2002) Polyamines and hyperhydricity in micropropagated carnation plants. Plant Sci 162:671–678
6. Watad AA, Ahroni A, Zuker A, Shejtman H, Nissim A, Vainstein A (1996) Adventitious shoot formation from carnation stem segments: a comparison of different culture procedures. Sci Hortic 65:313–320
7. Yadav MK, Gaur AK, Garg GK (2003) Development of suitable protocol to overcome hyperhydricity in carnation during micropropagation. Plant Cell Tissue Organ Cult 72:153–156
8. Van Altvorst AC, Koehorst HJJ, Bruinsma T, Dons JJM (1994) Improvement of adventitious shoot formation from carnation leaf explants. Plant Cell Tissue Organ Cult 37: 87–90
9. Karami O, Deljou A, Pour AM (2007) Repetitive somatic embryogenesis in carnation on picloram supplemented media. Plant Growth Regul 51:33–39
10. Karami O, Deljou A, Esna-Ashari M, Ostad-Ahmadi P (2006) Effect of sucrose concentrations on somatic embryogenesis in carnation (*Dianthus caryophyllus* L.). Sci Hortic 110:340–344

11. Saher S, Piqueras A, Hellin E, Olmos E (2005) Prevention of hyperhydricity in micropropagated carnation shoots by bottom cooling: implications of oxidative stress. Plant Cell Tissue Organ Cult 81:149–158
12. Jain A, Kantia A, Kothari SL (2001) De novo differentiation of shoot buds from leaf-callus of *Dianthus caryophyllus* L. and control of hyperhydricity. Sci Hortic 87:319–326
13. Frey L, Saranga Y, Janick J (1992) Somatic embryogenesis in carnation. HortScience 27:63–65
14. Crouch NR, Van Staden J (1993) In vitro culture of *Dianthus zeyheri* subsp. *natalensis*, a South African carnation. Plant Cell Tissue Organ Cult 35:81–85
15. Pâques M (1991) Vitrification and micropropagation: causes, remedies and prospects. Acta Hortic 289:283–290
16. Saher S, Piqueras A, Hellin E, Olmos E (2004) Hyperhydricity in micropropagated carnation shoots: the role of oxidative stress. Physiol Plant 120:152–161
17. Olmos E, Hellín E (1998) Ultrastructural differences of hyperhydric and normal leaves from regenerated carnation plants. Sci Hortic 1182:1–10
18. Fernández-García N, Piqueras A, Olmos E (2008) Sub-cellular location of H_2O_2, peroxidases and pectin epitopes in control and hyperhydric shoots of carnation. Environ Exp Bot 62:168–175

Chapter 12

In Vitro Propagation of *Jasminum officinale* L.: A Woody Ornamental Vine Yielding Aromatic Oil from Flowers

Sabita Bhattacharya and Sanghamitra Bhattacharyya

Abstract

The growing demand for flower extracts in perfume trade can primarily be met by increasing flower production and multiplying planting material. The major commercial aromatic flower yielding plants including *Jasminum officinale* L., a member of the Family Oleaceae have drawn the attention of a large section of the concerned sectors leading to a thrust upon developing advanced propagation technologies for these floral crops, in addition to conventional nature-dependent agro-techniques. This chapter describes concisely and critically, a protocol developed for in vitro propagation of *Jasminum officinale* by shoot regeneration from existing as well as newly developed adventitious axillary buds via proper phytohormonal stimulation. To start with nodal segments as explants, March–April is the most ideal time of the year when planting material suitable for in vitro multiplication is abundantly available. Prior to inoculation of explants in the culture medium, special care is needed to reduce microbial contamination by spraying on selected spots of the donor plant with anti-microbial agents 24 h prior to collection; treatment with antiseptic solution after final cleaning and surface sterilization by treating explants with mercuric chloride. Inoculated explants are free from brown leaching from cut ends by two consecutive subcultures within 48 h in MS basal medium. Multiplication of shoots, average 4–5 at each node, takes place in MS medium containing 4.0 mg/L BAP, 0.1 mg/L NAA, and 40 g/L sucrose over a period of 8 weeks. For elongation of regenerated shoots, cultures are transferred to MS medium, supplemented with a single growth hormone, kinetin at 2.0 mg/L. Emergence and elongation of roots from shoot base is facilitated by placing on the notch of a filter paper bridge. The hardened in vitro propagated plants are able to grow normally in soil like other conventionally propagated *Jasminum officinale*.

Key words: *Jasminum officinale*, Sterilization, Phenolic leaching, Axillary shoot, Rooting of shoots, Hardening

1. Introduction

Tissue culture has in recent years been recognized as an important tool in agricultural and horticultural development and most popular application of this technique is micropropagation. Ornamental

S.M. Jain and S.J. Ochatt (eds.), *Protocols for In Vitro Propagation of Ornamental Plants*, Methods in Molecular Biology, vol. 589
DOI 10.1007/978-1-60327-114-1_12, © Humana Press, a part of Springer Science+Business Media, LLC 2010

plants including aromatic species, occupying a significant position in global trade need increased rate of production to meet the ever-increasing demand from business. Along with the supplies from wild and cultivated sources (in some cases), tissue culture plants are accepted by commercial growers (1).

Jasmine flowers are well known to all perfume lovers throughout the world; its natural odor cannot be imitated. *Jasminum officinale*, the elite species used in the industry is a woody vine, perennial deciduous, 10–15 ft in fully grown state and need support to climb by means of twining. Besides utilization of its essential oil in perfumery, the flowers and other parts of *J. officinale*, are also used in traditional medicines for its various medicinal properties like antimicrobial (2), aphrodisiac (3), antifertility (4), analgesic, febrifuge, antidepressant (5). Recently, in aromatherapy it is used in relaxing nerves and reducing nervous tension.

The highly commercial jasmine, *Jasminum officinale* is one of the thrust area floral crops of India that has received larger attention of the Government departments of agriculture and horticulture. The thrust has been mainly on crop improvement and standardization of agro-techniques including mass propagation for ready planting materials required for cultivation in larger areas (1). According to Bajaj (6), one of the constraints to meet the demand of the perfumery industries is the nonavailability of sufficient planting material for large-scale cultivation.

Even though the techniques of tissue culture are known for a long time, limited work has been done on the woody, aromatic, and ornamental plants for enhancing propagation rate. Attempts through callus culture (7), protoplast culture (8), organogenesis and somatic embryogenesis (9) may be mentioned as few examples of micropropagation of woody ornamentals.

Jasminum officinale in vitro culture was made earlier by Khoder et al., (10) and Jonard (11) and the only report available on in vitro multiplication of the species using nodal explant till date is of Bhattacharya and Bhattacharyya (12). Axillary shoot proliferation by forcing dormant bud germination and subsequent adventitious bud development mediated through application of proper phytohormones was found to be an effective method for rapid multiplication of woody plants (13, 14). This tissue culture-based propagation method is accepted widely because stimulation of axillary branching is the most reliable method for maintaining the genetic integrity of clonal population (15); such integrity may be lacking in population raised by conventional seed germination method due to the heterogeneous nature of seeds.

In this chapter, we describe micropropagation protocol of jasmine (*Jasminum officinale*) developed in our laboratory (12).

2. Materials

2.1. Choice of Donor Plant

A 25-year-old jasmine plant (*Jasminum officinale*).

2.2. Collection of Explants

1. A solution of 1 g/L Bavistin (Carbendazim, BASF, Mumbai) and 200 mg/L Pantomycin 0.2 g/L (Streptomycin sulfate, Tetracyclin hydrochloride, ARIES, Mumbai) (*see* Note 1).

2.3. Cleansing of Explants

1. Running tap water.
2. Few drops of Tween 20 (HiMedia Laboratories Pvt. Ltd., Mumbai, India).
3. 5 g/L Cetrimide (Tetradecyl trimethyl ammonium bromide (SRL Pvt. Ltd, Mumbai) aqueous solution).
4. Distilled water.

2.4. Surface Sterilization of Cleaned Explants

1. Aqueous solution of 1 g/L mercuric chloride ($HgCl_2$ Qualigen Glaxo Fine Chemicals, India).
2. Sterile forceps, scalpels, and sterile filter papers.

2.5. Inoculation, Establishment of Explants in Media and Multiplication of Shoots

1. Sterilized MS basal medium (16) supplemented with 40 g/L sucrose, pH 5.7 (*see* Note 2), gelled with agar (HiMedia Laboratories, Mumbai, India), referred to as "establishment medium."
2. MS supplemented with 4.0 mg/L BAP (6-benzylaminopurin; Sigma), 0.1 mg/L NAA (Naphthalene acetic acid; Sigma) (*see* **Note 3**), and 45 g/L sucrose gelled with agar 8 g/L, adjust pH 5.7, referred to as "regeneration medium."

2.6. Elongation of Regenerated Shoots

1. MS medium or Miller's medium (17) containing 4.0 mg/L BAP, 2.0 mg/L Kin (Kinetin, 6-furfurylaminopurin; Sigma Chemicals), 0.1 mg/L NAA,45 g/L Sucrose, referred to as "shoot elongation medium."

2.7. Rooting of Shoots

1. IBA 10 mg/L (Indole butyric acid, Sigma Chemicals) (*see* Note 3) MS ½ (inorganic salts) medium or Miller's ½ (inorganic salts) medium (*see* Note 4), referred to as "rooting medium," fortified with 20 g/L in two different states – liquid and solid gelled with 8 g/L agar.
2. Filter Paper Bridge (*see* Note 5).

2.8. Acclimatization and Field Transfer of the Rooted Plants

1. Sterile soilrite (Technolab Scientific Company, Kolkata), sand, and soil.
2. Perforated polythene bags and plastic cups (*see* Note 6).
3. Earthen pots filled with soil.

Table 12.1
Media used at different steps of the protocol in micropropagating *Jasminum officinale* through various developmental stages

Steps of protocol	Medium + growth hormone + sucrose	Developmental stages of micropropagation
I	MS basal medium[a] + 40 g/L sucrose	Establishment of explant, resumption of growth
II	MS + 4.0 mg/L BAP + 0.1 mg/L NAA + 45 g/L sucrose	Regeneration of shoots at nodal axils
III	MS + 4.0 mg/L BAP + 2.0 mg/L kinetin + mg/L NAA + 45 g/L sucrose	Growth and elongation of regenerated shoots
IV	½ MS (liquid medium)[b] + 10.0 mg/L IBA + 20.0 g/L sucrose	Induction of roots from shoots
V	½ MS (solid medium)[c] + 20.0 g/L sucrose	Root growth and elongation

[a] All ingredients of MS without growth hormone
[b] Inorganic salts of MS at half concentration, other ingredients as usual, dissolved in distilled water, pH 5.7
[c] ½ MS, solidified with agar

Table 12.1 briefly presents the steps of the in vitro part of micropropagation protocol of *Jasminum officinale.*

3. Methods

Using the materials listed above and following the methods described in the following section, in vitro clonal propagation of *Jasminum officinale* (micropropagated jasmine plants) could be done.

3.1. Explant Collection

1. Collect explants in between March –April (*see* Note 7).
2. Spray a solution made of 1 g/L Bavistin and 0.2 mg/L Pantomycin on the leafy young branches of the source plant 24 h before excision of explants (Fig. 12.1a, b, c) (*see* Note 8).
3. Cut with the help of a sharp knife or a blade semi hard parts (*see* Note 9) of young branches into pieces 8–10 cm each containing 2–3 nodes with axillary buds.
4. Dip stem segments into a beaker filled with water immediately after removal from the branches.

3.2. Cleaning of Explants

1. Cut the stem segments into smaller pieces so that each piece contains 1 or two axillary buds at the axils of opposite leaves.
2. Remove leaf lamina, leaving the basal petiolar part only and rinse the explants under running tap water for 30 min to remove the traces of adhering Bavistin–Pantomycine.

Fig. 12.1. (**a–c**) A full grown *Jasminum officinale* plant: showing climbing nature and profusely growing young branches. (**b**) A close view of the plant at blooming condition. (**c**) Close view of a flowering branch.

3. Cleanse with a few drops of Tween 20 in water preferably by stirring in a magnetic stirrer to remove all trapped air bubbles, in particular on the surface of axillary buds.
4. Wash with Cetrimide solution for 1 min.
5. Wash thoroughly 5–6 times with distilled water.

3.3. Surface Sterilization Under Aseptic Condition

1. Treat explants with mercuric chloride solution for 2 min.
2. Wash 4 times with sterile distilled water.
3. Prepare nodal segments (1.0–1.5 cm) finally by trimming off the extra stem tissue.

3.4. Inoculation and Establishment of Explants

1. Soak adhering water from the surface of sterilized explants by pressing against sterile filter papers, then inoculate in the establishment medium in Erlenmayer flasks (*see* Note 10) by placing the explants vertically with the axillary buds facing upwards.
2. Incubate cultures in 16 h photoperiod (light intensity 41.76 μmol $m^{-2} s^{-1}$) at 25 ± 2°C.
3. Transfer the explants to a fresh medium of similar composition after 24 h when a light brown leachate is found to be stacked at the base of the explants (*see* Note 11).
4. Keep explants in this medium for 7 days (*see* Note 12).

3.5. Multiplication and Elongation of the Axillary Shoots

1. Transfer explants to MS regeneration medium preferably single explant/culture tube (*see* Note 13) and culture for 4 weeks.
2. Subculture the same explants with emerged axillary branches from preexisting buds (Fig. 12.2a, b).
3. Maintain culture for another 4 weeks in the same medium to stimulate initiation and growth of additional shoot buds.
4. Transfer cultures to shoot elongation medium and maintain for 6 weeks (Fig. 12.2c).

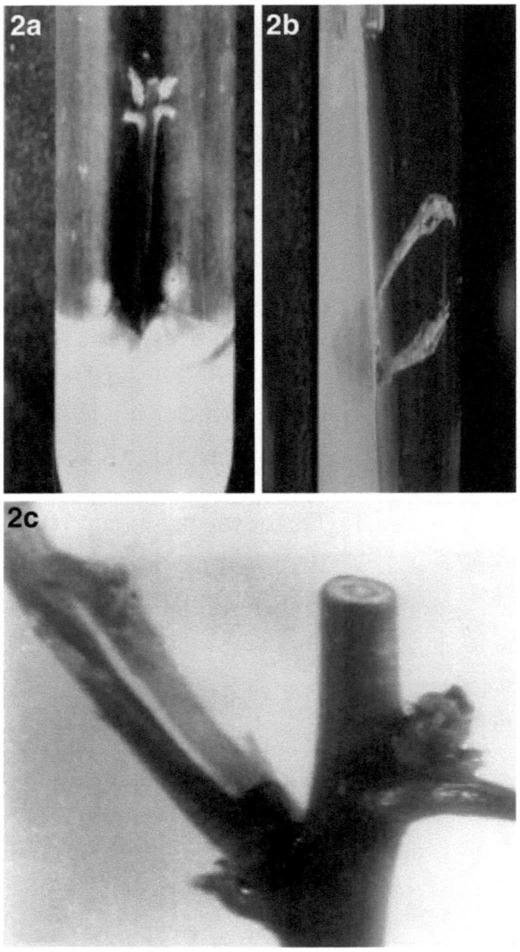

Fig. 12.2. (**a–c**) Regeneration of axillary shoots: (**a**) Enlarged axillary buds at the node after 1 week of culture. (**b**) Emergence of shoots by breaking buds after 2 weeks of culture. (**c**) Elongated shoots and multiple adventitious bud formation.

3.6. Rooting of Shoots

1. Place the erect, newly regenerated, elongated shoots separately in liquid rooting medium with the help of filter paper bridge.
2. Change medium when one or two stout roots are found to grow geotropically (Fig. 12.3a) from the stem base (*see* Note 14).
3. Transfer rooted shoots to solidified rooting medium (Fig. 12.3b) for further root growth.

3.7. Transfer of Well-Rooted Shoots from Gelled Substrate to Solid Substrate

1. Wash roots and shoot base thoroughly and carefully in tap water followed by sterile water to remove adhering agar (*see* Note 15).
2. Place the plantlets in plastic cups containing sterile soilrite, cover with perforated polythene bags and keep under tissue culture room condition for 2 weeks.

Fig. 12.3. (**a–b**) From root formation to field transfer of *J. officinale*: (**a**) Linear growth of root, initiated from the cut end of shoot. (**b**) Further development of roots in semisolid medium. (**c**) In vitro propagated *Jasminum officinale* grown in earthen pots.

3. Transfer the same sets from the laboratory environment to outside and keep in a shady place for another 7 days.
4. Scoop the whole content of the plastic cup along with the plantlet and place it in the groove made centrally in the earthen pots (quite larger than the plastic cups), filled with soilrite-sand-garden soil (1:1:1) and cover the plant base by layering. Wet the substrate with water as required, sprinkle water on leaves, and cover again with polythene bags as before.
5. Remove the polythene covers immediately after the young shoots resume apical growth by emerging new leaves.
6. Transfer young plants from small pots to larger ones containing soil following the technique of scooping and placing centrally (Fig. 12.3c), and water properly. Keep plants under sun after they attain a height of 15 cm or more.

4. Notes

1. Bavistin is insoluble in water, Pantomycine is soluble. To prepare a solution with these two antimicrobial compounds, weigh compounds separately according to the concentration and mix in a stirrer in 25–30 ml distilled water for 5 min. and make up the volume. The resultant solution is a turbid one.
2. Adjust pH of medium before adding agar to it, with 1 N NaOH or 1 N HCl in a pH meter, calibrate each time before adjusting new sample medium. Sterilize glass wares and culture media in autoclave at 121°C and 103 kPa for 20 min.
3. To prepare stock solutions of BAP and Kinetin, add few drops of 0.1 N HCL to dissolve the crystals, then fill up to the required volume in a volumetric flask. Prepare IBA and NAA in the same way by dissolving crystals with 0.1 N NaOH.
4. Prepare ½ strength of MS and Miller's medium by reducing the inorganic salts only by ½ of the normal concentration, keeping other ingredients at full strength.
5. Make Filter paper bridge by folding filter paper (Whatman No.1, 7.0 cm diameter) so that a 5-layered thick, 1 cm broad strip is developed, make a notch and a little damage at the middle part and place carefully at the base of the culture tube containing liquid rooting medium.
6. Perforate poly bags at different sites, and cover the plastic cups containing plantlets. The poly bags protect young plants from direct air flow outside, and thus minimize the rate of transpiration and allow the entry of atmospheric gasses through small pores.

7. Specifically, March–April is the best time for collecting explants because at this time *Jasminum officinale* start growing vigorously after spending a lag phase during winter. After this period (March–April) again, plants undergo important hormonal changes, and start flowering when most of the vegetative buds change to flowering buds, and therefore unsuitable for in vitro culture and forced branching.
8. Spraying with antimicrobial agents before in vitro use is a necessary step as it reduces the high risk of spoiling explants after inoculation.
9. Avoid selecting tender parts of young branches because cut ends of explants from these parts start developing callus at the early stage of culture.
10. To complete phenolic leaching from cut end of the explants place 10 segments/Earlenmayer flasks (100 ml volume) containing 200 ml medium.
11. Intensity of brown leachate indicates the time of subculture.
12. Two subsequent cultures are usually enough to stop leaching, but it is advisable further changes in this regard if needed (due to varietal changes or eco-geographic differences).
13. Use separate culture tubes for single explants in shoot multiplication experiment because loss due to contamination hamper less in single culture rather than that in multiple culture in flasks. Promising cultures may be transferred from culture tube to larger container.
14. Keep at least a small bit of hard tissue at the base of newly emerged regenerated shoots and place properly on the notch of the filter paper bridge to allow smooth penetration of emerging roots.
15. Take special care during cleaning and removing of agar from roots to keep root tips intact and during transferring to solid substratum for placing roots properly and not clinging to each other.

References

1. Dadlani NK (1998) Cut flower production in India. In: Cut flower production in Asia. RAP Publ.
2. Kumar GS, Jayaveera KN, Ashok Kumar CK, Sanjay UP, Vrusabendra Swamy BM, Kishore Kumar DV (2007) Antimicrobial effects of Indian medicinal plants against acne-inducing bacteria. Trop J Pharm Res 6:717–723
3. Brown D (1995) Encyclopedia of herbs and their uses. Dorling, Kindersley, London, ISBN 0-7513-020-31
4. Iqbal M, Ghosh AKM, Saluja AK (2006) Antifertility activity of floral buds of *Jasminum officinale* var. *grandiflorum* in rats. Phytotherapy Res 7:5–8

5. Duke JE, Ayensu ES (1985) *Jasminum officinalis.* In: Medicinal plants of China. Reference Publ.
6. Bajaj YPS (ed) (1089) Biotechnology in agriculture and forestry, Vol.7 Medicinal and aromatic plants II. Springer, Berlin, Heidelberg, New York
7. George D, Chenieux JC, Ochatt SJ (1993) Plant regeneration from aged callus cultures of woody ornamental species of *Lonicera japonica* cv. "Hall's Prolific". Plant Cell Rep 13:91–94
8. Ochatt SJ (1991) Requirement for plant regeneration from protoplasts of the shrubby ornamental honey suckle *Lonicera nitida* cv. Maigram. Plant Cell Tiss Org Cult 25:161–167
9. Pati PK, Rath SP, Sharma M, Sood A, Ahuja PS (2006) In vitro propagation of rose – a review. Biotech Adv 24:94–114
10. Khoder M, Villemur P, Jonard R (1979) Micropropagation et bouturage in vitro chez le Jasmine (*Jasminum officinale* L.). CR Acad Sci Paris Ser D 293:343–346
11. Jonard R (1989) *Jasminum* spp. (Jasmine): Micropropagation and the production of essential oils. In: Bajaj YPS (ed) Biotechnology in agriculture and forestry. medicinal and aromatic plants II, Vol. 7. Springer-Verlag, Berlin-Heidelberg, New York, pp 315–331
12. Bhattacharya S, Bhattacharyya S (1997) Rapid multiplication of *Jasminum officinale* L., in vitro culture of nodal explants. Plant Cell Tiss Org Cult 51:57–60
13. Sahoo Y, Chand PK (1998) Micropropagation of *Vitex negundo* L. woody aromatic medicinal shrub, through high frequency axillary shoot proliferation. Plant Cell Rep 18:301–307
14. Ruzic DjV, Vujovic TI (2008) The effect of cytokinin types and their concentration on in vitro multiplication of sweet cherry cv. Lapins (*Prunus avium* L.). Hort Sci (Prague) 35:12–21
15. George FE (1993) Plant propagation by tissue culture, Part-2. Exegetics Ltd, UK
16. Murashige T, Skoog F (1962) A revised medium for rapid growth and bioassays witVh tobacco tissue cultures. Phyisol Plant 15:473–497
17. Miller CO (1967) Cytokinins in *Zea mays.* Ann NY Acad Sci 144:250–257

Chapter 13

Micropropagation of *Lysionotus pauciflorus* Maxim. (Gesneriaceae)

Toshinari Godo, Yuanxue Lu, and Masahiro Mii

Abstract

Numerous shoots were directly regenerated from the leaf explants of *Lysionotus pauciflorus* on the MS medium containing 0.5–2 μM NAA with or without 1 μM BA. The calli were induced from the leaves on MS medium supplemented with 2 μM 2, 4-D. The calli proliferated about four times in fresh weight in the liquid medium of the same composition as the callus induction medium after 4 weeks of culture on a rotary shaker at 100 rpm. Shoots were induced from these calli on the regeneration medium amended with 32 μM BA or 0.5 μM zeatin. Regenerated shoots rooted easily on ½ MS medium without any plant growth regulators. Most of the regenerants from callus were diploid, whereas eight of 66 acclimatized plantlets were tetraploid determined by flow cytometric analysis. Furthermore, flow cytometric analysis of calli also revealed tetraploidy.

Key words: Shoot regeneration, Callus, *Lysionotus pauciflorus*, Micropropagation, Plant growth regulator

1. Introduction

The family Gesneriaceae consists of about 2,000 species mainly distributed in the tropical and subtropical areas worldwide. The family includes many species having high ornamental value with compact size and colorful flowers. Many wild species and cultivars in some genera such as *Achimenes*, *Aeschynanthus*, *Chirita*, *Gloxinia*, *Saintpaulia* and *Streptocarpus* have been cultivated as exotic indoor plants. Many wild species have become endangered due to the destruction of their natural habitat and over-collection for horticultural uses. Recently, conservation and sustainable utilization of endangered species have been recognized as global tasks. Therefore, the establishment of propagation and breeding systems for Gesneriaceae species conserved in botanical gardens or botanical institutes as genetic resources is urgently needed (1).

S.M. Jain and S.J. Ochatt (eds.), *Protocols for In Vitro Propagation of Ornamental Plants*, Methods in Molecular Biology, vol. 589
DOI 10.1007/978-1-60327-114-1_13, © Humana Press, a part of Springer Science+Business Media, LLC 2010

Many researchers have already reported micropropagation of plants of the Gesneriaceae family, mainly genus *Saintpaulia* (2–7). Although several species of the genus *Lysionotus* have high ornamental value, micropropagation has only been reported on *Lysionotus pauciflorus* Maxim., which was collected from Japan as an endangered species (8). Recently we also reported the establishment of two micropropagation systems, shoot regeneration from leaf directly and via callus formation, of *L. pauciflorus* native to Yunnan Province, China (1).

2. Materials

2.1. Preparation of In Vitro Plant

1. Sterilization solution: 1% (available chlorine) NaOCl solution containing 1% (w/v) Tween 20. Prepare just before use.
2. Half strength MS (9) medium: half strength MS mineral salts, full strength MS organic constituents and 20 g/L sucrose, adjust pH to 5.8 before autoclaving at 121°C for 15 min.
3. Solidifying agent: 2 g/L gellan gum (Phytagel: Sigma-Aldrich, Inc. St. Louis, MO, USA). Add to the medium after adjusting pH, before autoclaving.
4. Parafilm (Alcan Packaging, Neenah, WI, USA).
5. 70 μm nylon mesh of Falcon cell strainer (Becton Dickinson Labware, Franklin Lake, NJ, USA).
6. Falcon 50 mL conical tube (Becton Dickinson Labware).

2.2. Direct Shoot Regeneration and Callus Induction from Leaf

1. MS medium: MS mineral salts, MS organic constituents, several concentrations of plant growth regulators and 30 g/L sucrose, pH adjust to 5.8 before autoclaving at 121°C for 15 min.

2.3. Flow Cytometric Analysis of Regenerated Plants

1. Buffer A: Partec High Resolution Staining Kit for Plant DNA Analysis (Partec GmbH, Münster, Germany).
2. DAPI staining solution: 2.5 mg/L DAPI (4′, 6-diamidino-2-phenylindole), 10 mM Tris-HCl, 50 mM sodium citrate, 2 mM $MgCl_2$, 1% (w/v) PVP K-30 (Polyvinylpyrrolidone K-30), 0.1% (v/v) Triton X-100, pH 7.5. Store at –20°C.

3. Methods

3.1. Preparation of In Vitro Plant

1. Collect seeds of *Lysionotus pauciflorus* Maxim. (Gesneriaceae) plants. This plant is distributed in China, Vietnam and Japan.

It has high ornamental value with beautiful flowers (Fig. 13.1; *see* Note 1).

2. Soak seeds in a glass tube (20 × 150 mm) by adding 20 mL sterilization solution, seal the tube with parafilm and surface sterilize seeds while shaking slowly for 10 min.
3. Collect seeds on 70 μm nylon mesh of falcon cell strainer attached to a Falcon 50 mL conical tube, and rinse seeds on the mesh with sterilized water using a Pasteur pipette.
4. Sow 3–5 seeds in a glass tube (20 × 150 mm).containing 10 mL gellan gum-solidified ½ MS medium devoid of plant growth regulators.
5. For seed germination, incubate these cultures at 25 ± 2°C under the white fluorescent light, 40 μmol/m^2/s with a 16 h photoperiod (*see* Note 2).
6. Select the best growing plants among the seedlings for further plant propagation.
7. Cut the selected plants to single node segments and transfer 5 to 7 node segments in 500 mL glass bottles having 100 mL gellan gum-solidified ½ MS medium devoid of plant growth regulators.
8. Culture them at 25 ± 2°C under white fluorescent light, 40 μmol/m^2/s, with a 16 h photoperiod.

Fig. 13.1. *Lysionotus pauciflorus* Maxim. (Gesneriaceae) collected in the southeastern part of Yunnan province, China.

9. Maintain the cultures at 2–3 month-intervals by transferring single node segments (*see* Note 3).

3.2. Direct Shoot Regeneration and Callus Induction from Leaf

1. Excise leaves from in vitro plants. When using a large leaf, cut it perpendicularly into two explants at the midrib.
2. For direct shoot regeneration, place ten leaf explants in plastic Petri dish (90 × 20 mm) filled with 50 mL solidified MS medium containing 0.5–2 μM NAA with/without 1 μM BA.
3. For callus induction, place ten leaf explants in each plastic Petri dish (90 × 20 mm) containing 50 ml solidified MS medium containing 2 μM 2, 4-D.
4. Incubate both cultures at 25 ± 2°C under the white fluorescent light, 40 μmol/m^2/s, 16 h photoperiod. The results are shown in Table 13.1, Figs. 13.2 and 13.3 (*see* Notes 4 and 5).

3.3. Callus Proliferation and Establishment of Suspension Culture

1. After 4 months of culture, transfer calli induced on MS medium supplemented with 2 μM 2, 4-D onto 50 mL of the fresh solidified MS medium in plastic Petri dish (90 × 20 mm), and culture on the same condition for proliferation.
2. After 2 months, transfer 2 g (fresh weight) proliferated calli into 40 mL liquid MS medium with the same composition as callus induction medium in 100 mL Erlenmeyer flasks with polypropylene caps.
3. Culture on a rotary shaker, 100 rpm, 25°C, under 16 h dim light condition (ca. 3 μmol/m^2/s).
4. Subculture 2 g (fresh weight) calli to the same fresh liquid culture medium at 4-week-interval by chopping large callus clusters (over 1 cm diameter) to small pieces of ca. 0.5–1 cm in diameter using spatula.
5. After transfer, culture on a rotary shaker under the same conditions for proliferation and maintenance (Fig. 13.4).

3.4. Shoots Regeneration from Callus

1. Transfer ten callus clusters (ca. 5 mm in diameter, ca. 0.6 g of total weight) proliferated in the liquid MS medium amended with 2 μM 2, 4-D onto 50 mL solidified MS medium containing 32 μM BA or 0.5 μM zeatin in 200 mL glass bottle.
2. Culture at 25 ± 2°C under the white fluorescent light (40 μmol/m^2/s) with a 16 h photoperiod. The results are shown in Table 13.2 and Fig. 13.5 (*see* Note 6).

3.5. Rooting

1. Excise 1 cm long regenerated shoots from leaf segments or calli.
2. Transfer 10 to 15 shoots to 100 mL fresh solidified ½ MS medium without any plant growth regulators in each 500 mL glass bottle.

Table 13.1
Effect of plant growth regulators on leaf culture of *Lysionotus pauciflorus* (Reproduced from ref. 1 with permission from Sejani Ltd.)

Plant growth regulator (μM)			
BA	NAA	2,4-D	Response
0	0	0	S,R
0	0.25	0.25	S,R
1	0.25	0.25	S,R
0	0.5	0	S,R
0	0	0.5	S,R
0	1	1	R
1	1	1	R
0	2	0	S,R
1	2	0	S,R
0	0	2	C
1	0	2	C
0	4	4	C
1	4	4	C
0	8	0	S,R
0	0	8	C
0	16	16	-
1	16	16	-
0	32	0	S,R
0	0	32	-
0	64	64	-
1	64	64	-
0	128	0	R
0	0	128	-

All data were scored after 2 months of culture
S shoot, *R* root, *C* callus, - no response

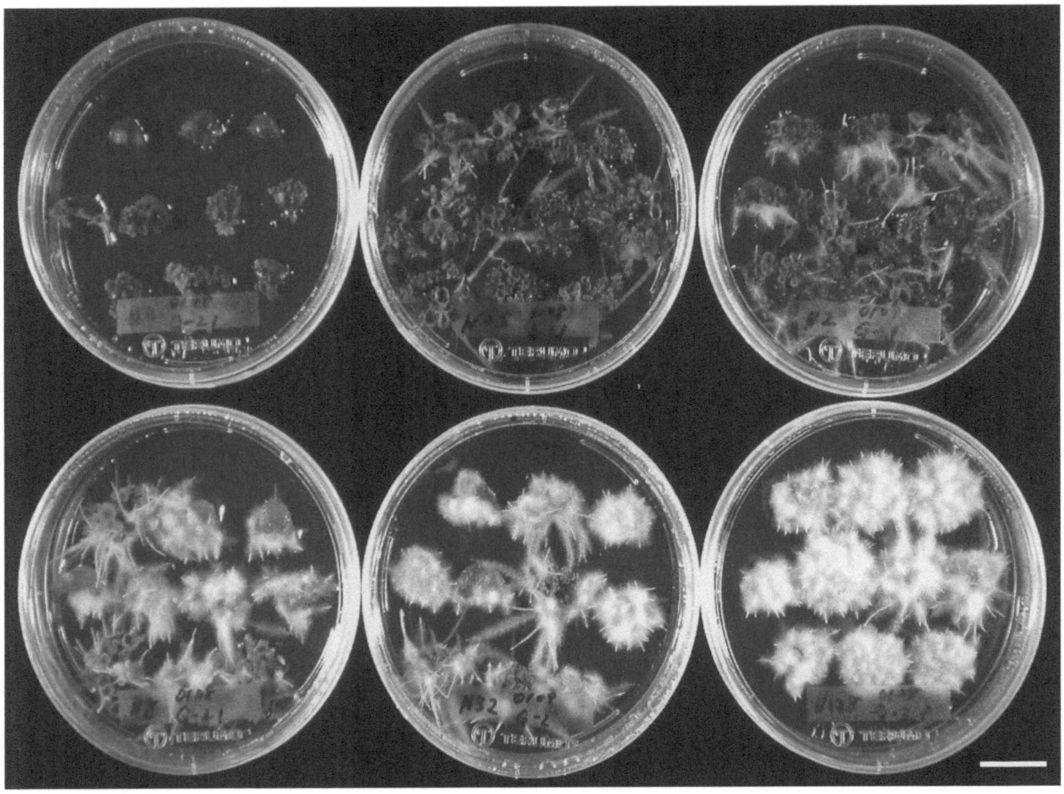

Fig. 13.2. Response of leaf explants of *L. pauciflorus* to different concentrations of NAA in culture medium. *Upper left*: 0 μM. *Upper center*: 0.5 μM. *Upper right*: 2 μM. *Bottom left*: 8 μM. *Bottom center*: 32 μM. *Bottom right*: 128 μM. Numerous shoots were induced from leaf explants on medium containing low concentrations (0.5–2 μM) of NAA. On the other hand, numerous adventitious roots were induced on medium with high concentration (32 or 128 μM) of NAA. The bar indicates 3 cm (Reproduced from ref. 1 with permission from Sejani Ltd.).

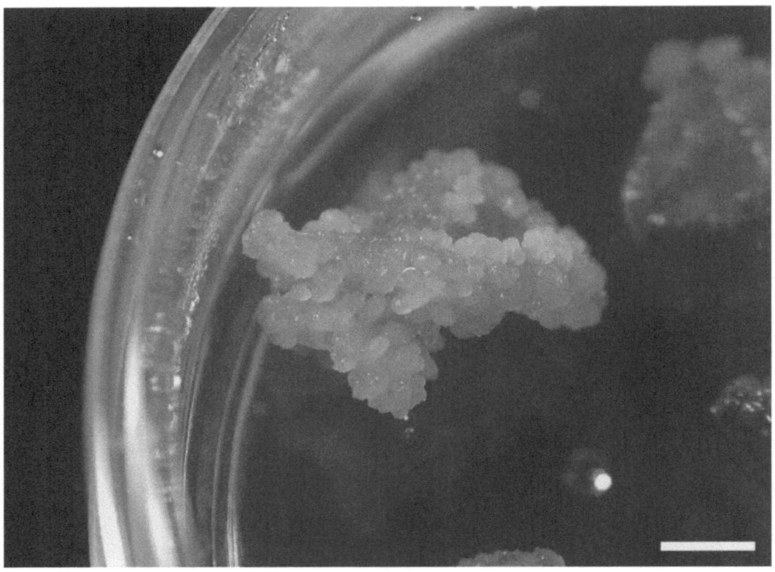

Fig. 13.3. Callus induced from leaf segment of *L. pauciflorus* after 2 months of culture on MS medium containing 2 μM 2,4-D. The callus formation was initiated within 2 weeks after initiation of leaf culture. Although calli were induced on MS medium containing 2–8 μM 2,4-D irrespective of the presence of 1 μM BA or 4 μM NAA, the medium with 2 μM 2,4-D alone showed good performance in induction and growth of callus. Consequently, it was used as a subculture medium of callus. The bar indicates 1 cm (Reproduced from ref. 1 with permission from Sejani Ltd.).

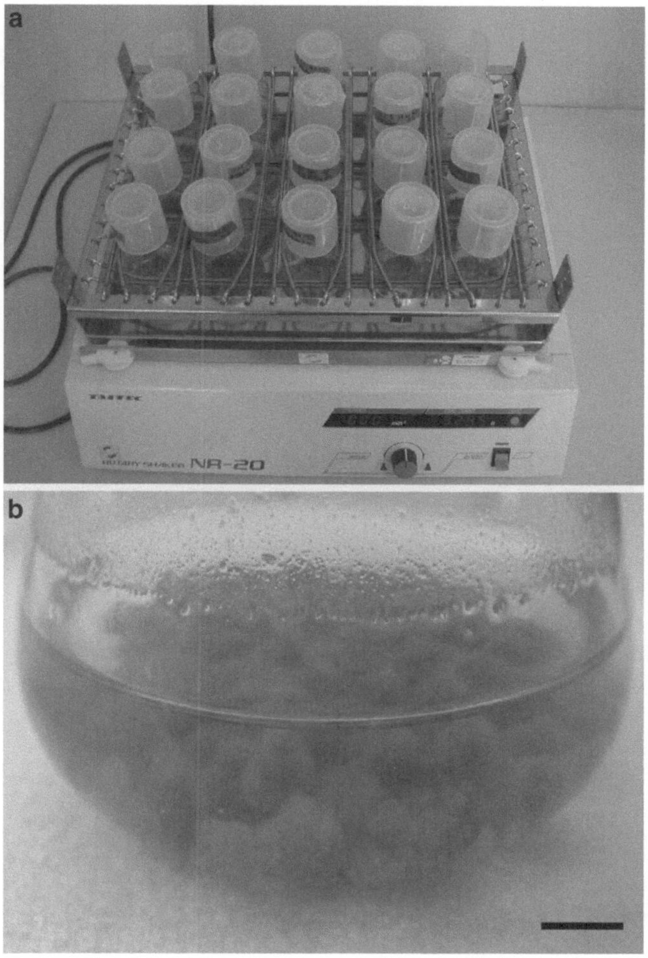

Fig. 13.4. Suspension culture of callus established in liquid medium containing 2 μM 2,4-D. After transfer to liquid medium supplemented with 2 μM 2,4-D and culture on a rotary shaker at 100 rpm, calli proliferated as clusters of ca. 0.5–1 cm in diameter. Four weeks after transferring, 2 g of callus proliferated about four times in fresh weight. (**a**) a rotary shaker used for suspension culture. (**b**) callus proliferated in liquid medium 4 months after transferring. The bar indicates 1 cm (Reproduced from ref. 1 with permission from Sejani Ltd.).

3. Culture at 25 ± 2°C under the white fluorescent light (40 μmol/m^2/s) with a 16 h photoperiod for rooting and further growth (*see* Note 7).

3.6. Acclimatization of Regenerated Plants

1. Gently pull out the rooted plantlets from the culture glass bottles.
2. Remove the culture medium adhering to roots very carefully without any damage in a water container.
3. Transfer the plantlets in 3 cm plastic pots filled with compost such as vermiculite without adding fertilizer.
4. Keep plantlets in a container covered with a plastic sheet to maintain over 90% relative humidity at 25 ± 2°C for 1–2 weeks.

Table 13.2
Effects of type and concentration of plant growth regulators on plant regeneration from callus of *Lysionotus pauciflorus* (Reproduced from ref. 1 with permission from Sejani Ltd.)

BA (μM)	TDZ (μM)	Zeatin (μM)	2,4-D (μM)	Shoot regeneration (%)[a]	Death (%)
2	–	–	–	5.0 ± 5.0 b	0
8	–	–	–	5.0 ± 5.0 b	0
32	–	–	–	30.0 ± 0.0 a	0
64	–	–	–	0 b	0
2	–	–	2	0 b	100[b]
8	–	–	2	0 b	100[b]
32	–	–	2	0 b	100[b]
–	2	–	–	15.0 ± 5.0 a	0
–	8	–	–	0 b	0
–	32	–	–	0 b	70
–	–	0.5	–	32.5 ± 7.5 a	0
–	–	2	–	17.5 ± 2.5 a	0
–	–	8	–	5.0 ± 5.0 b	0

All data were recorded after 4 months of culture

[a]Means in the column followed by the same letter are not significantly different at $p <$ the 0.05 level by Ryan's method

[b]Dead after callus proliferation

Plastic sheet can be removed after 2–3 weeks of acclimatization of plants.

5. Transfer the acclimatised plantlets to the greenhouse for further growth (Fig. 13.6) (*see* Note 8).

3.7. Flow Cytometric Analysis of Regenerated Plants

1. Take a leaf sample from each plantlet grown in the greenhouse, callus cultured in the liquid medium, and a leaf of in vitro mother plant as a control.
2. For the isolation of nuclei, chop the leaf tissues or 0.2 g callus with a razor blade in a plastic Petri dish containing 1 ml buffer A.
3. After 5 min incubation, add 1.5 ml DAPI staining solution.
4. Filtrate the nuclear suspension through a 40 μm nylon mesh to remove tissue segments and incubate for 10 min at 25°C.
5. Analyze all samples with a Partec PA flow cytometer. The results are shown in Fig. 13.7 (*see* Note 9).

Fig. 13.5. Shoot regeneration from callus after 4 months of culture. After transferring to regeneration medium containing cytokinin from liquid proliferation medium containing 2 μM 2,4-D, color of calli turned green and shoots were induced. A single application of 32 μM BA or 0.5 μM zeatin to the regeneration medium gave the highest shoot regeneration rate (about 30%) from calli, 4 months after transferring, but addition of 2,4-D inhibited shoot regeneration induced by BA. The bar indicates 5 mm (Reproduced from ref. 1 with permission from Sejani Ltd.).

Fig. 13.6. Acclimatised plants regenerated from calli of *L. pauciflorus* (Reproduced from ref. 1 with permission from Sejani Ltd.).

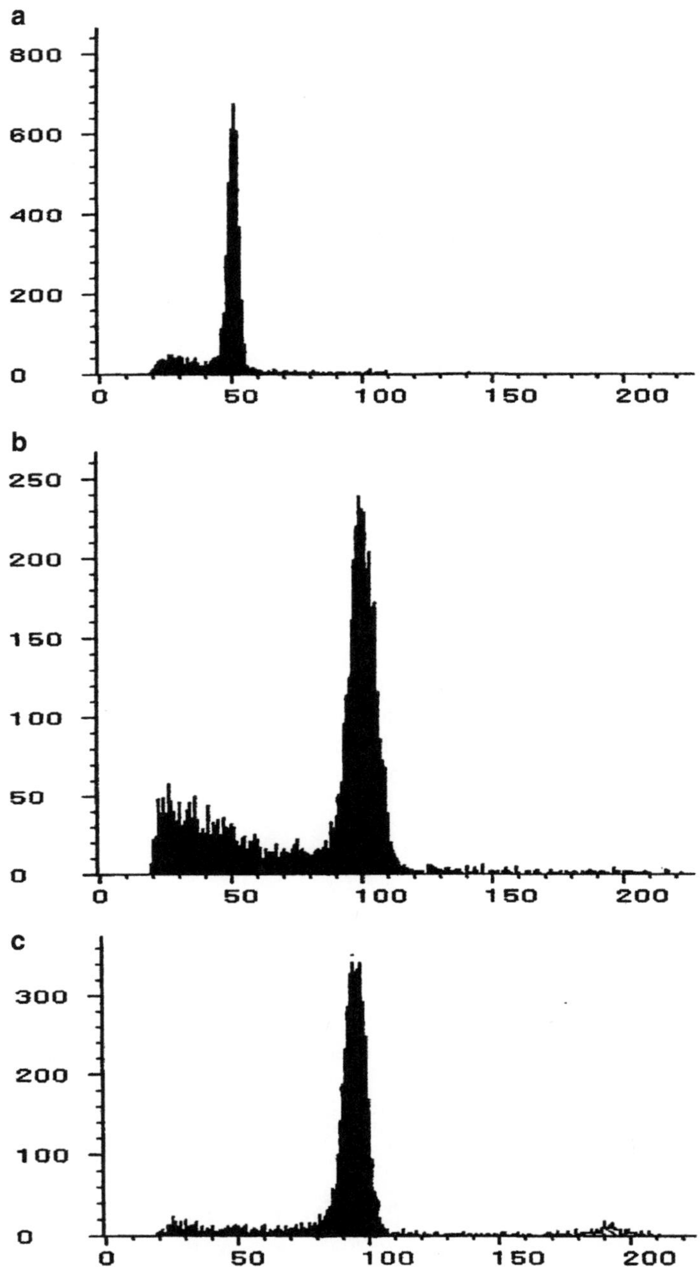

Fig. 13.7. Flow cytometric analysis for nuclear DNA contents of original in vitro plant (**a**), tetraploid regenerant (**b**) and tetraploid calli (**c**) (Reproduced from ref. 1 with permission from Sejani Ltd.).

4. Notes

1. Our previous results showed *Lysionotus pauciflorus* native to Yunnan, China is a diploid with 32 chromosomes (1).
2. Optimam temperature is 25°C and light are indispensable for seed germination of this species (8).

3. Plants propagated by a single-node culture were genetically stable as previously reported (10).
4. Embryo-like structures (Fig. 13.8a) were induced from leaf explants cultured on medium supplemented with NAA and 2,4-D, each 1 μM. Although they looked like somatic embryos initially, histological observations showed that these structures were monopolar (Fig. 13.8b) and developed into tiny roots with root hairs during culture on the same medium (Fig. 13.8c). Therefore, we concluded that these structures were not somatic embryos but root primordia.

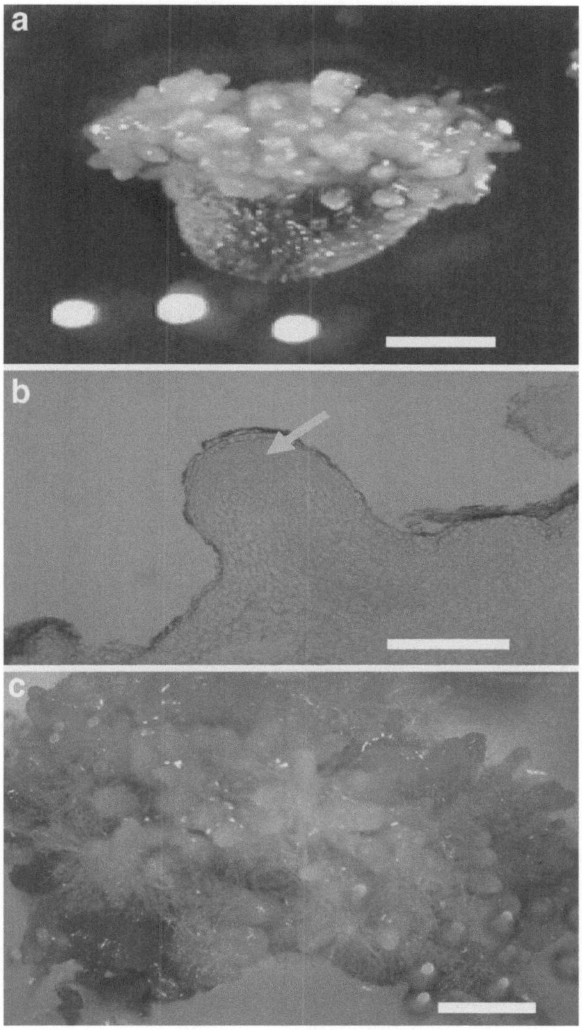

Fig. 13.8. Formation of root primordium-like structures on leaf explants. (**a**) numerous root primordium-like structures induced on the surface of leaf explants on medium containing both 1 μM NAA and 1 μM 2,4-D after 2 months of culture. The bar indicates 5 mm. (**b**) histological observation of root primordium-like structure. Arrow indicates meristem of adventitious root. The bar indicates 1 mm. (**c**) root primordium-like structures with root hairs. The bar indicates 3 mm (Reproduced from ref. 1 with permission from Sejani Ltd.).

5. High concentrations (16–128 μM) of 2,4-D inhibited both callus and shoot induction from leaf tissues (Table 13.1).
6. Since the cell culture system has retained high plant regeneration ability for more than 2 years, it might also be utilized for somatic hybridisation and genetic transformation studies. In Gesneriaceae plants, protoplast culture and genetic transformation are known to be recalcitrant and both could be achieved by utilising callus culture with high plant regeneration ability (11,12).
7. Regenerated shoots easily rooted on ½ MS medium without any plant growth regulators.
8. Shoots having a few variegated leaves were often regenerated from callus, but they did not produce variegated leaves continuously during subsequent in vitro and ex vitro cultures.
9. Most of the regenerants from callus were diploid, whereas 8 out of 66 acclimatised plantlets were tetraploid, shown by flow cytometric analysis (Fig. 13.6). Tetraploids are sometimes useful in ornamental plants due to alterations of morphological and physiological characteristics such as longer shoots, thicker stems, deeper green leaf color, larger and thicker leaves and larger flowers (13–18). They could become an excellent source of high value added ornamental plants for the floriculture industry.

Acknowledgments

We wish to thank Dr. Masashi Nakata of Botanic Gardens of Toyama for chromosome observations, Dr. Tohru. Ohmiya of Botanic Gardens of Toyama for cytological observation, Dr. Dong Poh Chin of Chiba University for flow cytometric analysis and Dr. Kaiyun Guan of Kunming Institute of Botany for useful information.

References

1. Lu YX, Godo T, Chin DP, Mii M, Guan KY (2006) Establishment of callus culture with high plant regeneration ability from leaf segments of *Lysionotus pauciflorus* Maxim. Propagation Ornament Plants 6:180–186
2. Norris R, Smith RH, Vaughn KC (1983) Plant chimeras used to establish de novo origin of shoots. Science 220:75–76
3. Lo KH, Giles KL, Sawhney VK (1997) Acquisition of competence for shoot regeneration in leaf discs of *Saintpaulia ionantha* × *confusa* hybrid (African violet) cultured in vitro. Plant Cell Rep 6:416–420
4. Lo KH, Giles KL, Sawhney VK (1997) Histological changes associated with acquisition of competence for shoot regeneration in leaf discs of *Saintpaulia ionantha* × *confusa* hybrid (African violet) cultured in vitro. Plant Cell Rep 6:421–425
5. Molgaard JP, Roulund N, Deichmann V, Irgens-Moller L, Andersen SB, Farestveit B (1991) In vitro multiplication of *Saintpaulia*

ionantha Wendl. by homogenization of tissue cultures. Sci Hortic 48:285–292

6. Ohki S (1994) Scanning electron microscopy of shot differentiation in vitro from leaf explants of the African violet. Plant Cell, Tissue Organ Cult 36:157–162
7. Smith RH, Norris RE (1983) In vitro propagation of African violet chimeras. HortScience 18:436–437
8. Haruki K, Inamura H (2003) Propagation and growing of *Lysionotus pauciflorus* Maxim. Bull Shimane Agric Experiment Station 34:61–70 (in Japanese)
9. Murashige T, Skoog F (1962) A revised medium for rapid growth and bioassays with tobacco tissue cultures. Physiologia Plantarum 15:473–497
10. Pierik RLM (1997) In vitro culture of higher plants. Kluwer Academic Publishers, Dordrecht, pp 190–192
11. Hoshino Y, Nakano M, Mii M (1995) Plant regeneration from cell suspension-derived protoplasts of *Saintpaulia ionantha* Wendl. Plant Cell Rep 14:341–344
12. Kushikawa S, Hoshino Y, Mii M (2001) *Agrobacterium*-mediated transformation of *Saintpaulia ionantha* Wendl. Plant Science 161:953–960
13. Chen CH, Goeden-Kallemeyn Y (1979) In vitro induction of tetraploid plants from colchicine-treated diploid daylily callus. Euphytica 28:705–709
14. Lindsay GC, Hopping ME, O'Brien IEW (1994) Detection of protoplast-derived DNA tetraploid lisianthus (*Eustoma grandiflorum*) plants by leaf and flower characteristics and by flow cytometry. Plant Cell, Tissue Organ Cult 38:53–55
15. Takamura T, Miyajima I (1996) Colchicine induced tetraploids in yellow-flowered cyclamens and their characteristics. Sci Hortic 65:305–312
16. Väinölä A (2000) Polyploidization and early screening of *Rhododendron* hybrids. Euphytica 112:239–244
17. Kermani MJ, Sarasan V, Roberts AV, Yokoya K, Wentworth J, Sieber VK (2003) Oryzalin-induced chromosome doubling in *Rosa* and its effect on plant morphology and pollen viability. Theor Appl Genet 107:1195–1200
18. Gu XF, Yang AF, Meng H, Zhang JR (2005) *In vitro* induction of tetraploid plants from diploid *Zizyphus jujuba* Mill. cv. Zhanhua. Plant Cell Rep 24:671–676

Chapter 14

Micropropagation of *Rhododendron*

Tom Eeckhaut, Kristien Janssens, Ellen De Keyser, and Jan De Riek

Abstract

Methods for in vitro initiation and multiplication and general culture practices of *Rhododendron* are presented. Also acclimatization procedures are described. Protocols for callus, shoot, and root induction are described. Several protocols for breeding applications are highlighted in more detail.

Key words: Azalea, Callus, Interspecific hybridization, In vitro, Micropropagation, Polyploidization, Regeneration, Rhododendron, Sport induction

1. Introduction

The genus *Rhododendron* (±1,000 species) is divided into 8 subgenera, the 4 most important subgenera being Tsutsusi (evergreen azaleas, 117 species), Pentanthera (deciduous azaleas, 30 species), Rhododendron (lepidote *Rhododendrons*, 542 species) and Hymenanthes (elepidote *Rhododendrons*, 302 species) (1). These subgenera are economically important due to their ornamental value: the latter two constitute what gardeners loosely refer to as "garden rhododendrons," whereas Tsutsusi and Pentanthera species are the origin of many hybrid groups of both indoor and outdoor azaleas. Tropical *Rhododendron* species (Vireya's, 310 species) are included in the subgenus Rhododendron. Traditionally, stem cuttings are used to propagate most *Rhododendron* species. However, in vitro micropropagation, that gradually became common after the development of shoot-tip culture, is becoming increasingly important in commercial production. Efforts to establish efficient protocols have been ongoing for some decades. Protocols for micropropagation, organogenesis, somaclonal variation, polyploidization and interspecific hybridization in *Rhododendron* are presented.

S.M. Jain and S.J. Ochatt (eds.), *Protocols for In Vitro Propagation of Ornamental Plants*, Methods in Molecular Biology, vol. 589
DOI 10.1007/978-1-60327-114-1_14, © Humana Press, a part of Springer Science+Business Media, LLC 2010

2. Materials

1. Prior to in vitro initiation, pot plants are kept in the greenhouse, under normal conditions, 18–20°C, 65% relative humidity.
2. The four main basal media that are described for in vitro culture of *Rhododendron* are Anderson medium (2), woody plant medium (3), Murashige and Skoog medium (4) and Economou and Read medium (5), henceforth respectively referred to as BMA, BMW, BMM and BME. They are purchased as a powder, dissolved in water and supplemented with a carbon source, phytohormones (cytokinins, auxins and/or gibberellins) and/or a gelling agent, pH-corrected, autoclaved, and poured into a suitable recipient and inoculated with plant explants.
3. Cytokinins 2iP (2-isopentenyladenine), Z (zeatin) and BAP (6-benzylaminopurine) are dissolved in a few drops of HCl and further diluted with H_2O; auxins IAA (indole-3-acetic acid), NAA (1-naphthalene acetic acid), IBA (indole-3-butyric acid) and 2,4-D (2,4-dichlorophenoxy acetic acid) and cytokinin TDZ (thidiazuron) are dissolved in a few drops of KOH and further diluted with H_2O (10 mg/mL). All solutions can be stored for several months (–20°C, darkness).
4. The standard carbon source for all media described in the manuscript is 30 g/L sucrose, and the gelling agent is 6–8 g/L agarose. The pH of all media is set at 5.0 (deciduous azaleas) – 5.8 (evergreen azaleas) by means of small aliquots of 0.1 N KOH or HCl. Media are autoclaved (121°C, 30 min, 500 hPa) and can be subsequently stored (4°C, darkness) for 1 month.
5. Recipients used are Meli-jars (100 mL), Weck pots (100 mL), petri dishes of either 5 or 9 cm diameter (respectively 10 and 25 mL), and glass tubes (20–40 mL). They are usually sealed with low density polyethylene tape.
6. All sterile handlings (filling of the recipients, filter sterilizations, explant sterilization and handling) were performed under a laminar flow cabinet that was surface cleaned with 70% ethanol. Paper was sterilized (220°C, 3 h) to provide a sterile working surface. Scalpels and pincets are sterilized in a glass bead sterilizer (250°C).
7. Explants are grown in an environmentally controlled room with temperatures of 21–25°C and 16 h PAR (photosynthetic active radiation) of 40–75 $\mu mol/m^2/s$, supplied by cool white fluorescent lamps.
8. All ingredients are added prior to autoclaving, except GA_3, trifluralin, oryzalin and colchicine. Solutions (various concentrations) of the latter 3 are made in H_2O enriched with 0.5% DMSO and

0.05% Teepol. They are cell mitosis inhibitors and should be handled carefully. GA_3 is dissolved in a few drops of KOH and further diluted with H_2O (10 mg/mL). Non autoclavable solutions are sterilized by filtration (0.22 μm) and stored (−20°C, darkness).

9. The mineral, sucrose and agar levels and all environmental circumstances mentioned in this section are used for all applications mentioned in this manuscript and for any genotype, except when specifically mentioned otherwise.

3. Methods

3.1. Initiation

3.1.1. Vegetative Shoot Tips

1. Plants cultivated in pots in greenhouse conditions are periodically fertilized and pruned for good vegetative growth. This ensures supply of 1 cm long unlignified shoots that can be used as explant material.
2. A young shoot with approximately 15 pale green leaves is cut from the mother plant. The majority of the leaf is cut off, only the petiole and the leaf base are maintained onto the shoot to protect the axillary buds during sterilization.
3. Surface sterilization is performed by rinsing in 70% ethanol for 30 s and subsequent sterilization for 15–20 min in 1% NaOCl, along with a few drops of detergent (e.g., Tween20). Finally, explants are rinsed three times in sterile water (Note 1).
4. On sterile paper, the base of the sterilized shoot is removed because it is contaminated with NaOCl. The remaining part is cut into pieces of approximately 1.5 cm. These cuttings are planted horizontally on multiplication medium. Sterile Weck pots or Meli-jars are the most common culture containers. Axillary buds will sprout and generate young shoots (1–4 months).

3.1.2. Floral Structures and Seeds

1. Ovaries, ovules or seeds are dissected from pot plants kept in greenhouse conditions. With a fine scalpel, hairs are removed from the ovary before sterilization.
2. Sterilization procedures: see above (*see* Note 2).
3. After rinsing (see above), ovaries can be directly inoculated. For the inoculation of seeds or ovules, ovaries are cut longitudinally on sterile paper into five parts.
4. Explants are usually initiated in small petri dishes in basal or regeneration medium.

3.1.3. Meristems

1. Sterilization procedures: see above.
2. Using a binocular microscope and fine forceps, the vegetative buds are peeled down and the meristem is placed on 20–40 ml basal or regeneration medium in a small tube.

3.2. Multiplication (Table 14.1)

1. On sterile paper, shoot bundles are dissected into individual shoots, which are transferred to fresh medium and vertically planted (*see* Note 3).
2. For standard shoot multiplication, 10 g/L bacto-agar, 15 mg/l 2iP and 4 mg/L IAA (indole-3-acetic acid) are added to BMA (*see* Note 4). Adventitious shoots are induced at the base of the explant.
3. Axillary shoots are induced at lower cytokinin concentrations, down to 0.1 mg/L 2iP.
4. This is periodically repeated after every subculture cycle (usually 6–10 weeks) (Fig. 14.1). Prior to eventual transfer to soil, plantlets must be rooted.

Table 14.1
Protocols for in vitro multiplication of various *Rhododendron* genotypes

Genotype	Explant	Medium	Reference
VGE	Meristem	1/10 BMM + 2–5 mg/L BAP + 20 g/L sucrose	(11)
VGE	Shoots	1/2 BMA + 0.05 mg/L TDZ + 0.5 mg/L Z	(12)
VGE	Shoots	1/10 BMM + 245 mg/L KH_2PO_4 + 250 mg/L KNO_3 + 600 mg/L $MgSO_4$ + 20 mg/L sucrose + 2.25 mg/L BAP	(13)
R. "Petrick," *R.* "Rex"	Shoots	BME (sequestrene replaced by $Na_2EDTA + FeSO_4$) + 10 mg/L 2iP + 0.9 mg/L IAA	(14)
Menziesia × VGE	Seedlings	BMA + 10 mg/L 2iP + 3 g/L gellan gum	(15)
VGD	Shoots	BME + 1 mg/L IAA + 5 mg/L 2iP	(5)
R. prinophyllum	Shoots	BME + 1 mg/L IAA + 5 mg/L 2iP	(16)
R. "Montego"	Shoots	BMW + 2–10 mg/L 2iP	(17)
R. "PJM" hybrids	Shoots	BMA + 1–2 mg/L 2iP (pH 6.0)	(18)
R. "President Roosevelt"	Vegetative and flower buds, shoots	BMW + 8.12 mg/L 2iP + 1.5 g/L gelrite	(19)
R. maddenii	Shoots	BMA + 2 mg/L 2iP + 0.1 mg/L NAA	(20)
R. laetum	Shoots	BMA + 15 mg/L 2iP	(21)
Various genotypes	Shoots	BMW + 1.6–3.2 mg/L 2iP	(22)

VGD various genotypes of deciduous azalea, *VGE* various genotypes of evergreen azalea

Fig. 14.1. *Rhododendron simsii* "Inga" on BMA + 5 mg/L 2iP + 1 mg/L IAA: typical shoot bundle (*left*). *Rhododendron simsii* (unnamed seedling) on BMA + 15 mg/L 2iP: shoot multiplication in Meli jars (*right*). Bars = 1 cm.

Table 14.2
Protocols for in vitro rooting of various *Rhododendron* genotypes

Genotype	Medium	Reference
VGE	1/10 BMM + 2 mg/L IAA + 2.5 g/L charcoal	(11)
VGE	1/10 BMM + 20 g/L sucrose + 2 mg/L IAA + 2.5 g/L charcoal	(23)
VGE	1/10 BMM + 245 mg/L KH_2PO_4 + 250 mg/L KNO_3 + 600 mg/L $MgSO_4$ + 20 g/L sucrose + 1.75 mg/L IAA + 0–3 g/L charcoal	(13)
R. "Petrick," *R.* "Rex"	BME (sequestrene replaced by Na_2EDTA + $FeSO_4$) + 20 g/L sucrose	(14)
Menziesia × VGE	BMA + 3 g/L gellan gum	(15)
VGD	1/2 BME + 1–4 mg/L IAA or IBA	(5)
R. prinophyllum	1/2 BME + 20 g/L sucrose + 1–4 mg/L IAA or IBA + 1 g/L charcoal	(16)
R. catawbiense	1/2 BMA + 20 g/L sucrose + 2–4 mg/L IAA or IBA + 1 g/L charcoal	(24)
R. ponticum	BMA (1/2 macrosalts)	(25)

VGD various genotypes of deciduous azalea, *VGE* various genotypes of evergreen azalea

3.3. Rooting (Table 14.2)

1. Shoot bundles are dissected into individual shoots, on sterile paper.
2. Those shoots are transferred to a new medium promoting rhizogenesis (*see* Note 5). Rooting can take up to 2–3 months.
3. Rooted plantlets (Fig. 14.2) can be acclimatized.

Fig. 14.2. Root induction in *Rhododendron simsii* on BMA + 20 g/L sucrose + 2.5 g/L charcoal + 0.8 mg/L IBA. Bar = 1 cm.

3.4. Acclimatization

1. Remove agar carefully from roots of plantlets (single shoots).
2. After 3 weeks rooting in plastic seedling trays, in a fog unit of the greenhouse (18–20°C, relative humidity 95–100%, 16 h photoperiod), plants are transferred to a multi-layer growing room (21 ± 1°C, relative humidity 75–80%, 16 h photoperiod, CO_2-concentration 350–450 ppm).
3. During the following 9 weeks and after acclimatization, plantlets are fertilized regularly. Recommended substrates are sphagnum peat, pH 4.0–4.5, containing 60% organic matter (acclimatization) or sphagnum peat 4: coconut fiber 1, pH 4 (subsequent cultivation).

3.5. Callus Induction (Table 14.3)

1. The top 5 leaves of an in vitro plantlet are cut and placed on the medium in petri dishes with the adaxial side up.
2. For both induction and regeneration of callus (*see* Note 6), transversally cutting the leaves is preferable over other wounding types.
3. If necessary, callus burning can be prevented by lower light intensities or covering the recipients with 1 layer of paper.

Table 14.3
Protocols for in vitro callus induction in various *Rhododendron* genotypes

Origin	Medium	Reference
VGE, meristem	1/10 BMM + 20 g/L sucrose + 0.2–2 mg/L IAA	(11)
R. simsii leaves	BMA + 20 g/L sucrose + 2–4 mg/L 2,4-D + 1.2 mg/L Z	(23)
R. uwaense 5 × 5 mm leaf sections	1/2 BMM + 0.5 mg/L NAA + 0.5–1 mg/L 2iP or TDZ + 2.5 g/L gelrite	(26)
R. "PJM" hybrids stamina	BMA - 1.75 mg/L IAA + 5 mg/L 2iP + 0–0.2 mg/L TDZ	(27)
R. catawbiense "Grandiflorum" leaves	BMM + 60 g/L sucrose + 40 mg/L adenine hemisulphate + 100 mg/L casein hydrolysate + 2.5 mg/L TDZ	(28)
R. laetum × *R. aurigeranum* axillary shoot tips	BMA + 4 mg/L IAA + 15 mg/L 2iP	(21)

VGE various genotypes of evergreen azalea

Fig. 14.3. Callus induction on *Rhododendron simsii* (unnamed seedling) leaves (*left*) and subsequent regeneration of shootlets (*right*), respectively on BMA + 1 mg/L Z + 2 mg/L 2,4-D and BMA + 2 mg/L TDZ + 0.5 mg/L NAA. Bars = 1 mm.

4. Callus is formed from 2 weeks onwards (Fig. 14.3), and cultures generally last 1–2 months. Mostly regeneration through shoots is attempted (Fig. 14.3).

3.6. Shoot Induction (Table 14.4)

1. Callus is cut off from the explants and pressed firmly on the surface of the new medium (*see* Note 7), in a petri dish or a tube. Alternatively, an intact explant is cultured, with or without callus, dependent on the genotype.

Table 14.4
Protocols for in vitro shoot induction in various *Rhododendron* genotypes

Origin	Medium	Reference
VGE leaves	BMA + 5 mg/L TDZ + 4 mg/L IAA	(29)
R. simsii leaf callus	BMA + 160 mg/L adenine sulfate + either 2 mg/L TDZ or 24 mg/L Z	(23)
R. "Petrick," *R.* "Rex" leaves	BME (sequestrene replaced by $Na_2EDTA + FeSO_4$) + 10 mg/L 2iP + 0.9 mg/LIAA	(14)
R. "Fuchsia," *R.* "Hino Crimson" leaves	1/2 BMA + 0.5 mg/L Z or 0.25–0.5 mg/L TDZ or 0.25 mg/L TDZ + 0.5 mg/L Z	(12)
R. uwaense leaf callus	1/2 BMM + 2.5 g/L gelrite; after 3 months transfer to 1/2 BMM + 0.5 mg/L BA	(26)
R. "PJM" leaves	BMA + 10 mg/l 2iP + 2 mg/l IBA; after 2 weeks transfer to BMA + 0.2 mg/l IBA + 5–15 mg/l 2iP or 0.02 mg/l TDZ	(30)
R. laetum × *R. aurigeranum* leaf callus	BMA + 4 mg/l IAA + 30 mg/l 2iP	(21)
Various genotypes leaves	BMA + 1 mg/L IBA + 15 mg/L 2iP	(31)
R. prinophyllum ovaries	BMA + 4 mg/L IAA + 15 mg/L 2iP (darkness); after 1 month 35–50 μmol/ m^2/s	(16)
R. "Irina" ovary and pedicels	BMA + 20 g/L sucrose + 10 g/L glucose + 15 mg/L 2iP + 3 mg/L IBA + 0.05–0.1 mg/L TDZ (PAR 35–50 μmol/m^2/s)	(32)
R. "PJM" hybrids stamina	BMA + 1.75 mg/L IAA + 5 mg/L 2iP + 0–0.2 mg/L TDZ	(27)
R. catawbiense ovary and pedicels	BMA + 1–4 mg/L IAA + 5–15 mg/L 2iP (darkness)	(24)
R. maddenii cotyledonary nodal parts	BMA + 2 mg/L 2iP + 0.1 mg/L NAA	(20)

VGE various genotypes of evergreen azalea

2. Cultures are incubated in the dark for 1 week followed by low light intensity culture (30 μmol/m^2/s), or 2 weeks–1 month incubation in the dark followed by culture in 30–90 μmol/m^2/s, which usually stimulates final shoot induction efficiency.
3. Shoots are usually harvested at 0.5–1 cm size, after 6–12 weeks, and rooted.

3.7. Sport Regeneration (6)

1. Transversally cut leaves, abaxial side down, were used as explants and put on callus induction medium consisting of BMA + 1 mg/L IAA + 2 mg/L Z.

2. After 2 weeks, the explants were transferred to regeneration medium (BMA + 2 mg/L TDZ).
3. For cultivars resulting in excessive callus formation, entire leaves including the petiole are placed on 1/2 BMM + 25 mg/L NaFeEDTA (i.s.o. 36.7 mg/L) + 0.37 mg/L NAA + 2 mg/L TDZ.
4. Shoots are harvested at 0.5–1 cm size, rooted, acclimatized, and evaluated in vivo.

3.8. Polyploidy Induction

3.8.1. Regeneration of Tetraploid Petal Margins (7)

1. *R. simsii* "Fabiola," "Koningin Marcella," "Starlight" and "Gerda Keessen" petals (Note 8) were initiated in vitro.
2. Shoots were initiated from callus derived from these petals that are cultured on BMA + 2 mg/L 2,4-D + 0.75 mg/L Z and after 4 weeks transferred to BMA + 2 mg/L TDZ.
3. Tetraploids were regenerated from the flower margins of these picotee chimeras.

3.8.2. Induction of Polyploids from Seed (8)

1. Normal, mature seed is sterilized and sown in vitro on BMA (see above).
2. 1–2 µl of a solution of trifluralin, oryzalin (10–100 mg/L) or colchicin (100–1,000 mg/L) is carefully administered between freshly opened cotyledons on BMA, on a daily basis during 3–7 consecutive days.
3. Tetraploids were recovered among surviving seedlings; 100 mg/L trifluralin is the most efficient treatment.

3.8.3. Induction of Polyploids from Microshoots (9)

1. Microshoots of evergreen rhododendron and deciduous azalea are cut in 2-node sections and placed on BMW.
2. Oryzalin or trifluralin are dissolved in DMSO and added to liquid BMW to result in a final concentration of 2% DMSO.
3. The 2-node sections are placed in Erlenmeyer flasks with 20 ml treatment solution, and shaken (60 rpm) for 24–48 h.
4. Polyploid plants grow out of the sections; 52 mg/L oryzalin is the most efficient treatment.

3.9. Interspecific Hybridization ((10), Table 14.5)

1. Interspecific pollinations (*see* Note 9) are performed in vivo, analogous to normal pollinations. Fresh pollen is put on a moist stigma of the seed parent. The seed parent is always emasculated.
2. The ovaries are left on the plant as long as possible before spontaneous abortion occurs, typically 4–5 months, except 1.5–2 months for Vireya seed parents.
3. Ovaries are sterilized and ovules are dissected. Separate ovules are initiated on "embryo rescue" media in small petri dishes, on "embryo rescue" medium that is based on BMA + 80 mg/L adenine sulfate + 50 mg/L GA_3 + 0.4 mg/L thiamine HCl.

Table 14.5
Protocols for in vitro "embryo rescue" after various interspecific *Rhododendrvon* crosses

Cross	Medium	Reference
Evergreen × decidous azaleas	BMA + 40 g/L sucrose + 0.1 mg/L NAA	(33)
Menziesia × *Rhododendron*	BMA + 50 mg/L GA_3 + 3 g/L gellan gum	(15)
Various *Rhododendron* genotypes × *R. simsii*	BMW + 50 mg/L GA_3, followed by BMW + 1 mg/L 2iP	(10)

4. Seedlings that develop are screened with both morphological and molecular tools to confirm their hybrid status.
5. Multiplication and rooting are attempted according to previous protocols (see above).

4. Notes

1. An alternative procedure for heavily contaminated material is a sequential treatment with 30% H_2O_2 (15 min), 7% NaOCl (10 min), 0.1% $HgCl_2$ (10 min) with 0.1% Tween 20 added to every solution.
2. Ultrasonication (1 min) can be applied prior to NaOCl treatment.
3. Multiplication can also be started from in vitro meristem, floral structures or seedling cultures.
4. The most widely applied cytokinin is 2iP. It is cheaper than Z and more efficient than BAP, yielding shoot multiplication rates of 7–20. TDZ strongly promotes shoot proliferation, but shoots are often distorted, compact, hyperhydric and unsuitable for rooting in vivo; therefore it is rarely used for multiplication purposes, but rather applied for regeneration. Auxins like IAA or NAA are occasionally used to counter excessive cytokinin effects and to render more vigorous shoots.
5. Another option is the insertion of a 10 days pre-treatment period of shoots in BMA + 5–15 mg/L IBA, before in vivo explant.
6. Although multiple callus types occur, hard and compact callus is generally preferable over soft and friable callus that seldom regenerates.

7. Different cytokinins can be used. TDZ favors shoot induction rates whereas others (mainly Z or 2iP) generate higher quality shoots with higher rooting percentages.
8. Only these cultivars are suitable as their flowers have a broad edge that is differently colored compared to the rest of the petal and is composed of tetraploid tissue, as opposed to the diploid remains of the plant.
9. Interspecific hybridization can be performed between *Rhododendron* genotypes that belong to the same subgenus. Inter-subgeneric crossing is normally prohibited by incongruity barriers.

Acknowledgments

The authors wish to thank the staff of ILVO-Plant Applied Genetics and Breeding, especially Ronald Van den Oord, Pepijn De Raeymaecker and Romain Uytterhaegen for their indispensable technical support.

References

1. Chamberlain D, Hyam G, Argent G, Fairweather G, Walter KS (1996) The genus *Rhododendron*. Its classification & synonymy. Royal Botanical Garden Edinburgh, Edinburgh 181pp
2. Anderson WC (1984) A revised tissue culture medium for shoot multiplication of Rhododendron. J Amer Soc Hort Sci 109:343–347
3. Lloyd GB, McCown BH (1980) Commercially feasible micropropagation of mountain laurel, Kalmia latifolia, by use of shoot tip culture. Comb Proc Int Plant Prop Soc 30:421–426
4. Murashige T, Skoog TA (1962) A revised medium for rapid growth and bioassays with tobacco tissue cultures. Plant Physiol 15:473–497
5. Economou AS, Read PE (1984) In vitro shoot proliferation of Minnesota deciduous azaleas. Hort Sci 19:60–61
6. Samyn G, De Schepper S, Van Bockstaele E (2002) Adventitious shoot regeneration and appearance of sports in several azalea cultivars. Plant Cell Tissue Organ Cult 70:223–227
7. De Schepper S, Leus L, Eeckhaut T, Van Bockstaele E, Debergh P, De Loose M (2004) Somatic polyploid petals: regeneration offers new roads for breeding Belgian pot azaleas. Plant Cell Tissue Organ Cult 76:183–188
8. Eeckhaut T, Van Huylenbroeck J, De Schepper S, Van Labeke M (2006) Breeding for polyploidy in Belgian azalea (*Rhododendron simsii* hybrids). Acta Hort 714:113–118
9. Vaïnola A (2000) Polyploidization and early screening of Rhododendron hybrids. Euphytica 112:239–244
10. Eeckhaut T, De Keyser E, Van Huylenbroeck J, De Riek J, Van Bockstaele E (2007) Application of embryo rescue after interspecific crosses in the genus *Rhododendron*. Plant Cell Tissue Organ Cult 89:29–35
11. Preil W, Engelhardt M (1977) Meristem culture of azaleas (*Rhododendron simsii*). Acta Hort 78:203–208
12. Hsia CN, Korban SS (1997) The influence of cytokinins and ionic strength of Anderson's medium on shoot establishment and proliferation of evergreen azalea. Euphytica 93:11–17
13. Dabin P, Bouharmont J (1983) Application of in vitro cultures in azalea (*Rhododendron simsii* Planch). Acta Hort 131:89–93
14. Radice S, Caso OH (1990) In vitro organogenesis in leaves of azaleas 'Petrick' and 'Rex'. Sci Hort 41:343–347

15. Kita K, Kurashige Y, Yukawa T, Nishimura S, Handa T (2005) Intergeneric hybridization between Menziesia and Rhododendron based on molecular phylogenetic data. J Japan Soc Hort Sci 74:51–56
16. Dai C, Lambeth VN, Taven R, Mertz D (1987) Micropropagation of Rhododendron prinophyllum by ovary culture. Hort Sci 22:491–493
17. Brand MH, Kiyomoto R (1997) The inducion of tissue proliferation-like characteristics in in vitro cultures of Rhododendron 'Montego'. Hort Sci 32:989–994
18. Ettinger TL, Preece JE (1985) Aseptic micropropagation of Rhododendron P.J.M. hybrids. J Hort Sci 60:269–274
19. Pogany MF, Lineberger RD (1990) Phenotypic variation during micropropagation of the chimeral Rhododendron 'President Roosevelt'. Plant Cell Tissue Organ Cult 21:201–209
20. Tiwari ON, Chauhan UK (2006) Rhododendron conservation in Sikkim Himalaya. Current Sci 90:532–541
21. Iapichino G, Chen TH, Fuchigami LH (1991) Adventitious shoot production from a vireya hybrid of rhododendron. Hort Sci 26:594–596
22. McCown BH, Lloyd GB (1983) A survey of the response of Rhododendron to in vitro culture. Plant Cell Tissue Organ Cult 2:77–85
23. Mertens M, Werbrouck S, Samyn G, Botelho H, Debergh P (1996) In vitro regeneration of evergreen azalea from leaves. Plant Cell Tissue Organ Cult 45:231–236
24. Meyer MM (1982) In vitro propagation of *Rhododendron catawbiense* from flower buds. Hort Sci 17:891–892
25. Almeida R, Goncalves S, Romano A (2005) In vitro micropropagation of endangered *Rhododendron ponticum* L. subsp. Baeticum (Boissier & Reuter) Handel-Mazzetti. Biodivers Conserv 14:1059–1069
26. Yamaguchi S (2004) Micropropagation of *Rhododendron uwaense*, an endemic and red-data (endangered/rare) species in Japan with commercial production potential. Acta Hort 630:337–346
27. Shevade A, Preece JE (1993) In vitro shoot and floral organogenesis from stamen explants from Rhododendron PJM group clone. Sci Hort 56:163–170
28. Vejsadova H, Pretova A (2003) Somatic embryogenesis in Rhododendron catawbiense 'Grandiflorum'. Acta Hort 616:467–470
29. Hsia CN, Korban SS (1998) Effect of growth regulators, dark treatment and light intensity on shoot organogenesis from leaf tissues of evergreen azalea. J Hort Biotechnol 73:56–60
30. Preece JE, Imel MR (1991) Plant regeneration from leaf explants of Rhododendron PJM hybrids. Sci Hort 47:68–71
31. Iapichino G, McCulloch S, Chen TH (1992) Adventitious shoot formation from leaf explants of Rhododendron. Plant Cell Tissue Organ Cult 30:237–241
32. Tomsone S, Gertnere D, Novikova D (2004) The influence of thidiazuron on shoot regeneration and proliferation of rhododendrons in vitro. Acta Universitatis Latviensis 676:239–242
33. Yamaguchi S (1986) In vitro culture of remote hybrid seedlings aiming to breed new yellow flowered evergreen azalea. Plant Cell Incompatibility Newslett 18:50–51

Chapter 15

In Vitro Regeneration and Multiplication of *Passiflora* Hybrid "Guglielmo Betto"

Luca Pipino, Luca Braglia, Annalisa Giovannini, Giancarlo Fascella, and Antonio Mercuri

Abstract

With more than 450 species, *Passiflora* is the most important genus of the family *Passifloraceae*. It comprises many species grown for their edible fruits, for their high ornamental value, and further for the therapeutic properties. With their striking exotic flowers, they are of particular interest for the floriculture market. With the aim of evaluating the in vitro propagation of an Italian ornamental hybrid, axillary tendrils of *Passiflora* "Guglielmo Betto" M. Vecchia (*P. incarnata* L. × *P. tucumanensis* L.) were sterilized and placed in vitro. Direct shoot regeneration was achieved from young tendrils cultivated on MS medium containing, either 4.43 μM 6-benzylaminopurine (BAP) and 11.41 μM indoleacetic acid (IAA), or 49.20 μM 6-γ-γ-dimethylallylaminopurine (2iP) and 2.68 μM α–naphthalene acetic acid (NAA), respectively. In vitro shoot multiplication, rooting, and regenerated plant acclimatization protocols were established.

Key words: Auxin, Garden pot plant, Passion flower, Shoot regeneration, Tendril

1. Introduction

Micropropagation has been an important tool that has improved research and development in many fields of plant science, allowing fundamental research to be conducted on the physiology of morphogenesis and differentiation. In the field of ornamental plants, tissue culture has allowed mass propagation of superior genotypes, thus enabling the commercialization of healthy and uniform planting material (1). Furthermore, in combination with advanced

S.M. Jain and S.J. Ochatt (eds.), *Protocols for In Vitro Propagation of Ornamental Plants*, Methods in Molecular Biology, vol. 589
DOI 10.1007/978-1-60327-114-1_15, © Humana Press, a part of Springer Science+Business Media, LLC 2010

technologies, in vitro culture offers new opportunities for plant improvement (2).

The genus Passiflora, comprising about 500 species of vines, lianas, and small trees, is the largest in the Passion flower family (*Passifloraceae*). With the exception of a few Malayan, Chinese, North America, and Australian species, the genus is distributed in the warm temperate and tropical regions of the New World. Several species are grown in the tropics for their edible fruits (*Passiflora edulis* Sims.) and many others are grown either outdoors, in the warmer parts of the world, or in glasshouses. Some Passiflora species exhibit several unique floral features, including multiple series of brightly colored coronal filaments and elaborate floral nectary structures (3) that are of particular interest for the floriculture market.

Tissue culture protocols have been established for some *P. edulis* varieties (4), for *P. alata, P. caerulea, P. mollissima, P. coccinea, P. herbertiana,* and *P. suberosa* (5), as well as for the medicinal *P. incarnata* (6). In vitro shoot regeneration was achieved from *P. coerulea* leaves (7), from *P. edulis f. flavicarpa* leaf protoplasts (8), from tendrils and immature flower buds in *P. trifasciata, P. manta,* and *P. foetida* (9).

In the Italian floricultural market, Passiflora species and hybrids are commercialized as garden pot plants, propagated with in vitro culture techniques. In this work, we consider the Passiflora hybrid "Guglielmo Betto" due to its interesting ornamental traits: floral size (diameter >8 cm), double series of colored filaments, rapid growing and abundant flowering, and high hardiness and resistance at low temperatures (–7/–8°C). Efficient shoot regeneration and multiplication systems were established.

2. Materials

2.1. Plant Material

Plants of Passiflora hybrid "Guglielmo Betto" (*P. incarnata* L. × *P. tucumanensis* L.), supplied by the Italian breeder Maurizio Vecchia (Fig. 15.1), were cultivated in the farm of the C.R.A.-FSO (Research Unit for Floriculture and Ornamental Species), in Sanremo (Italy). In this work, 1-year-old greenhouse-grown plants were used as source of explants.

2.2. Sterilization

1. Liquid soap (dishwashing liquid soap Palmolive®, USA).
2. Ethanol absolute ≥99.8 (v/v) (Sigma-Aldrich, USA)
3. Sodium hypochlorite (Commercial bleach).
4. Tween 20 – Polyethylene glycol sorbitan monolaurate – Nonionic detergent (Sigma-Aldrich).
5. Petri dishes – 90 mm × 15 mm (Barloworld, USA).

Fig. 15.1. Flower of Passiflora hybrid "Guglielmo Betto".

2.3. Shoot Induction Media

1. MS medium including basal salts and vitamins (10), powder for plant cell culture (Duchefa, The Netherlands). Prepare medium by dissolving the powder stock into distilled deionised water before use.
2. Plant growth regulators, powder (Sigma-Aldrich): α-naphthalene acetic acid (NAA); 6-γ-γ-dimethylallylaminopurine (2iP); indole-3-acetic acid (IAA); 6-benzylaminopurine (BAP). Stock solution of 1 mg/mL for each growth regulator is prepared by using few drops of 1 N NaOH (Sigma-Aldrich) as solvent, dilute with water to final volume of 100 mL and store for up to 6 months at –20°C, except for BAP stored at 4°C.
3. Pure sucrose crystallized (Duchefa).
4. Plant agar powder (Duchefa).

2.4. Shoot Multiplication Media

1. MS-Shoot medium including basal salts and vitamins (11), powder for plant cell culture (Shoot multiplication medium B – Duchefa). Medium is prepared by dissolving the powder stock into distilled deionised water before use.
2. Plant growth regulator, powder (Sigma-Aldrich): 6-benzylaminopurine (BAP). Stock solution of 1 mg/mL is prepared by using few drops of 1 N NaOH (Sigma-Aldrich) as solvent, dilute with water to final volume of 100 mL and then store for up to 6 months at 4°C.
3. Pure sucrose crystallized (Duchefa).
4. Plant agar powder (Duchefa).

2.5. Shoot Elongation and Root Induction Media and Culture Conditions

1. MS-Shoot medium including basal salts and vitamins, powder for plant cell culture (Shoot multiplication medium B – Duchefa). Dissolve powder stock in distilled deionised water before use while preparing the medium.
2. Pure sucrose crystallized (Duchefa).
3. Plant agar powder (Duchefa).
4. All in vitro cultures are kept in a growth chamber at 24 ± 1°C, 16 h photoperiod with 40 W Osram cool-white fluorescent tubes providing a quantum flux density of 28 μE/m^2/s.

2.6. Hardening

1. Neutral Peat (20% total Carbon – 0.5% Nitrogen – 40% total organic matter) Tercomposti (Italy); autoclaved for 20 min at 121°C.
2. Common silica sand (Asplanato Vittorio, Italy); autoclaved for 20 min at 121°C.

3. Methods

3.1. Plant Material

Young axillary tendrils, filamentary in shape without coil structures (Fig. 15.2) are cut with sterile blade from the apical area of the plant and placed in a glass jar (*see* Note 1).

Fig. 15.2. Apical young tendrils used as source of explants.

3.2. Sterilization

1. Tendril explants are washed with a solution of water and liquid soap (10%, v/v) for 15 min and rinsed with sterile distilled water for 10 min.
2. Rinsed explants are surface sterilized by immersion in ethanol 70% for 60 s and then soaked for 20 min in a solution containing 1% sodium hypochlorite and two drops of Tween 20. Finally, the explants are rinsed three times in sterile distilled water for 5 min each (Fig. 15.3).
3. After the removal of chlorine-damage area, each tendril should be cut into a segment approximately 1 cm long and should be placed on Petri dishes containing shoot induction media (SIM).

3.3. Shoot Induction Media

1. SIM0 – dissolve 4.3 g MS powder stock into 1 L distilled deionised water, without growth regulators.
2. SIM1 – dissolve 4.3 g of MS powder stock into 1 L distilled deionised water supplemented with 0.5 mL NAA and 10 mL 2ip stock solutions. Final concentrations of growth regulators in the medium are 2.68 μM NAA and 49.20 μM 2iP.
3. SIM2 – dissolve 4.3 g of MS powder stock into 1 L distilled deionised water containing 1 mL IAA and 2 mL BAP stock solutions. Final concentrations of growth regulators in the medium are 11.41 μM IAA and 4.43 μM BAP.
4. For all media, add sucrose (30 g/L) and growth regulators, adjust pH to 5.7–5.8 with 1 N NaOH or 1 M HCl, and mix plant agar (8 g/L). Autoclave the final solution for 20 min at

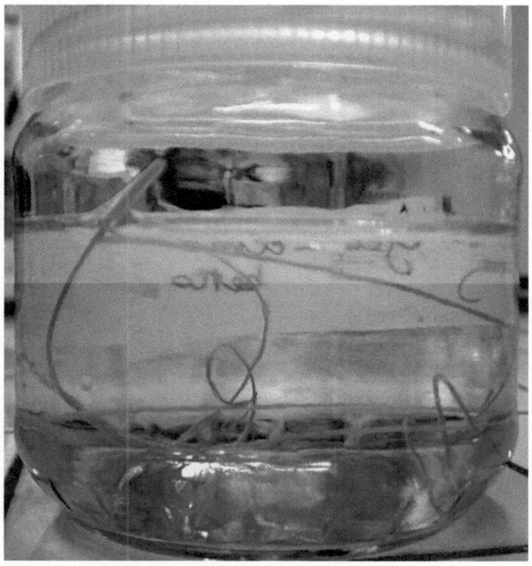

Fig. 15.3. Sterilized tendrils rinsed in water.

121°C. After cooling, under aseptic conditions, pour medium into Petri dishes (25 mL each).

5. Culture the explants (three repetitions with 5 explants) on each medium.
6. Subculture the explant cultures two times in the same media, approximately every 4 weeks (*see* Note 2).
7. After 2 months of culture, calculate the shoot regeneration efficiency from the total number of explants cultured for each treatment (*see* Note 3) (Fig. 15.4) (Table 15.1).

Fig. 15.4. Shoot regeneration from tendrils after 2 months of culture on SIM2.

Table 15.1
Shoot induction from young tendril cuttings

Medium	No of shoots[a]	Shoot regeneration efficiency (%)[b]
SIM0	0 a	0 a
SIM1	11.21 ± 0.29 c	90 ± 0 c
SIM2	2.45 ± 0.38 b	64.22 ± 14.66 b

Different letter indicates significant differences among treatments at $P \leq 0.05$ according to one-way ANOVA and Tukey test. Percentage values were subjected to arcsin transformation before analysis

[a]Mean number ± Standard Error

[b]Transformed percentages ± Standard Error

3.4 Shoot Multiplication Media (MM)

1. After 2 months, isolate regenerated shoots (20–30 mm long) obtained from young tendrils and multiply through nodal cuttings.
2. MM0 – dissolve 4.5 g MS-Shoot powder stock into 1 L distilled deionised water, without growth regulators.
3. Other multiplication media are prepared with MM0 amended with 0.1, 0.3, 0.5, 0.8 mL BAP stock solution. Final concentrations of growth regulator into the medium are 0.02, 0.07, 0.11 and 0.18 μM BAP (MM1, MM2, MM3, and MM4 respectively).
4. Add 30 g/L sucrose and growth regulator to all media, adjust pH to 5.7–5.8 with 1 N NaOH or 1 M HCl, and mix 8 g/L plant agar. The final solution is autoclaved for 20 min at 121°C. After cooling, under aseptic conditions, pour medium into Petri dishes (25 mL each).
5. Five repetitions with 6 explants are cultured for each treatment. Evaluate shoot formation after 2 months of culture (Fig. 15.5) (Table 15.2) (*see* Note 4).

3.5. Shoot Elongation and Root Induction Medium (ERM)

1. Culture single shoots separated from multiple shoot clumps in a plant growth regulator- free medium.
2. ERM – dissolve 4.5 g MS-Shoot powder stock into 1 L distilled deionised water, without growth regulators.
3. Add 30 g/L sucrose, adjust pH to 5.7–5.8 with 1 N NaOH or 1 M HCl, add 8 g/L plant agar and mix very well. Autoclave the final solution for 20 min at 121°C and cool it under aseptic conditions. Pour the medium into Petri dishes (25 mL each).

Fig. 15.5. Multiplication stage on MM3.

Table 15.2
Multiplication of Passiflora on different media

Medium	BAP (μM)	No of shoots/exp[a]
MM0	0	0 a
MM1	0.02	4.52 ± 0.37 b
MM2	0.07	5.60 ± 0.78 b
MM3	0.11	8.52 ± 0.66 c
MM4	0.18	4.24 ± 0.55 b

Different letter indicates significant differences among treatments at $P \leq 0.05$ according to one-way ANOVA and Tukey test
[a]Mean number ± Standard Error

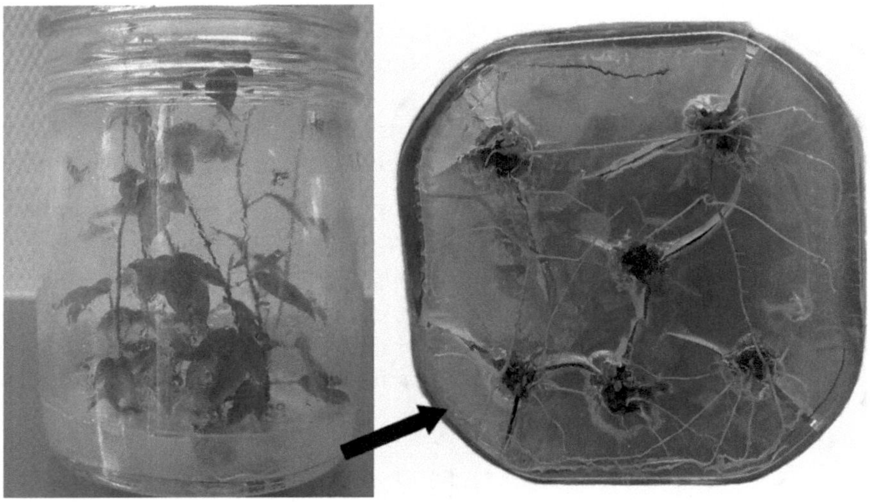

Fig. 15.6. Elongation and rooting in vitro on ERM.

4. Shoot elongation and root induction are achieved simultaneously without subculture. About 96% regenerated shoots on this medium elongate and form roots in vitro within 5 weeks (Fig. 15.6).

3.6. Hardening

1. In vitro-rooted plantlets are transferred to sterilized sand: soil (1:1 v/v) mixture in filled perforated potting bags and harden under mist conditions (10 s nebulization every 30 min) for 15 days.
2. After hardening, plants are maintained in the greenhouse at 23 ± 2°C/15 ± 2°C, with natural day length. The percentage of acclimatization is about 70%.

3.7. Statistical Analysis

1. Data are subjected to one-way ANOVA with Tukey test for mean comparison ($P \leq 0.05$). Percentage values are subjected to arcsin transformation before analysis. For statistical analysis use software SPSS release 13.0 for Windows.

4. Notes

1. In this study, direct adventitious shoot proliferation was observed at the edges of P. "Guglielmo Betto" tendril cuttings. Early physiological studies on *Cucurbita moschata* and *P. coerulea* tendrils by immersing the tip in a solution containing different auxins, concluded that IAA, in particular, acts to induce coiling by immediately interacting with auxin-specific receptors in the tip of the tendril (12). More, Bai et al. (13) have isolated two early auxin response genes in pea, homologous to the *Arabidopsis* auxin efflux carrier *PINOID* (*PID*). The *PsPK2* mRNA is more abundant in tendrils than in leaflets of pea leaves, irrespective of genotype, suggesting that it might play a role in the thigmotropic response. The authors suggested that tendrils transported auxin much more efficiently than other leaf parts. The efficient auxin transport in *Passiflora* tendril tips could be responsible for their strong morphogenic behavior, in our regeneration protocols. This hypothesis could be confirmed by an experiment performed in our laboratory: adding to SIM an auxin carrier inhibitor, such as 2,3,5-triiodobenzoic acid, no shoot regeneration from tendril cuttings was achieved.
2. Little callus formation was observed in any explant on SIM1 and SIM2. After the first subculture, tendrils became slightly swollen giving rise to new shoot formation localized mainly in cut areas.
3. On SIM1 the highest number of regenerated shoots and regeneration percentage were achieved, but shoots were small in size and showed vitrification. Shoots regenerated on SIM2 were well developed and dark green in color, and thus they were selected for the multiplication step. On hormone-free medium (SIM0) no regenerations were observed.
4. Different concentrations of BAP were used. MM3 (0.11 μM BAP) gave the highest number of new shoots per explant. After 2 months, shoots were well developed and elongated, and were ready for being separated and transferred on elongation and rooting medium. MM1, MM2, and MM4 showed similar results concerning the number of shoots, but with MM4, where the concentration of BAP was the highest, shoots were chlorotic and reduced in size.

Acknowledgments

The authors would like to thank Dr. Maurizio Vecchia for providing plant material and Mr Cesare Bianchini for technical help. This research was financed by the Italian Ministry of Agriculture and Forestry, Project 192/7303/2006 "Individuazione, caratterizzazione e valorizzazione di specie dotate di caratteristiche mediterranee (VIVAFLOR)." Publication no 7.

References

1. Winkelmann T, Geier T, Preil W (2006) Commercial in vitro plant production in Germany in 1985–2004. Plant Cell Tissue Organ Cult 86:319–327
2. Nhut DT, Don NT, Vu NH, Thien NQ, Thuy DTT, Duy N, Teixeira da Silva JA (2006) Advanced technology in micropropagation of some important plants. In: Teixeira da Silva JA (ed) Floriculture ornamental and plant biotechnology, vol II. Global Science Books, UK, pp 325–335
3. Dornelas MC, Tsai SM, Martinelli Rodriguez AP (2006) Expressed sequence tags of genes involved in the flowering process of *Passiflora* species. In: Teixeira da Silva JA (ed) Floriculture ornamental and plant biotechnology, vol I. Global Science Books, UK, pp 483–488
4. Isutsa DK (2004) Rapid micropropagation of passion fruit (*Passiflora edulis* Sims.) varieties. Sci Hortic 99:395–400
5. Drew RA (1991) In vitro culture of adult and juvenile bud explants of Passiflora species. Plant Cell Tissue Organ Cult 26:23–27
6. Mingozzi M, Lucchesini M, Mensuali-Sodi A (2003) La propagazione in vitro della *Passiflora incarnata*. Colture Protette 9:139–144
7. Vestri F, Schiff S, Bennici A (1990) In vitro shoot regeneration in *Passiflora coerulea*. Acta Hortic Wageningen 280:105–107
8. Davey MR, Anthony P, Power JB, Lowe KC (2005) Plant protoplasts: status and biotechnological perspectives. Biotechnol Adv 23:131–171
9. Pipino L, Braglia L, Giovannini A, Fascella G, Mercuri A (2008) In vitro regeneration of Passiflora species with ornamental value. Propagation Ornament Plants 8:47–49
10. Murashige T, Skoog F (1962) A revised medium for rapid growth and bioassays with tobacco tissue cultures. Physiol Plant 15:473–497
11. Huang LC, Murashige T (1976) Plant tissue culture media: major constituents, their preparation and some applications. TCA Manual 3(1):539–548
12. Jaffe MJ (1975) The role of auxin in the early events of the contact coiling of tendrils. Plant Sci Lett 5:217–225
13. Bai F, Watson JC, Walling J, Weeden N, Santner AA, DeMason DA (2005) Molecular characterization and expression of PsPK2, a PINOID-like gene from pea (*Pisum sativum*). Plant Sci 168:1281–1291

Chapter 16

In Vitro Propagation of Rose

Pratap Kumar Pati, Navtej Kaur, Madhu Sharma, and Paramvir Singh Ahuja

Abstract

In vitro propagation of rose is an important tool for rapid multiplication and development of new cultivars with desirable traits. However, successful in vitro propagation requires an understanding of specific requirements and precise manipulation of various factors. Efficient protocols for different stages of micropropagation using apical buds or nodal segments are currently available. Recently, new challenges for refinements of protocols for high rate of shoot multiplication and development of cost effective methods has gained importance. Significance of the liquid static culture for shoot proliferation and root induction for rose has also gained prominence. Other distinct approaches of in vitro propagation include organogenesis and embryogenesis. These approaches are important for the successful implementation of various biotechnological techniques used for rose improvement programmes. Types of explants, media and optimization of conditions are major factors for successful regeneration of rose.

Key words: Rose, Micropropagation, In vitro propagation, Shoot proliferation, Root initiation, Rooting, Shoot regeneration, Somatic embryogenesis

1. Introduction

Roses are among the most important ornamental plants worldwide. Besides being grown as a garden plant, it is also cultivated for cut flower production, root stock and as a source of rose oil. It assumes different forms like bushes, shrubs and climbers and may be either deciduous or evergreen. They are also adapted to varied agroclimatic conditions and exhibit a wide range of variation in flower shape, size, colour, fragrance, etc.

Roses are generally propagated by vegetative methods like cutting, layering, budding and grafting. Seeds are used for propagation of species, new cultivars and for production of rootstocks (1). Although propagation by vegetative ways is a predominant

S.M. Jain and S.J. Ochatt (eds.), *Protocols for In Vitro Propagation of Ornamental Plants*, Methods in Molecular Biology, vol. 589
DOI 10.1007/978-1-60327-114-1_16, © Humana Press, a part of Springer Science+Business Media, LLC 2010

technique in roses, it does not ensure healthy and disease-free plants. Moreover, dependence on season and slow multiplication rates are some of other major limiting factors in conventional propagation (2). In vitro culture technique, an alternative method for plant propagation, offers unique features of rapid multiplication in a relatively short span of time; production of healthy and disease free plants; ability to generate propagules around the year. Micropropagation of roses has been reported by various researchers using cultures of axillary buds and apical meristems (2–5). The successful micropropagation protocol of rose proceeds through a series of stages, each with a specific set of requirements. The stages commonly involve (1) initiation of aseptic cultures, (2) shoot multiplication, (3) rooting of microshoots, and (4) hardening and field transfer of tissue culture raised plants.

In roses, there are many reports on rapid regeneration and multiplication through organogenesis via callus formation. However, only a few deal with direct regeneration from different explants. Lloyd et al. (6) reported, for the first time, direct shoot regeneration using leaf and root explants in *R. persica* × *xanthina*, *R. laevigata* and *R. wichuriana*. In vitro leaf explants were also used for direct shoot regeneration in ornamental rose such as *R. hybrida* (7–10). Dubois and de Vries (11) used leaf explants of in vivo grown glasshouse rose cultivars for direct regeneration. However, all these observations have been on ornamental roses. We report that similar regeneration response can be evinced on leaf petioles for *R. damascena* (12).

Furthermore, this method has considerable implications for the rose breeder as it facilitates the process of creation and rapid multiplication of new varieties. In vitro propagated plant also proved to be beneficial due to the change in the growth characteristics (13–15).

2. Materials

2.1. Chemicals, Glassware/Plasticware

1. All chemicals used are of s.d. fine or Qualigens brand (AR grade). Plant growth regulators (PGRs) were either procured from Hi-media laboratories, Mumbai or Sigma Chemicals, USA; Bavistin (BASF, India Ltd., Mumbai); streptomycin sulphate (HiMedia Laboratories, Mumbai, India).
2. All glassware used are Borosilicate glass (Borosil works Ltd., Mumbai) or Duran (Germany). Nodal segments are grown in culture tubes (8.9 in. × 1.0 in.) and later subcultured in 250-ml Erlenmeyer flasks or 350-ml jars. However, leaf petiole segments and for initiation of somatic embryo, Petri dishes (size 9.0 cm) can be used.

2.2. Washing of Glassware/ Plasticware

All laboratory glassware are cleaned by soaking in chromic acid solution (for preparation of chromic acid solution, 6.8 g potassium dichromate is dissolved in 100-ml sulphuric acid and the total volume is made up to 1,000 ml) followed by vigorous tap water rinses and a final rinse with double distilled water (16). Dry washed glassware in oven at 150°C. The reusable plastic lab ware are washed using mild non-abrasive detergent followed by a rinse in tap water and then finally with distilled water.

2.3. Preparation of MS medium

1. For most of the tissue culture work in rose, MS medium (17) is most suitable. The MS medium stock solutions and the medium are prepared as given in the Table 16.1. Stock solutions and medium are prepared in double distilled water. Stock solutions are stored at 4°C in the refrigerator.
2. The different plant growth regulators are prepared as given in Table 16.2, store them at 4°C and add desired amount to the tissue culture medium.
3. The solid media are prepared with 0.8% agar (Bacteriological grade; Qualigens, India Ltd, Mumbai). For liquid media no gelling agent is used.

Table 16.1
MS medium stock solution preparation (17)

	Constituents	Concentration (mg/l)	Concentration of stock solution (mg/l)	Volume of stock/ liter of medium (ml)
Macroelements (stock-I)	NH_4NO_3	1,650	33,000	50
	KNO_3	1,900	38,000	
	$CaCl_2$ $2H_2O$	440	8,800	
	$MgSO_4$ $7H_2O$	370	7,400	
	KH_2PO_4	170	3,400	
Microelements (stock-II)	KI	0.83	166	5
	H_3BO_3	6.2	1,240	
	$MnSO_4$ $4H_2O$	22.3	4,460	
	$ZnSO_4$ $7H_2O$	8.60	1,720	
	Na_2MoO_4 $2H_2O$	0.25	50	
	$CuSO_4$ $5H_2O$	0.025	5	
	$CoCl_2$ $6H_2O$	0.025	5	
Iron (stock-III)	$FeSO_4$ $7H_2O$	27.85	5,560	5
	Na_2EDTA	37.25	7,460	
Vitamins (stock-IV)	Meso inositol	100	20,000	5
	Glycine	2.0	400	
	Nicotinic acid	0.5	100	
	Pyridoxine–HCl	0.5	100	
	Thiamine–HCl	0.1	20	

Table 16.2
Preparation and storage of different plant growth regulators used for in vitro propagation of rose

Growth regulator	Molecular weight	Preparation and storage		
		Solvent	Diluent	Storage
2,4-D	221.0	EtOH/1 N NaOH	Water	0–5°C
IBA	203.2	EtOH/1 N NaOH	Water	0–5°C
NAA	186.2	1 N NaOH	Water	0–5°C
BAP	225.3	1 N NaOH	Water	0–5°C
TDZ	220.2	DMSO	Water	0–5°C

4. The pH should be adjusted to 5.8 with either 0.1 N HCl or 0.1 N NaOH.
5. The media are autoclaved for 20 min at 1.1 kg/cm^2 pressure and 121°C.

3. Methods

3.1. Choice of Explant

1. For micropropagation of rose, the most commonly used explant is a nodal stem segment, wherein the axillary bud is made to proliferate from multiple shoots. The performance of nodal segments as explants is much better than the shoot tips as shown by Horn (1).
2. Nodal segments (9.0–10.0-mm long; 3.0–4.0-mm diameter) from actively growing branches of different age, i.e., from freshly sprouted to 1-year old are selected. The nodal explants are normally from the middle part of the branches.

3.2. Surface Sterilization and Inoculation

1. Nodal explants taken from field grown plants are properly cleaned with TWEEN-80 using a sable hairbrush. Prickles are removed before cleaning.
2. Pre-sterilization is carried out with 0.04% (v/v) solution of sodium hypochlorite containing 0.2% (v/v) TWEEN-80 and 0.001% (w/v) Tetracycline (Hi-media, Mumbai) for 25 min on a gyratory shaker.
3. Finally, surface sterilization of explants is done in a laminar flow cabinet with 0.05% (w/v) mercuric chloride solution containing 0.2% (v/v) TWEEN-80 for 4–5 min and thereafter,

washed repeatedly with sterile distilled water to remove all the traces of sterilizing agents.

4. About 1–2 mm of the exposed parts of cut surfaces is removed before the explants are inoculated onto a culture medium.
5. Nodal explants are inoculated in test tubes containing MS medium supplemented with BAP (5.0 μM), sucrose (3.0%) and agar (0.8%) (*see* Note 1).
6. Incubate cultures in the light, 70 ± 5 μmol/m²/s¹ from cool white fluorescent lamps at 25 ± 2°C. Maintain day length of 14 h on a 24-h light/dark cycle.

3.3. Shoot Proliferation

1. After 4 weeks, the proliferating shoots are excised from the explants and are transferred to either agar gelled (0.8%) or to a static liquid MS medium supplemented with BAP (5.0 μM) and sucrose (3.0%) (Fig. 16.1).
2. The rate of proliferation of shoot is higher in liquid (eight-folds) compared to agar-gelled medium (fourfolds) (18).
3. For optimum proliferation of shoots, 20 ml of liquid medium can be used in 350-ml capacity jars. (*see* Note 2).
4. For agar gelled medium the subculture period is 4 weeks, whereas for static liquid medium it can be extended up to 6 weeks.
5. Incubate cultures in light, 70 ± 5 μmol/m²/s¹ from cool white fluorescent lamps at 25 ± 2°C. Maintain day length of 14 h on a 24-h light/dark cycle.

3.4. Rooting of Microshoots

1. Individual and bunches of microshoots (each containing 5–6 shoots with an average length of 4.0 cm and 2.0 mm diameter) can be transferred to half strength MS liquid medium containing 10 μM IBA and 3% sucrose and incubated in dark for 1 week (11).
2. After 1 week, the medium is replaced by MS liquid medium without PGRs and with 3% sucrose.
3. The survival of rooted cultures is found maximum when they are maintained for 6 weeks before transfer to a green house for hardening.
4. Subsequent to the development of good root systems (3 weeks), they are transferred to 6-in. earthen pots containing sterile mixture of sand, soil and cow-dung manure in the ratio of 1:1:1 (w/w/w). The pots were kept in the greenhouse for acclimatization.

3.4.1. Culture Vessel for Efficient Rooting of Microshoots

1. Bulk rooting of microshoots can be done using specialized culture vessel (11). It comprises of an outer transparent

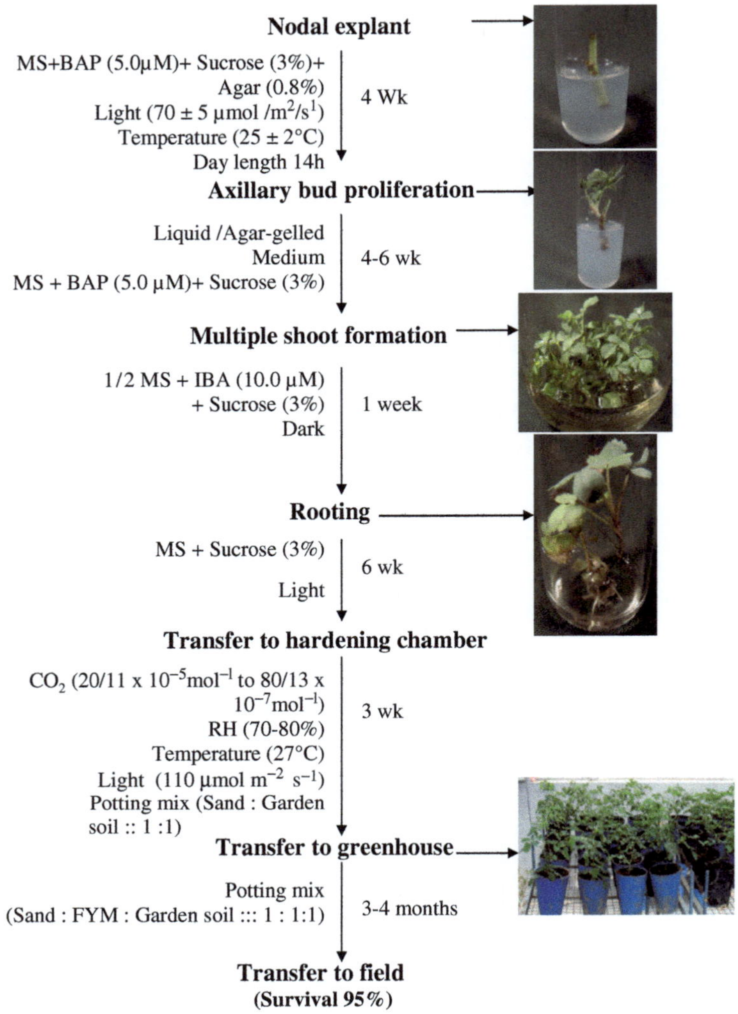

Fig. 16.1. Protocol for micropropagation of *Rosa damascena*, *R. bourboniana*, and *R. hybrida*

autoclavable container with a tight lid in which a perforated platform with holes of suitable sizes is placed.

2. The excised shoots are inserted in the holes ensuring that their cut ends constantly touch the rooting medium in the container (Fig. 16.2).

3. Different steps involved in the process are (1) preparation of a perforated platform (A) in order to provide a support system for the microshoots, (2) placement of the platform in a container (B), (3) pouring of rooting medium in the box and sterilization, (4) inserting microshoots in the holes of perforated platform, (5) sealing the lid (C) of the container with parafilm, and (6) incubating the container under appropriate conditions of temperature and light.

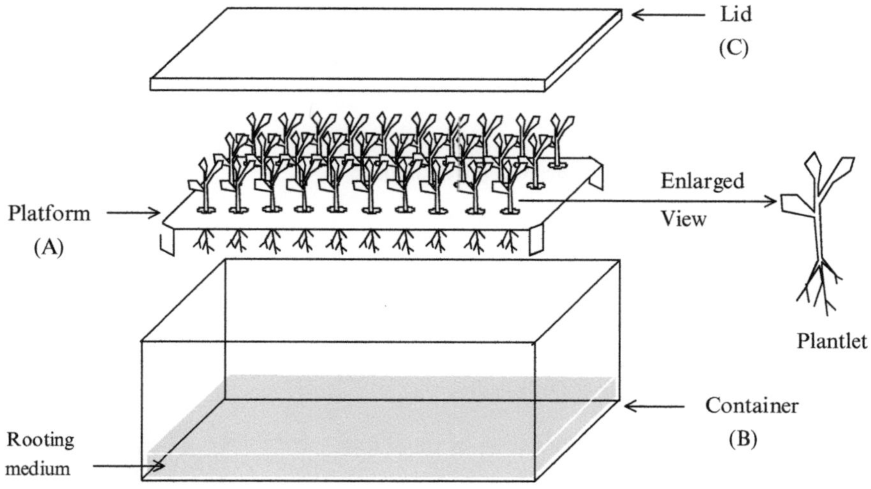

Fig. 16.2. Polypropylene box used for rooting of microshoots

3.5. Hardening and Transfer to Pots

1. The rooted microshoots are washed in water and transferred to Hikko trays/pots.
2. The Hikko trays are placed in hardening chambers/green house with RH maintained at 70–80% and provided with light intensity of 35 μmol/m^2/s^1 from the top. CO_2 (20/11 × 10^{-5} mol l^{-1} to 80/13 × 10^{-7} mol l^{-1}) enrichment also favours better hardening (19).
3. If microshoots are rooted in bunches they can be separated at a later stage i.e., after 3 weeks of hardening.
4. After developing good root systems (3 weeks), they are subsequently transferred to 6-in. earthen pots containing sterile mixture of sand, soil and cow-dung manure in the ratio of 1:1:1 (w/w/w).

3.6. Regeneration Studies

In vitro plant regeneration is often the most important step for successful implementation of various biotechnological tools used for plant improvement programmes.

3.6.1. Protocol for Direct Regeneration from Leaf Petioles of *Rosa damascena* (12)

1. The old shoots are pruned and grown in liquid MS medium supplemented with BAP (5.0 μM) and sucrose (3.0%).
2. Fully developed young leaves, obtained after 4 weeks of pruning of old shoots, are excised and used as explants (Fig. 16.3). These are placed with their abaxial sides touching the medium.
3. Initially, the explants are cultured on induction medium comprising half-strength MS medium containing 6.8 μM thidiazuron (TDZ), 0.27 μM α-naphthalene acetic acid (NAA), and 17.7 μM $AgNO_3$ for 3 weeks (*see* Notes 3).
4. The cultures are incubated at 25°C in the dark initially for 3 days before transferring to light.

In vitro raised microshoots from nodal explant

↓ MS+BAP (5µM) + Sucrose (3%)+0.8% Agar | Subculture cycle 4 wk

pruned microshoots

↓ MS+BAP (5µM) + Sucrose (3%)+0.8% Agar | 2 wk

Petiole explant from newly proliferated microshoots

↓ ½MS+TDZ (6.8 µM)+ NAA (0.27 µM)+Sucrose (3%)+ $AgNO_3$ (3mg/l)+ 0.8% Agar

Induction of shoot bud →

↓ MS+BAP (2.25 µM)+ NAA (0.054 µM)+Sucrose (3%)+ 0.8% Agar | 4 wk

Shoot bud regeneration →

↓ MS+BAP (5µM)+ Sucrose (3%)+ 0.8% Agar | 4 wk

Shoot proliferation

↓ 1/2 MS + IBA (10.0 µM) + Sucrose (3%)+ 0.8% Agar | 1 week Dark

Root induction

↓ MS + Sucrose (3%)+ 0.8% Agar | 4 wk Light

Transfer to hardening chamber

Fig. 16.3. Direct regeneration protocol from leaf explants of *Rosa damascena*.

5. After 3 weeks, the cultures are transferred to the regeneration medium consisting of full-strength basal MS medium supplemented with 2.25 µM BA and 0.054 µM NAA (12).
6. For induction and regeneration medium, agar (0.75%) is used as a gelling agent and explants are inoculated on 25-ml medium in a 90-mm Petri dish.

3.6.2. Histological Studies

In order to ascertain the origin of these shoot buds, histological studies are essential for which purpose the following protocol is to be employed.

Killing, Fixing and Dehydration of Material

Petiole explants with emerging primordia are fixed in FAA (Formalin Acetic Acid: Ethanol 50%::5:5:90) for 1 week. The material can be preserved in 70% ethanol until use after which the tissue is dehydrated in the following TBA (*t*-butyl alcohol) series:

	Rect. alcohol (ml)	TBA (ml)
(a)	30	20
(b)	50	20
(c)	50	35
(d)	45	55
(e)	25 (Ethanol)	75
(f)	–	100

Water is added to each grade (a–c) to make total volume up to 100 ml. The material is kept in each grade for 2 h and is left overnight in grade "f," i.e., pure TBA (*see* Notes 4).

Waxing

The temperature of the oven is set at 60°C. The vial containing tissue in TBA is placed in the oven and waited for about half an hour so that temperature of TBA also comes to the same temperature as that of the oven. The cap of the vial is removed and a few paraffin wax flakes are added after every 15–20 min. The whole process is carried out for a minimum of 3–4 h after which it is left overnight in the oven. During whole process the temperature of oven is maintained at 60°C, which is the melting point of paraffin wax. The material becomes ready for block making only when there is no smell of TBA.

Block Making and Section Cutting

Blocks are made with molten paraffin wax. If the wax is not clean and is showing some suspended particles, it may be filtered inside the oven. Use this clear molten wax for block making. Orientation of material in the wax is very important, otherwise sections will be cut in different plane and wrongly interpreted. The shape of the block has to be square or rectangular. Sections (10–12 μm) thick are cut with the help of a rotary microtome.

Mounting and Stretching

Fresh egg albumen is used as adhesive. Alternatively gelatin 1% (adhesive prepared from gelatin powder) can also be used. For this purpose 1.0-g gelatin powder is dissolved in 100-ml boiling distilled water. Cool down and add 2–3 crystals of phenol for long term storage. Store at 4°C(*see* Notes 5). A drop of egg albumen is applied on a clean slide, 2–3 drops of distilled water is added and a strip of sections is placed on it. The sections are allowed to stretch on a hot plate at 60°C and then kept overnight at room temperature in a dust free environment.

Dewaxing

From this point onwards, staining jars or staining dishes are used. Slides are put in pure xylol for 1–2 h and pass subsequently through each of the following grades for 2–3 min unless otherwise specified:

	Xylol (ml)	Ethanol (ml)
(a)	75	25
(b)	50	50
(c)	25	75
(d)	Rectified alcohol	
(e)	25 Water	75 Ethanol
(f)	50 Water	50 Ethanol
(g)	75 Water	25 Ethanol
(h)	Safranin (6–24 h)	
(i)	75 Water	25 Ethanol
(j)	50 Water	50 Ethanol
(k)	25 Water	75 Ethanol
(l)	Rectified alcohol	
(m)	Ethanol	
(n)	Ethanol	
(o)	Clove oil 25% in ethanol	
(p)	Clove oil 50% in ethanol	
(q)	Fast green (prepared in 50% clove oil)	
(r)	Clove oil 50% in Xylol	
(s)	Clove oil 25% in Xylol	
(t)	Xylol (30 min)	
(u)	Xylol (30 min)	

Mounting of Slides

The slides are mounted in D.P.X. mountant.

3.6.3. Shoot proliferation

1. Grow newly induced shoot buds in the shoot multiplication MS medium containing 5.0 μM BAP and 3% sucrose by two subcultures, each at 4-week interval.
2. Multiple shoots obtained in static liquid culture are subjected to rooting.

3.6.4. Rooting of Micro-Shoots and Hardening

1. Induce rooting in vitro on half-strength MS liquid medium supplemented with 10.0 μM IBA and 3% sucrose. Cultures are initially kept in the dark for 1 week (18).
2. The micro-shoots are later transferred to the hormone-free MS liquid medium without changing sucrose concentration.
3. Transfer rooted micro-shoots to Hikko trays/pots containing sand:garden soil (1:1) in the specially-designed hardening chamber for 3 weeks and later transfer to the greenhouse in larger pots (20-cm diameter) with sand:garden soil:farmyard manure (1:1:1) for further growth.

3.7. Protocol for Somatic Embryogenesis in *R. bourboniana* (20)

3.7.1. Explant Sterilization and Initiation of Embryogenic Callus

1. Prepare MS (17) supplemented with 15 μM 2,4-D, 3% sucrose and 0.8% agar.
2. The pH of the medium is adjusted to 5.7 prior to autoclaving 1.1 kg/cm^2 at 121°C for 20 min.
3. 25-ml medium is poured in a petriplate (90 mm) under aseptic conditions.
4. Immature fruits (hips) are collected which are green in colour (Fig. 16.4).
5. Dissect seeds from hips and washed with TEEPOL solution for 5 min and later with running tap water for 2–3 min. Thereafter, they are treated with 0.1% Bavistin and 0.1% streptomycin sulphate for 25–30 min and wash with distilled water.
6. The next step of disinfection is the treatment with 70% (v/v) ethanol for 30 s followed by 0.04% mercuric chloride (w/v) with 1–2 drops of TWEEN-20 for 8–10 min. TWEEN-20 is added as a wetting agent.
7. After 3–4 washings with autoclaved distilled water, remove the testa, taking care that the embryo is not injured.
8. Embryos are excised aseptically in the laminar flow cabinet and inoculate on the medium poured in petri dish. Inoculate at least five embryos per petri dish.
9. Cultures are incubated initially in dark for 3–4 weeks at 25 ± 2°C and later on shifted to light conditions with photosynthetic photon flux density (PPFD) of 70 ± 5 μmol/m^2/s^1 from cool white fluorescent lamps and temperature at 25 ± 2°C. The day length is maintained at 14 h in a 24h light/dark cycle.

3.7.2. Proliferation and Maturation of Somatic Embryos

1. The embryogenic callus with fully developed somatic embryos is transferred to MS medium supplemented with 5 μM 2,4-D and 600 mg/l l-Proline to promote embryo proliferation and maturation (20).
2. For maturation of somatic embryos, subculture embryogenic tissues on PGR free MS medium along with 3% sucrose (Fig. 16.4).

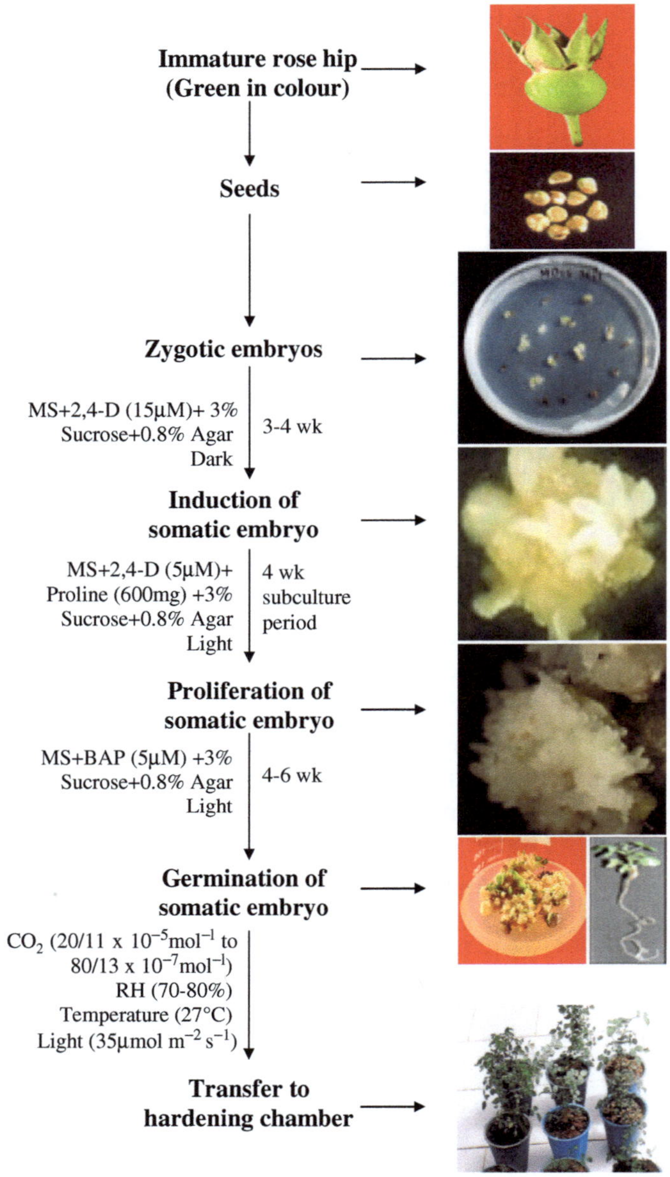

Fig. 16.4. Protocol for somatic embryogenesis in *R. bourboniana* and *R. hybrida*.

3.7.3. Somatic Embryo Germination

1. For germination, somatic embryos at cotyledonary stage are transferred to MS medium with 5 μM 6-Benzyl aminopurine with 3% sucrose and 0.8% agar.
2. Sometimes germinated embryo produce shoots without roots. In such case shoots are transferred to liquid medium containing 10 μM IBA (single-shoot/culture tube). The quantity of liquid medium used is 5 ml per culture tube.

3. Initially the cultures are kept in dark for 1 week and later on, transferred to liquid MS medium without PGRs and with the same concentration of sucrose for rooting. The cultures are maintained in this medium for 4 weeks at 25 ± 2°C.

3.7.4. Histological Studies

The procedure followed is the same as described earlier in Subheading 3.6.2.

37.5. Hardening and Acclimatization

1. Take out the rooted plants from the culture vessels and wash thoroughly in water.
2. Transplant rooted microshoots initially in plastic pots, 10-cm diameter, containing sand and keep in greenhouse covered with plastic jars for 3–4 weeks and subsequently transfer to bigger pots, 20-cm diameter, filled with sand:soil (1:1) and maintain under the greenhouse conditions

4. Notes

1. 15–20-ml medium is poured in culture tubes (8.9 in. × 1.0 in.). In case the culture medium turns brown within 72 h then transfer it to a fresh medium of same composition.
2. In liquid culture system it is optimized that 20 ml should be used in 350-ml jar. If more medium is used it leads to hypehydricity and the multiplication rate decreases (18).
3. $AgNO_3$ is filter sterilized and added to the medium after autoclaving, as suggested by Escalettes and Dosba (21).
4. While transferring from one grade to another, care must be taken that the whole of the solution is not decanted, otherwise the tissue will directly come in contact with next grade solution. Due to sudden change, the soft tissue may be damaged.
5. The advantage of using adhesive prepared from gelatin powder is that it can be used for longer duration whereas egg albumen has to be used fresh, otherwise it starts giving fowl smell.

Acknowledgments

Financial assistance from the Council of Scientific and Industrial Research, New Delhi and the University Grants Commission, New Delhi is gratefully acknowledged.

References

1. Horn WAH (1992) Micropropagation of rose (*Rosa* L.). In: Bajaj YPS (ed) Biotechnology in agriculture and forestry, vol 20, High-tech and micropropagation IV. Springer, Germany, pp 320–342
2. Pati PK, Rath SP, Sharma M, Sood A, Ahuja PS (2006) In vitro propagation of rose – a review. Biotechnol Adv 24:94–114
3. Skirvin RM, Chu MC, Young HJ (1990) Rose. In: Ammirato PV, Evans DR, Sharp WR, Bajaj YPS (eds) Handbook of plant cell culture. Springer, New York, pp 716–743
4. Yan M, Byrne DH, Jing C (1996) Propagation of rose species in vitro. In Vitro Cell Dev Biol Plant 32:103–108
5. Rout GR, Samantaray S, Mottley J, Das P (1999) Biotechnology of the rose: A review of recent progress. Sci Hortic 81:201–228
6. Lloyd D, Roberts AV, Short KC (1988) The induction in vitro of adventitious shoots in *Rosa*. Euphytica 37:31–36
7. Leffering L, Kok E (1990) Regeneration of rose via leaf segments. Prophyta 8:244
8. Ibrahim R, Mondelaers W, Debergh PC (1998) Effects of X-irradiation on adventitious bud regeneration from in vitro leaf explants of *Rosa hybrida*. Plant Cell Tissue Organ Cult. 54:37–44
9. Ibrahim R, Debergh PC (2000) Improvement of adventitious bud formation and plantlet regeneration from in vitro leaflet explants of roses (*Rosa hybrida* L.). Acta Hortic 520:271–280
10. Ibrahim R, Debergh PC (2001) Factors controlling high efficiency adventitious bud formation and plant regeneration from in vitro leaf explants of roses (*Rosa hybrida* L.). Sci Hortic 88:41–57
11. Dubois LAM, de Vries DP (1995) Preliminary report on direct regeneration of adventitious buds on leaf explants of in vivo grown glass house rose cultivars. Gartenbauwissenschaft 60:249–253
12. Pati PK, Sharma M, Sood A, Ahuja PS (2004) Direct shoot regeneration from leaf explants of *Rosa damascena* Mill. In Vitro Cell Dev Biol Plant 40:192–195
13. Onesto JP, Poupet R, Julien P (1985) Production de potees Fleuries de rosier a partir d plantules obtenus par multiplication in vitro conforme automme 1983-printemps 1984. Horticulture 176:3–10
14. Reist A (1985) Culture in vitro en pipiniere de rosiers: une alternative au bouturage ou au greffage des varietes? Rev Suisse Vitic Arboric Hortic 17:361–364
15. Dubois LAM, Roggemans J, Soyeurt G, De Vries DP (1988) Comparison of the growth and development of dwarf rose cultivars propagated in vitro and in vivo by softwood cuttings. Sci Hortic 35:293–299
16. Street HE (1973) Laboratory organization. In: Street HE (ed) Plant tissue and cell culture. University of California Press, Berkeley, CA, pp 11–30
17. Murashige T, Skoog F (1962) A revised medium for rapid growth and bioassays with tobacco tissue cultures. Physiol Plant 15:473–497
18. Pati PK, Sharma M, Sood A, Ahuja PS (2005) Micropropagation of *Rosa damascena* and *R. bourboniana* in liquid cultures. In: Hvoslef-Eide AK, Preil W (eds) Liquid systems for in vitro plant propagation. Springer, The Netherlands, pp 373–385
19. Sharma M, Sood A, Nagar PK, Prakash O, Ahuja PS (1999) Direct rooting and hardening of tea (*Camellia sinensis* L.) O. Kuntze microshoots in the field. Plant Cell Tissue Org Cult 58:111–118
20. Kaur N, Pati PK, Sharma M, Ahuja PS (2006) Somatic embryogenesis from immature zygotic embryos of *Rosa bourboniana* Desp. In Vitro Cell Dev Biol Plant 42:124–127
21. Escalettes V, Dosba F (1993) In vitro adventitious shoot regeneration from leaves of *Prunus* spp. Plant Sci Limerick 90:201–209

Chapter 17

In Vitro Propagation of Chrysanthemum

D. Nencheva

Abstract

Chrysanthemum flowers *Chrysanthemum x grandiflorum* (Ramat.) Kitam., are commercially significant worldwide as there are large number of cultivars for cut flowers, pot flowers, and garden flowers. Commercial in vitro multiplication of chrysanthemum is often based on stem nodal explants with lateral meristems. This chapter describes a protocol for in vitro propagation from stem nodal explants and by direct organogenesis from pedicel explants producing large number of true-to-type plantlets in 4–8 week on Murashige and Skoog (MS) based media. Also, true mutants with changed flower color are obtained without producing chimeras after gamma-irradiation in mutation breeding.

Key words: Chrysanthemum, *Chrysanthemum x grandiflorum* (Ramat.) Kitam, Adventitious buds, Micropropagation, Pedicel, Stem node, Gamma-rays, Mutation

1. Introduction

Chrysanthemum x grandiflorum (Ramat.) Kitam. (synonyms: *Chrysanthemum x morifolium* Ramat., *Dendranthema x morifolium* Tzvelev., *Dendranthema x grandiflorum*, *Chrysanthemum sinense*) belongs to the Asteraceae family and is also known as the "florist's chrysanthemum" or "mum." There are many different types of chrysanthemums with different flower colors and styles. They are economically significant flowers worldwide as there are a great number of cultivars for cut flowers, pot flowers, and garden flowers (1). The chrysanthemum is a typical vegetatively propagated plant, e.g., by rooting of cuttings in vivo or by in vitro techniques. The techniques for in vitro propagation of ornamental plants and tissue culture laboratory equipment are being continuously improved to meet the demand of the floriculture breeding

S.M. Jain and S.J. Ochatt (eds.), *Protocols for In Vitro Propagation of Ornamental Plants*, Methods in Molecular Biology, vol. 589
DOI 10.1007/978-1-60327-114-1_17, © Humana Press, a part of Springer Science+Business Media, LLC 2010

and industry (2). Commercial in vitro multiplication of chrysanthemum is often based on stem node explants with lateral meristems (3, 4).

In this chapter, protocol for a complete cycle of in vitro propagation of chrysanthemum is described. It includes the steps beginning with shoot initiation based on direct organogenesis from pedicels as explants or from stem nod explants, multiplication, cool storage, and root initiation. An example for their application in mutation breeding is also given.

In vitro propagation by pedicel explants is based on the *adventitious buds technique.* In herbaceous plants, adventitious buds develop from leaves or pedicels, i.e., places other than the leaf axil or the shoot-tip. Some flowers produce adventitious buds in vivo, e.g., Begonia leaves, etc. Such buds are often used for conventional propagation of species such as Begonia, Saintpaulia, etc. (5, 6). Adventitious buds can also be used for micropropagation of many plant species when the specifically required conditions are provided and a suitable part of the plant is used for propagation. The in vitro cultivation of stem, leaf, or other explants yields a very large number of adventitious shoots. In some cases, the adventitious buds technique for in vitro propagation of chrysanthemum by leaf explants may be a source of new variability as a result of either chimera structure or somaclonal variation of the plants (7). We have successfully used the protocol for pedicel explants described here to obtain nonchimera mutants with changed flower color after gamma-irradiation in mutation breeding. Nonirradiated pedicel explants produced true-to-type plantlets in vitro.

In vitro propagation by stem node explants is based on the *proliferation of axillary buds.* Axillary buds develop at the junction of the stem and the petiole. The proliferation of axillary buds is the most common method for in vitro propagation (multiplication) of chrysanthemum and many other plant species (8–10). Activation of the growth of axillary buds is achieved by removing the apical part of the stem, which eliminates the dominance of the apical meristem and results in the development of lateral meristems. This approach, called "*pinching*," is applied to stimulate the branching especially of pot chrysanthemum in vivo. Stem node explants readily develop axillary shoots in vitro.

2. Materials

2.1. Plant Material

Commercially, glasshouse-grown chrysanthemum plants – vigorous and healthy-looking without damage caused by insects, fungal, bacterial, or viral diseases – are chosen (*see* Note 1).

2.2. Surface Sterilization Solution

1. Choose one of the following surface sterilization agents:
   - 4% calcium hypochlorite ($Ca(OCl)_2$, Sigma-Aldrich), or
   - 0.1% mercuric chloride ($HgCl_2$, Sigma-Aldrich), or
   - 0.8% silver nitrate ($AgNO_3$, Sigma-Aldrich).
2. For 250 mL sterilization solution, dissolve the following amounts in distilled water by stirring: 10 g $Ca(OCl)_2$, or 250 mg $HgCl_2$, or 2 g $AgNO_3$. Store the solution in a 500-mL dark wide-necked screw cap vessel at room temperature; keep it tightly closed.

2.3. Media

1. Four types of solid Murashige and Skoog (MS) (11) based media are used in the in vitro propagation of chrysanthemum. It is convenient to use commercial powder salts of MS macro nutrients and minor salts. Stock solutions of plant growth regulators are easily prepared.
2. In general, to prepare 1 L medium, dissolve the required quantity of MS salts and other dry substances in 300 mL distilled water, following the recipe given below. Unfreeze the growth regulators and add them (if required), then add 400 mL hot water, stir well, and adjust pH to 5.7 with 0.1 N NaOH. In another beaker, put agar and 300 mL of distilled water and heat up until the agar melts. Mix the two solutions, and stir very well. Dispense the necessary quantity of medium in culture tubes, close the caps, and sterilize by autoclaving for 15 min at 121°C.

2.3.1. Stock Solutions of Growth Regulators

0.1% N^6-Benzil adenine (BAP): dissolve 20 mg BAP (Duchefa B.V.) in 5 mL 0.1 N NaOH, add 15 mL distilled water, stir well, and freeze 2 mL aliquots for single use at –20°C.

0.1% 3-Indoleacetic acid (IAA): dissolve 20 mg IAA (Duchefa B.V.) in 5 mL ethanol and 2–3 drops 0.1 N NaOH, add 15 mL distilled water, stir well, and freeze 1 mL aliquots for single use at –20°C.

2.3.2. Medium for Shoot Initiation (MS Macro and Minor Salts and Vitamins in Full Strength Supplemented with Cytokinin and Auxin)

4.4 g/L powder of MS macro and minor salts including vitamins (Duchefa B.V.), 100 mg/L myoinositol (Sigma-Aldrich), 2 mg/L BAP (2 mL stock solution), 1 mg/L IAA (1 mL stock solution), 20 g/L sucrose, and 8 g/L agar (Duchefa B.V), pH 5.7.

2.3.3. Medium for Multiplication (MS Macro and Minor Salts and Vitamins in Full Strength)

4.4 g/L pvowder of MS macro and minor salts including vitamins (Duchefa B.V.), 20 g/L sucrose, and 6.5 g/L agar (Duchefa B.V.), pH 5.7.

2.3.4. Medium for Cool Storage (MS Macro and Minor Salts at Half Strength)

2.15 g/L powder of MS macro and minor salts (Duchefa B.V.), 20 g/L sucrose, and 6.5 g/L agar (Duchefa B.V.), pH 5.7.

2.3.5. Medium for Root Induction (MS Macro and Minor Salts at Half Strength Supplemented with Auxin)

2.15 g/L MS macro and minor salts (Duchefa B.V.), 2 mg/L IAA (2 mL stock solution), 20 g/L sucrose, and 6 g/L agar (Duchefa B.V.), pH 5.7.

2.3.6. Substrate for acclimatization

Mix of peat and pearlite in a 3:1ratio

2.4. Other Supplies

1. Glass cultural tubes with compatible autoclavable plastic caps (20 mm diameter, 18 cm high) (Danspharma, Sofia, Bulgaria).
2. Beakers (4 × 500 mL) sterilized in a hot air oven at 175°C for 2 h.
3. Filter paper (20 × 15 cm) sterilized in a hot air oven at 175°C for 2 h.
4. Distilled water in bottles (3 × 250 mL) sterilized by autoclaving for 1 h at 121°C.

3. Methods

3.1. Preparation of Explants

3.2. Pedicel Explants

1. Take pedicels with a diameter of 2–2.5 mm at a stage before flowering when the plants have already formed well-shaped green flower buds.
2. Cut the top part of the stems (approx. 10 cm long) including the green flower bud, pedicel, and the highest leaf (*see* Note 2).
3. Back in the laboratory, remove the flower buds and the bottom part of the cuttings (Fig. 17.1a) and use only pedicels without any leaves or lateral buds. Depending on the cultivar, the usable part (pedicel) is 2–4 cm long.

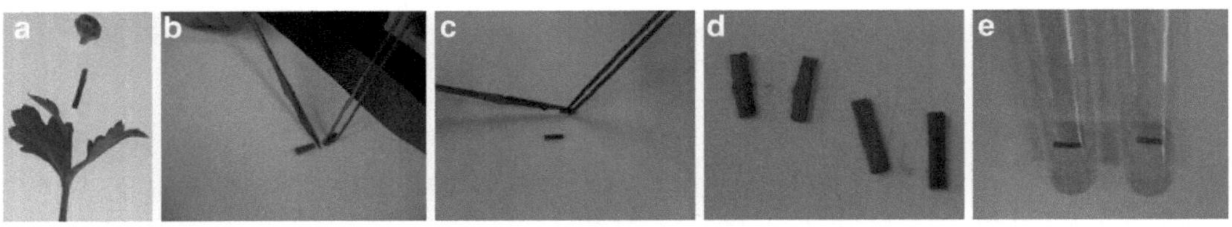

Fig. 17.1. Preparation of pedicel explants. Steps: Remove the flower buds and the bottom part of the cutting and sterilize the pedicel (**a**). Cut the sterilized pedicel into two or three fragments 8–10 mm long (**b**). Cut each fragment along its length to obtain four or six fragments from each pedicel (**c**, **d**). Place fragments onto medium for shoot initiation with the cut side facing downwards (**e**).

3.2.1. Nodal Explants

1. Use chrysanthemums grown in long-day conditions when flower buds have not yet formed (*see* Note 2).
2. Take the soft part of the stems (approx. 15 cm long).
3. Remove the top and the leaves with petioles.
4. Divide into cuttings with 3–4 nodes, approx. 5–6 cm long.

3.3. Surface Sterilization

1. Rinse the pedicels in 200 mL tap water with 1 or 2 drops of colorless liquid soap for 2 or 3 min and drain them on filter paper.
2. Drop 10–15 explants in the sterilization solution, close the screw cap, and swirl gently. The duration of soaking depends on the type of chemical: 12–15 min in 4% $Ca(OCl)_2$, or 4–5 min in 0.1% $HgCl_2$, or 7–8 min in 0.8% $AgNO_3$.
3. The following steps should be performed in sterile conditions in a laminar flow hood. Rinse the explants with sterile distilled water. Swirl gently for 2 min, then change the water, and rinse two or three more times. Pour out the last rinsing water.

3.4. Preparation of Fragments from Explants

1. Take out an explant, and lay it on a piece of filter paper.
2. Prune the outermost 1 mm from both ends to remove the parts damaged by the sterilization procedure.
3. *Pedicel explants*: Cut the pedicel into pieces 8–10 mm long (Fig. 17.1b). Then, cut each piece along its length. In this way, you finally obtain four or six semicylindrical fragments from each pedicel (Fig. 17.1c, d).
4. *Nodal explants:* Cut the stem explants into fragments containing one node (1.5–2.0 cm long) so that the bud stays at the upper 1/3 portion of the explant.

3.5. Induction of Shoots

1. Incubate each pedicel fragment (one per tube) on 15 mL medium for shoot initiation (*see* Point 2.3.2) with the cut side facing downwards (Fig. 17.1e).
2. Incubate each one-node fragment in a vertical position on 15 mL medium for shoot initiation with the bottom 1/3 portion of the fragment submerged in the medium. Be careful not to plant the explants up-side-down.
3. Cultivate at 24°C, 80 $\mu E/m^2/s$ and 16:8 hours day-night photoperiod. In 4–6 weeks, the shoots reach a height of 5–65 mm or more and are ready for micropropagation (*see* **Note 3**).

3.6. Micropropagation (Multiplication)

This process is based on the growth of axillary buds (lateral meristems) developing in the base of the leaves. Following the procedure described here, usually one shoot will develop from each one-node-microcutting.

1. Use plantlets from culture tubes with regenerated shoots.
2. Gently take out the explant with shoots from the test tube, and place it on a piece of filter paper (Fig. 17.2).
3. Cut off the shoots from the explant.
4. Cut the shoots into microcuttings bearing one node, i.e., one leaf, so that the larger part of the internode remains below the leaf (Fig. 17.3a, b).
5. Incubate the microcuttings on 10 mL medium for multiplication (*see* Point 2.3.3) in a vertical position with the bottom ½ of the internode submerged (Fig. 17.3c). Handle each explant individually, and label the vessels with the name of the cultivar and the date.

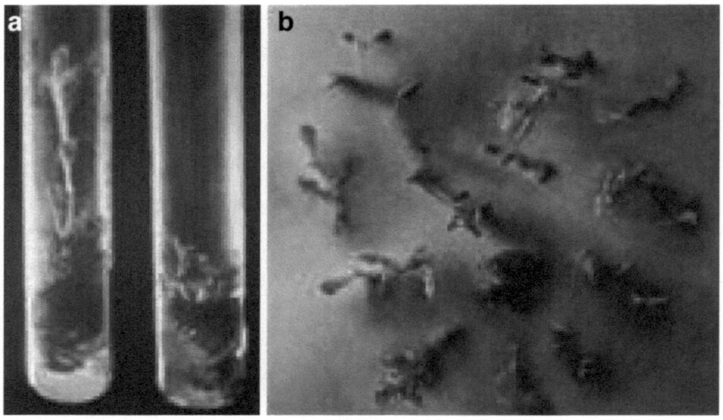

Fig. 17.2. Adventitious shoots formation. Pedicel explants with adventitious shoots in test tubes (**a**). Remove the explant, and cut off the shoots from the explant (**b**).

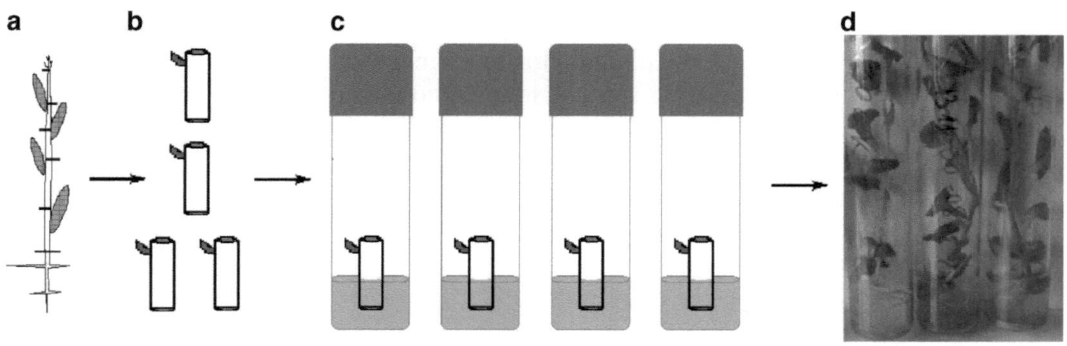

Fig. 17.3. Steps in micropropagation: Cut the shoots into one-node fragments (**a**, **b**). Plant the microcuttings in a vertical position with the bottom ½ of the internode submerged in the medium (**c**). Plantlets ready for a next cycle of micropropagation (**d**).

6. Cultivate at 24°C, 80 µE/m²/s and 16:8 hour day-night photoperiod for 3–4 weeks.
   When the shoots have grown to reach almost the top of the cultural vessel (Fig. 17.3d), they are ready for a next cycle of micropropagation (*see* Note 4).

3.7. Cool Storage

1. Prepare the microcuttings the same way as in the case of multiplication (Fig. 17.3a–c), but cultivate them on 15 mL medium for cool storage (*see* Point 2.3.4). Seal the caps with micropore tape.
2. Incubate at 24°C, 80 µE/m²/s and 16:8 hour day-night photoperiod for 10 days or until the plantlets develop three or four leaves (a height of 3–5 cm).
3. Transfer the plantlets into a cultivation room at 6°C, 50 µE/m²/s and 12:12 hour day-night photoperiod.

In such "poor" in vitro conditions, chrysanthemums grow very slowly and can be stored successfully up to 1 year without any handling. These plantlets could be successfully used as a resource for multiplication at any moment of their cool storage (*see* Note 5).

3.8. Root Induction

1. Prepare the microcuttings the same way as in the case of multiplication (Fig. 17.3a–c), but cultivate them vertically on 10 mL medium for root induction (*see* Point 2.3.5) at 24°C, 80 µmol/m²/s and 16:8 hour day-night photoperiod.
2. After 6 or 7 days, plantlets develop three to five short (4–6 mm) strong white roots on the basal part of the stems. At this stage, they are ready for acclimatization (Fig. 17.4a).

3.9. Acclimatization to In Vivo Conditions and Planting in Soil

1. Take out the rooted plantlets from the medium and rinse gently in water to remove the agar.

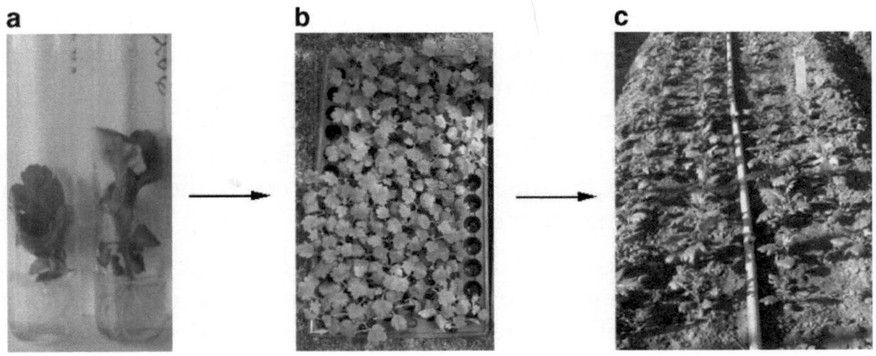

Fig. 17.4. Acclimatization to *in vivo* conditions. Steps: *In vitro* rooting (**a**), acclimatization in cell packs (**b**), and planting in soil/greenhouse (**c**).

2. Plant them in cell packs with a substrate of peat and pearlite (*see* Point 2.3.6) with moderate moisture. Cultivate at high air humidity (mist) and 22°C (Fig. 17.4b). Decrease the temperature to 17°C at the end of the acclimatization period. The air humidity has to be decreased gradually, too. Shading is necessary in hot sunny days. Following this procedure in 3 weeks, the plants are ready for planting in glasshouse (Fig. 17.4c) and grow at standard conditions for chrysanthemum.

4. Notes

1. When you work with more than one cultivar, pick the cuttings separately and label them punctually. If it is not possible to take the cuttings immediately before incubation, you can store them in plastic bags at 4–6°C for 3–4 days, but storage may increase the risk of contamination.
2. For in vitro propagation with pedicel explants, it is suitable to use chrysanthemum cultivars with big or middle-sized flowers whose pedicels are thick enough (at le st 2 mm). For propagation of cultivars with small flowers and thin pedicels, use stem node explants.
3. The first adventitious shoots usually appear on the surface of pedicel explants after 3 or 4 weeks of cultivation. Our experience shows that after 7 or 8 weeks of in vitro cultivation, most of the shoots reach a height of 5 mm and are ready for micropropagation. The number of shoots per explant varies from 4 or 5 to 50, and their height varies from 2 to 60–65 mm depending on the genotype (12). The general rule is: the greater the number of adventitious shoots from an explant, the smaller their size.
4. Depending on the genotype and the height of the vessel used, it takes a different time [12). For example, most chrysanthemums need about 4 weeks of in vitro cultivation to reach 10–12 cm of height and 8–12 leaves. Such plantlets can be cut into 7–10 microcuttings each. These microcuttings can be cultured the same way, and so on and so on. In commercial micropropagation, this makes it relatively easy to precisely predict the time it would take to obtain the necessary number of plants from each cultivar.
5. In our practice for 1 year in cool storage conditions, the plantlets grow up to 7.5–10.5 cm in height depending on the cultivar. When necessary, these plantlets can be used for micropropagation following the procedure described above. It is crucial to maintain the cool storage room faultlessly clean in order to avoid the risk of contamination with air-borne spores, which could germinate after the transfer to 24°C.

References

1. Bakker M, Van Dijk JM (eds) (1995) Fides mum manual. Fides, Holland BV, De Lier
2. Rout GR, Mohapatra A, Mohan Jain S (2006) Tissue culture of ornamental pot plant: a critical review on present scenario and future prospects. Biotechnol Adv 24(6):531–560
3. Broertjes C, Van Harten AM (1978) Adventitious bud techniques and other *in vivo* or *in vitro* methods of asexual propagation. In: Van Harten AM (ed) Developments in crop science (2). Elsevier, Amsterdam, pp 33–49
4. Himstedt JP, Jacobsen HJ, Fisher-Klüver G (2001) Shoot regeneration from stem and leaf explants of chrysanthemum (*Dendranthema x grandiflorum*). Acta Hortic 560:421–424
5. Broertjes C, Van Harten AM (1985) Single cell origin of adventitious buds. Euphytica 34:93–95
6. Chen CC, Chen SJ, Sagare A, Tsay HS (2001) Adventitious shoot regeneration from stem internode explants of *Adenophora triphylla* (Thunb.) A. DC. (Campanulaceae) – an important medicinal herb. Bot Bull Acad Sin 42:1–7
7. Zalewska M, Lema-Rumińska J, Miler N (2007) *In vitro* propagation using adventitious buds technique as a source of new variability in chrysanthemum. Sci Hortic 113(1):70–73
8. Read PE (2007) Micropropagation: past, present and future. Acta Hortic 748:17–27
9. Karim MZ, Amin MN, Azad MAK, Begum F, Rahman MM, Ahmad S, Alam R (2003) *In vitro* shoot multiplication of *Chrysanthemum morifolium* as affected by sucrose, agar and pH. Biotechnology 2(2):115–120
10. Karim MZ, Amin MN, Azad MAK, Begum F, Rahman MM, Islam MM, Alam R (2003) Effects of different plant growth regulator on in vitro shoot multiplication of *Chrysanthemum morifolium.* On Line J Biol Sci 3(6):553–560
11. Murashige T, Skoog F (1962) A revised medium for rapid growth and bioassays with tobacco tissue cultures. Physiol Plant 15:473–497
12. Nencheva D (2006) *In vitro* prediction of plant height for *Chrysanthemum x grandiflorum* (Ramat.) Kitam. J Fruit Ornam Plant Res 14:223–232

Chapter 18

Micropropagation of *Codiaeum variegatum* (L.) Blume and Regeneration Induction via Adventitious Buds and Somatic Embryogenesis

Silvia Radice

Abstract

Codiaeum variegatum (L) Blume cv. "Corazon de oro" and cv. "Norma" are successfully micropropagated when culture are initiated with explants taken from newly sprouted shoots. The establishment and multiplication steps are possible when 1 mg/L BA or 1 mg/L IAA and 3 mg/L 2iP are added to MS medium, according to the cultivar respectively selected.

Adventive organogenesis and somatic embryogenesis are induced from leaf explants taken from in vitro buds of croton. On leaf-sectioned of "Corazon de oro" cultured in vitro, 1 mg/L BA stimulates continuous somatic embryos development and induces some shoots too. Replacing BA with 1 mg/L TDZ induces up to 100% bud regeneration in the same explants. On the other hand, leaf-sectioned of *C. variegatum* cv. Norma does not start somatic embryo differentiation if 1 mg/L TDZ is not added to the MS basal medium. Incipient callus is observed after 30 days of culture, and then, subculture to MS with 1 mg/L BA allows the same process to show on the "Corazon de oro" cultivar. Somatic embryos show growth arrest that is partially overcome by transfer to hormone-free basal medium with activated charcoal. Root induction is possible on basal medium plus 1 mg/L IBA. Plantlets in the greenhouse have variegated leaves true-to-type.

Key words: Croton, In vitro, Multiplication, Ornamental, Regeneration, Somatic embryogenesis

1. Introduction

Codiaeum variegatum (L.) Blume is a member of the family *Euphorbiaceae*. It is commonly called "croton" or "variegated croton". It is native of East Asia, and is one of about 200 species in this genus. Garden croton occurs naturally in southern Asia,

S.M. Jain and S.J. Ochatt (eds.), *Protocols for In Vitro Propagation of Ornamental Plants*, Methods in Molecular Biology, vol. 589
DOI 10.1007/978-1-60327-114-1_18, © Humana Press, a part of Springer Science+Business Media, LLC 2010

Indonesia, and other Eastern Pacific islands where it grows in open forests and scrub.

Croton is an evergreen shrub with alternate, simple leaves mottled with white, yellow, or red flowers. Hence, this species has been selected for the different morphology and color combination of leaves, as an ornamental foliage plant. Color patterns range from multicolored spots to irregular color patches or solid-colored leaves with contrasting veins. Consequently, the apparition to off type could not be a problem. Many asexually propagated floriculture crop varieties are spontaneous somaclonal variants and chimeras (1).

Propagation of *Codiaeum spp.* by rooting of softwood cuttings has been a good development. Some authors investigated how different compounds of the substrate can improve root induction (2–4). On the other hand, Hata et al. (5) observed that a treatment combination of hot water and indole-3-butyric acid is beneficial to stimulate rooting and shoot development of cuttings of this species, although it has been observed as the temperature can be a restrictive factor to the rooting promotion. In fact, 27–30°C was determined that the optimum temperature for rooting *C. variegatum* cv. Aucubafolia (6). Nevertheless, terminal cuttings of *C. variegatum*, taken in August, soaked for 24 h in 10–20 ppm. IBA and planted in a 1:1 peat+sand mixture were 100% rooted (7).

Micropropagation would provide an alternative method to replace mother plant stocks with microcuttings maintained in vitro. The problems of rejuvenation, dormancy, or delayed flowering sometimes observed when hardy ornamental plants are produced in vitro (8) do not apply to *Codiaeum* spp., whose appearance lies in its foliage.

There are a few studies referred to *C. variegatum* (L.) Blume in vitro culture cited in the literature. It includes two reports on endosperm and embryo culture (9, 10). Protocols for micropropagation of two cultivars of *C variegatum* were developed (11, 12). In addition, Orlikowska et al. (13) improved rooting in vitro of defoliated micropropagated shoots of *C. variegatum* cv. Excellent.

The protocols presented afterwards have proven useful not only for in vitro clonally propagation, but also for adventitious buds and somatic embryogenesis induction. In this case, the somaclonal variation could provide an increment of leaf variegation that would enhance the possibilities of selection of new cultivars. The opportunities offered by variation can be interesting not only from the scientific point of view but also from applied consideration. In this, cultivar variations could offer a possibility to select plants with different color-leaf patterns.

2. Material

2.1. Donor Plant Preparation

Mother plants of different cultivars of *C. variegatum* (L) Blume were cultivated in pots and maintained in greenhouse conditions. The plants were pruned and fertilized to encourage new shoots and sprayed with a mix of Benlate and Agrimicine to minimize disease and contamination problems.

2.2. In vitro Culture

1. Plant material: Healthy and viable nodal segments of *C. variegatum* (L) Blume cv. "Corazon de oro" and "Norma" of approximately 20 mm long.
2. Water solution of alcohol, 70% (v/v); water solution of commercial bleach (NaClO) 25% (v/v); 0.1% Mercuric chloride ($HgCl_2$); 0.1% (v/v) liquid detergent (Tween-20).
3. Sterile deionized water.
4. Solution of inorganic and organic components of Murashige and Skoog's (14) medium modified according to Table 18.1.
5. Growth regulators: indole 3-acetic acid (IAA); N6 benzyladenine (BA); 6-(γ, γ-dimethilallylamino) purine (2iP); N-phenyl-N′-1, 2, 3-thidiazol-5-ylurea (TDZ) (Sigma Aldrich, USA) (Table 18.2).
6. Activated charcoal (Table 18.2).
7. 0.1 N NaOH and/or 0.1 N HCl.
8. Plastic packages, test tubes 2.5 cm in diameter, Petri dishes 10 cm in diameter, and glass flasks.

2.3. Hardiness of Propagules

Greenhouse equipped with mist system and running shade cloth. Pots, soil, crushed pine leaves and peat.

3. Methods

3.1. Preparation of Medium Stock Solution

1. Sucrose is dissolved in 200 mL distilled water.
2. Every compound described on Table 18.1 is a solution in a minimum quantity of water separately and then added to sucrose solution and strongly agitated to prevent precipitation.
3. Volume of stock solution is complete to 1,000 mL with distilled water.
4. The stock solution is dispensed in plastic packages containing 10 mL medium and kept in the freezer.

Table 18.1
Medium composition for in vitro culture *C. variegatum* (L) Blume based on Murashige and Skoog (14) medium

Compounds	Stock solution (100×)
Macroelements	(g/L)
KNO_3	190
NH_4NO_3	165
KH_2PO_4	17
$MgSO_4$ $7H_2O$	37
$CaCl_2$ $2H_2O$	44
NaEDTA	3,723
$FeSO_4$ $7H_2O$	2,795
Microelements	
KI	0.083
H_3BO_3	0.62
$MnSO_4$ $4H_2O$	2.23
$ZnSO_4$ $7H_2O$	0.86
$CuSO_4$ $5H_2O$	0.0025
$CoCl_2$ $6H_2O$	0.0025
Na_2MoO_4 $2H_2O$	0.025
Sucrose	3,000
Organic compounds	(mg/L)
Glycine	200
Mio-inositol	10,000
Thiamine	100

3.2. Media Preparation

1. Frozen basal medium is isolated to the freezer and defrosts. The amount will be the proportional one to the volume of culture medium to prepare (10 mL/L culture medium).
2. Growth regulators are added to basal medium according to the step to accomplish (Table 18.2), and then, pH of the medium is adjusted to 5.7 ± 0.1 with 0.1 N NaOH and/or 0.1 N HCl droplets.

Table 18.2
Addition of appropriate plant growth regulators in MS basal culture medium (see Table 18.1) for in vitro culture of *C. variegatum* (L) Blume cv. "Corazon de oro" and "Norma". Values are expressed in mg/L

Growth regulators	Micropropagation		SE induction on leaves sections		SE proliferation	Bud elongation	Rooting medium
	C. de oro cv.	Norma cv	C. de oro cv.	Norma cv			
BA	1		1		1		
IAA		1	1				
2iP		3					
IBA							1
TDZ				1			
Activated charcoal						1,000	

3. Semi-solid medium is achieved by 0.8% (w/v) of agar added to the culture medium and heated to microwave. Time was variable to the mix volume (100 mL = 1 m), and then, it is dispensed after a strong agitation about 2 m.
4. Media cultures, according to the morphogenetic induction step, are dispensed into 10 mL test tubes 2.5 cm in diameter or Petri dishes 10 cm in diameter. Fifty milliliter media are employed for propagation and rooting media into glass flasks.
5. Media are autoclaved at 0.1 Mpa for 15 min.

3.3. Micropropagation Culture Set Up

1. Nodal segment taken from mother plants are disinfected with 70% alcohol for a few minutes, 25% commercial bleach (2% chlorine) for 30 min, and 0.1% mercuric chloride for 5 min. Finally, they are then rinsed thoroughly with sterile deionized water.
2. Explants are made to remain in the last water bath until use to prevent tissue oxidation.
3. Every mini cutting is cut at the ends to take off damaged tissue and then sown into de semisolid medium according to the cultivar employed (Table 18.2).
4. Explants are subcultured every 35–42 days to the same medium (*see* Notes 1 and 2).

5. Culture vessels are maintained at 25 ± 2°C under a 16 h light photoperiod provided by cool white fluorescent light (57 μE/m^2/s) TLT 110 W/54 RS Philips day-light tubes.
6. Shoots with 2–3 leaves and 50 mm high are suitable to growth roots in vitro (Fig. 18.1d).
7. Plantlets with roots are transferred to pots containing a mixture of 1:2:1 soil: crushed pine leaves: peat, and they were maintained in a greenhouse for 2 weeks under mist (up to 85% RH), wherefore humidity conditions are gradually reduced to normal by 30 days from transfer.

3.4. Induction of Somatic Embryo

Excised leaves of sprouts growth in vitro are selected to promote somatic embryo induction. Leaves are cut in halves, and only the basal halves of leaves with the petiole portion are used. Leaves are placed in flask containing 50 mL MS medium supplemented with 1 mg/L IAA + 1 mg/L BA for "Corazón de oro" (Fig. 18.2) (*see* Note 3) cultivar, and only 1 mg/L TDZ is added when "Norma" cultivar is employed (Fig. 18.1) (Table 18.2). Flasks are covered by polycarbonate translucent top (Fig. 18.1d). Explants of two cultivars are subcultured every 30 days to the MS medium with 1 mg/LBA only. They show somatic embryo development directly without an intervening callus phase (Fig. 18.1). This morphogenetic response is continued in time, and it is noticeable that the repetitive somatic embryogenesis is at a high frequency.

3.5. Growth of Somatic Embryo

Clusters of somatic embryo are subculture to MS basal medium supplemented by activated charcoal only, twice or more time. Bud elongation is induced, but rooting remained unaltered. Embryo derived shoots from charcoal medium could be successfully rooted when they are transferred to MS with 1 mg/L IBA (*see* Note 4).

4. Notes

1. The multiplication step in *C. variegatum* cv. "Corazon de oro" was accomplished with axillary and adventitious buds that formed in the leaf axils. The subculture promoted a 3:1 multiplication rate every 35–40 days. Multiplication of "Norma" plants started after three subcultures on MS with 1 mg/L IAA + 3 mg/L 2iP, whereas the average bud propagation ratio was between 2.5-fold every 6 weeks. Explants were subcultured every 42 days without any deviation in phenotype even after 24 months of continuous subculturing. When BA was used in this cultivar, callus growth occurred after 10 days of culture, arresting the growth of buds (12).

Fig. 18.1. Morphogenetic response in *Codiaeum variegatum* (L.) Blume cv. "Norma". Somatic embryos developed on micropropagated leaves (**a**). Buds growth on micropropagated leaves (**b**). Micropropagated shoots (**c**). Rooted plants standing by to transfer to soil (**d**). Bars represented (**a**) = 5 mm; (**b–d**) = 10 mm.

2. Adventitious buds and somatic embryos in *C. variegatum* cv. "Corazon de oro" were observed after the fourth subculture on multiplication medium from expanded leaves attached to the stem but not in contact with the culture medium. The frequency of this phenomenon increased with time in culture, from 20% of explants by the fourth subculture to 90% after 12 months of culture.

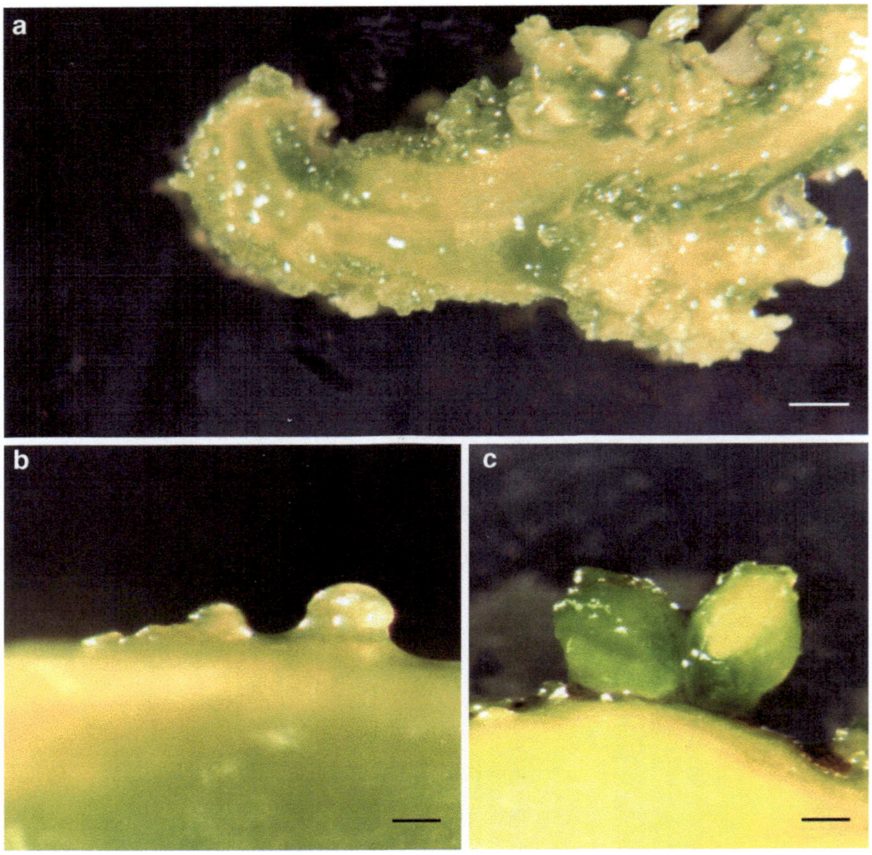

Fig. 18.2. Embryogenesis response in *Codiaeum variegatum* (L.) Blume cv. "Corazón de oro". Leaf explants after 30 days on MS medium with 1 mg/L indole-3-acetic acid (IAA) +1 mg/L N6 benzyladenin (BA) (**a**). Globular embryos (**b**). Development embryos (**c**). Bars = 1 mm.

3. Adventitious shoots and somatic embryos were simultaneously induced on the same explant for *C. variegatum* cv. "Corazon de oro". Nevertheless, when 1 mg/L TDZ was added to MS basal medium, in vitro leaf-sections showed bud induction after 30 days of culture. Bud initiation was observed on 77% explants, and it was possible to continue the micropropagation of these propagules when the 1 mg/L BA was employed.
4. Embryo germination in croton was poor. The removal of growth regulators from the transfer medium did not improve results, but it was observed that some germinated embryos in cultures remained in the same medium during 2–3 month (11).

Acknowledgments

Author thanks the Consejo Nacional de Investigaciones Científicas y Técnicas (CONICET) for supporting the researches presented.

References

1. Sagawa Y, Kunisaki JT (1990) Micropropagation of floriculture crops. In: Ammirato PV, Evans DA, Sharp WR, Bajaj Y (eds) Handbook of plant cell culture, vol 5, Ornamental species. New York, McGraw-Hill Publishing Company, pp 25–56
2. Tillmann MAA, Cavariani C, Piana Z, Minami K (1994) Comparacao entre diversos substratos no enraizamento de estacas de croton (*Codiaeum variegatum* L.). Sci Agric 51(1): 17–20
3. Chen JJ, Robinson CA, Caldwell RD, McConnell DB (2000) Waste composts as components of container substrates for rooting foliage plant cuttings. Proc Fla State Hort Soc 112:272–274
4. Dai-Bisheng (2007) Effect of carbendazim plus thiram and triadimefon plus ethylicin on the survival rate of three kind softwood cuttings. Huazhong Shifan Daxue Xuebao (Ziran Kexue Ban) 41(1):111–116
5. Hata TY, Hara AH, Nagao MA, Hu BKS (1994) Hot-water treatment and indole-3-butyric acid stimulates rooting and shoot development of tropical ornamental cuttings. Horttechnology 4(2):159–162
6. Hentig WU, von and Fischer M (1994) Bewurzelungstemperaturen von Zierpflanzenstecklingen fur Hydrokultur. Deutscher Gartenbau 41(9):522–525
7. Khattab M, Kamel H, Yacout M (1987) The effect of indolebutyric acid on rooting promotion in croton cutting. Alex J Agric Res 32(1):287–295
8. Capellades Queralt M, Beruto M, Vandeschaeghe A, Debergh PC (1991) Micropropagation. In: Debergh PC, Zimmerman RH (eds) Technology and application. Kluwer Academic, The Netherlands, pp 215–229
9. Chikkannaiah PS, Gayatri MC (1974) Organogenesis in endosperm tissue culture of *Codiaeum variegatum* Blume. Curr Sci 43:23–24
10. Chikkannaiah PS, Nataraja K, Gayatri MC (1976) Response of seed and isolated embryos of *Codiaeum variegatum* in vitro. Curr Sci 45:196–197
11. Marconi PL, Radice S (1997) Organogenesis and embryogenesis in *Codiaeum variegatum* (L.) Blume cv. Corazón de oro. In vitro Cell Dev Biol Plant 33:258–262
12. Radice S (2000) Micropropagation of *Codiaeum variegatum* (L) Blume cv "Norma". Phyton 69:143–146
13. Orlikowska T, Sabala I, Kucharska D, Cassells AC, Doyle BM, Curry RF (2000) Rooting of axillary shoots of *Codiaeum variegatum* Blume cv. Excellent obtained in vitro from defoliated shoot explants. Acta Hortic 530:253–256
14. Murashige T, Skoog F (1962) A revised medium for rapid growth and bioassays with tobacco tissue cultures. Physiol Plant 15:473–497

Chapter 19

Methods for In Vitro Propagation of *Pelargonium* x *hortorum* and Others: From Meristems to Protoplasts

Noëlle Dorion, Hatem Ben Jouira, Anthony Gallard, Anber Hassanein, Mazen Nassour, and Agnès Grapin

Abstract

Geraniums (*Pelargonium* spp.) are among the most popular bedding and pot plants (25 % of the French domestic market). On one hand, as vegetatively propagated plants, *Pelargonium* are submitted to pathogen pressure. On the other hand, innovation via interspecific hybridisation faces some difficulties. In this chapter, the two first protocols (from seeds and meristems) explain how in vitro plants free of virus could be obtained. The development of this technique is the long-term preservation of genetic resources via meristem cryopreservation. The third protocol describes propagation of *Pelargonium* with limited risks of variation. This technique also allows the constitution and the maintenance of a plant-stock from which explants can be taken for other studies. The two last protocols describe plant regenerations from leaf discs and mesophyll protoplasts, used for gene transfer and somatic hybridisation. These protocols were established mainly with *Pelargonium* x *hortorum* cultivars, but we propose possible solutions for the other species: *P.* x *peltatum*, *P.* x *domesticum*, *P. capitatum* and *P. graveolens*.

Key words: Leaf disc, Meristem, Micropropagation, Plant regeneration, *Pelargonium* spp., Protoplast

1. Introduction

Pelargonium spp. (Geranium) belongs to the family *Geraniaceae*. There are about 280 species in the genus *Pelargonium*, mostly native of South Africa. Geraniums are among the most popular bedding and pot plants in North America and Europe, e.g., 25% of the French national market, more than 150 million plants yearly sold. The widely cultivated *Pelargonium* falls into four major groups, *P.* x *hortorum* (zonal geranium), *P.* x *peltatum*

S.M. Jain and S.J. Ochatt (eds.), *Protocols for In Vitro Propagation of Ornamental Plants*, Methods in Molecular Biology, vol. 589
DOI 10.1007/978-1-60327-114-1_19, © Humana Press, a part of Springer Science+Business Media, LLC 2010

(ivy-leaved geranium), *P.* x *domesticum* (florist geranium), and fragrant leaf species and cultivars. In the context of sustainable agriculture, the use of *Pelargonium* in floricultural industry should rise again since good resistance to drought was found among them (1). However, genetic improvement is still needed both for agronomic, i.e. virus and/or bacterium resistance, superior drought resistance, and aesthetic (i.e. yellow flowers) traits. Improvement of geranium by traditional breeding faces some difficulties such as low fertility or incompatibility among species. Therefore, the use of non-conventional biotechnology methods could open new perspectives to geranium improvement provided that efficient plant regeneration systems exist. Also, in vitro culture provides means for the preservation, free of pathogen pressure, of this vegetatively propagated plant. These means involve classical micropropagation process but also meristem cryopreservation (2). Many groups have already worked with in vitro culture of *Pelargonium* (3–11). We summarise in this paper our 9-year experience with this plant. This includes introduction and propagation in vitro of virus-free plants (2, 12, 13). This step is needed because cultivars are mostly vegetatively propagated and thus often contaminated by virus, which interacts with growth, flowering, and consequently physiological state of the plant. It is also needed because it is easier to control environmental parameters in a growth chamber than in the greenhouse and therefore repeatedly have plants and explants in a quite similar physiological state. We also report plant regeneration from leaf discs (14) and mesophyll protoplasts (13, 15, 16). This crucial step is a prerequisite, whatever be the method used for plant improvement using somaclonal variation, mutagenesis, genetic transformation (17, 18), or somatic hybridisation (19).

2. Materials

2.1. General Materials

1. A greenhouse covered with glass or polycarbonate is needed to cultivate *Pelargonium* plants before in vitro introduction from meristems and for plant acclimatisation after in vitro culture. During winter, temperature should be at least around 18–20°C, and a supplementary photosynthetic illumination until 16 h can be added to ensure growth, especially during acclimatisation of in vitro plants. In summer, the greenhouse should be naturally ventilated (roof and aisle) and covered with white painting when internal temperatures reach 25–27°C. Cooling and/or mist system must be avoided. Acclimatisation needs the use of small plastic greenhouses

(1,500 cm^2, HR > 80%) for 3–4 weeks before setting the plants under normal greenhouse conditions.

2. Plants are potted in a *Pelargonium* specific substrate (i.e. Klasmann RHP15) containing mix-fertilised peat (120 mg/L N, 140 mg/L P_2O_5, 240 mg/L K_2O). For young acclimatized plants the substrate is first screened to eliminate the biggest fragments and mixed (1/3) with vermiculite (2/3).
3. Potted plants (pot diameter 12–13 cm) are irrigated (flux-reflux or dripping systems) with a nutrient solution containing a 3-2-6 equilibrium of NPK plus 0.6% (w/v) of MgO, plus 2.5% (v/v) microelements (Sevital) once a week in winter, twice a week in summer, or more precisely up to a level of –5 kPa, when soil water potential in the pot has reached –20 kPa.
4. A growth chamber is also needed for in vitro culture where temperature (24°C day/22°C night), photoperiod (16 h day / 8 h night), and intensity given by fluorescent light (70–120 μmol/m^2/s) can be controlled, as well as a laminar flow hood for aseptic manipulations and all current equipments and chemicals useful in an in vitro culture laboratory.
5. The protocols reported below are based on results obtained with several *Pelargonium* x *hortorum* genotypes, mainly PH-PaS ("Panaché Sud"), PH-DS ("Deep Salmon"), PH-A ("Alain"), PH-RB ("Renard Bleu"), and PH-PuS ("Pulsar Salmon"). However, when necessary, we shall explain, in Subheading 19.4, protocol variations adapted to *Pelargonium* x *peltatum* (PP-VP) "Ville de Paris" (R rose or L lilac), *Pelargonium graveolens* (PG-GLP "Grey Lady Plymouth"), *Pelargonium capitatum* (PC-BJ "Bois Joly"), and *Pelargonium* x *domesticum* (PD-AH "Autumn Haze").
6. Basal medium (BM) used for in vitro culture, contains MS macronutrients (20) sometimes diluted (*see* Subheading 19.3), micronutrient of MS (20), MW vitamins (21) (exception will be mentioned in Subheading 19.3), Fe EDTA (37.3 mg/L Na_2EDTA + 27.8 mg/L $FeSO_4{\cdot}7H_2O$) (19), sucrose (*see* Subheading 19.3), and agar depending on the step (*see* Subheading 19.3). The pH was adjusted to 5.8–6.1 with KOH or HCl (1 or 0.1N) before autoclaving (113°C, 20 min, 5 MPa). Growth substances, auxins, cytokinins, and gibberellic acid are sterilized with sterile filters (0.22 μm) and are added after autoclaving in the cooling medium (exception will be mentioned in Subheading 19.3). All media are prepared just before use, from stock solutions (MS macronutrients × 20; micronutrients × 1,000, vitamins × 500; Fe EDTA × 200), which can all be stocked for 2 months at 4°C except vitamins

(–18°C). In contrast, culture medium should not be stocked (2–3 days max at 4°C) except when growth substances are not included (1–2 weeks at 4°C).

7. Stock solutions (1 mg/10 mL) of auxins (IAA, IBA, and NAA) and GA_3 are prepared adding a few drops of ethanol (96°C) to the powder and adjusting the required volume with warm water (40°C). Stock solutions (1 mg/10 mL) of cytokinins (BAP, Zeatin) are prepared adding 1% (v/v) of HCl (1N) to the powder, adjusting the required volume with warm water (40°C) and shaking. TDZ stock solution (1 mg/10 mL) is prepared adding a few drops of DMSO to the powder until dissolution and adjusting the required volume with hot water. All solutions can be stocked at –18°C for several weeks.
8. If culture medium contains 2 g/L active charcoal, at the end of the autoclaving period, tubes or glass jars must be shaken and then quickly cooled in cold water to avoid charcoal sedimentation.

2.2. Specific Materials

2.2.1. In Vitro Mother Plant Establishment

1. A tea ball is useful for seed disinfestations.
2. A stereomicroscope with a zoom allowing high magnification (at least 50×) and also an optical fibre device that avoids explant warming up is needed for meristem extraction.
3. Glass jars (1,000 mL) with polyethylene cups are needed for propagation process (*see* Note 1).

2.2.2. Explants Culture

A puncher is used to design discs in the leaves used as explants (0.6–0.8 mm). First steps of culture need Petri dishes (9–10 cm diameter), the other ones being performed in test tubes and glass jars (*see* Subheading 19.3) as for meristem culture. A stereomicroscope (supplier name) is used to observe developing explants.

2.2.3. Protoplast Culture

1. Cellulase Onozuka RS (Yakult Biochemical, Tokyo, Japan)
2. Pectinase (Sigma).
3. Hematocytometer
4. Centrifuge with a horizontal rotor, low speed capacities (50–100 *g*)
5. Stainless steel sieves
6. Inverted microscope
7. Filters (0.22 μm)
8. Sucrose
9. Mannitol
10. Petri dishes

3. Methods

3.1. In Vitro Mother Plant Establishment

3.1.1. From Seeds

1. Put the seeds in a tea ball, and dip it in 70% ethanol for 20 s.
2. Then, dip the tea ball with seeds, for 20–25 min, in sodium hypochlorite solution (2.6% active chlorine) containing one drop Tween 20 per 200 mL (*see* Note 3).
3. Finally, under the laminar flow hood, rinse the device 3–4 times with sterile deionised water.
4. Bring out the seeds of the tea ball; drain them on a sterile filter paper.
5. Cut off a small part of the seed integument to help germination (scarification).
6. Sow the seeds, individually in tubes (25 × 3 cm) or by 5–6 in Petri dishes (9 cm diam) containing 20 mL of modified BM (Fig. 19.1).
7. Use BM with diluted macro-nutrients (½× or ¼), 15–20 g/L sucrose, 3 g/L Phytagel (SIGMA), or 7 g/L Bacto agar (DIFCO). If germination is less than 90 %, add 0.5 mg/L IAA to the medium.
8. Place the seeds at 22–24°C under 15–16 h fluorescent light (*see* Note 4).
9. After 6 weeks in tubes, treat the plant as mother plant (2.2.13) and introduce it in the propagation process.

3.1.2. From Meristems

1. In the greenhouse, harvest pelargonium branches with a scalpel or pruning scissors disinfested by 70% ethanol.
2. Cut into small pieces (5 cm) with at least one axillary bud in the middle.

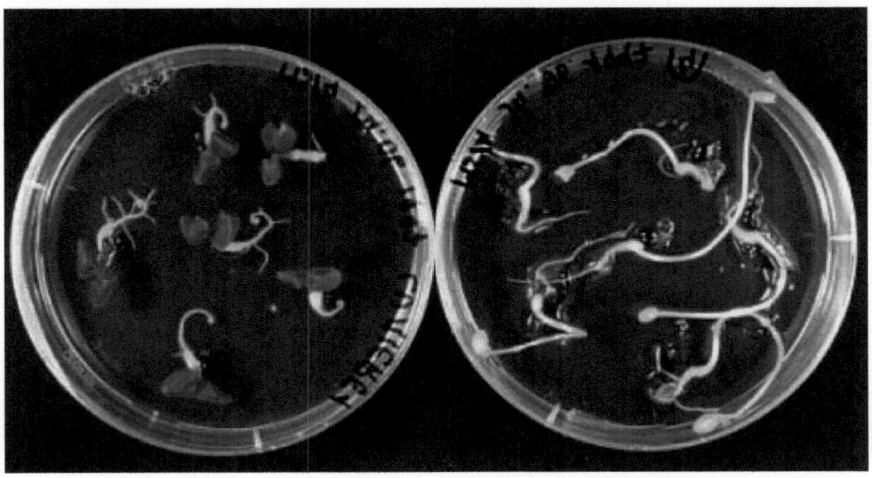

Fig. 19.1. In vitro germination (9-cm diameter Petri dishes), of *Pelargonium* x *hortorum* "Pulsar Salmon" seeds, after 1 week in light (left) and in darkness (right).

3. Cut off the leaves without damaging the axillary buds.
4. Put explants in a plastic bag and keep in an icebox until use, but avoid direct contact with ice.
5. Under laminar flow hood, take one explant, withdraw the stipules manually, and spray it with ethanol (70%).
6. Rub the explant with a cotton plug wetted with ethanol (70%).
7. Put the explant in a sterile Petri dish under the stereomicroscope and begin the dissection.
8. Remove successively the stipules and the leaves with a sterile scalpel (blade no. 11).
9. Use a freshly sterilized scalpel at each level of leaf.
10. When the meristem is visible, remove it with one or two foliar primordium (Fig. 19.2a). Do not forget to use a freshly sterilized scalpel.
11. Put the meristem on the culture medium in a Petri dish (i.e., 10 meristems/9 cm diam dish and 15 mL medium). Seal the Petri dish with Parafilm.
12. Use BM with 30 g/L sucrose, 0.5 mg/L IAA, 0.1 mg/L BAP, and 2.5 g/L Gelrite. Hormones are added before autoclaving (*see* Note 5 and 6).
13. Place the meristems in the growth chamber (24/22°C) in darkness for 1 week.
14. After 1 week, put the Petri dishes containing the meristems under light 70 μmol/m²/s.
15. Three weeks later, transfer the growing meristems into small tubes (14 × 1.5 cm) containing 10 mL of the previous medium where 0.5 mg/L IPA was added (see Note 6).

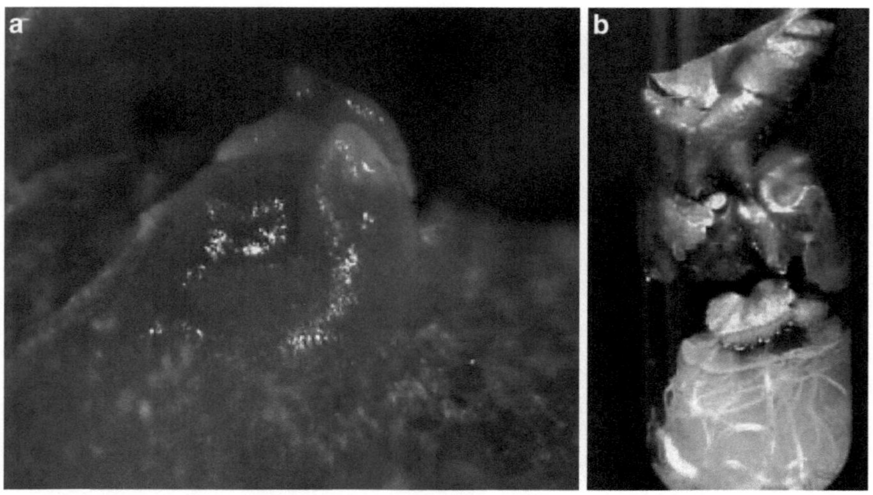

Fig. 19.2. Meristem culture. (**a**) meristem of *Pelargonium* x *hortorum* "Cahors" with one leaf primordium. (bar: 400–500 μm); (**b**) Rooted plant of *Pelargonium* x *hortorum* "Springtime" after meristem culture.

16. Repeat the step every 4 weeks until the shoot is around 2 cm long (2–3 times).
17. Isolate 2 cm-long shoots and subculture them for rooting on the previous medium without BAP, and with 2.5 g/L Phytagel instead of Gelrite.
18. Introduce rooting plants in the propagation cycle to obtain mother stock plants (Fig. 19.2b).

3.2. In Vitro Mother Plant Propagation

The method reported below cannot be considered efficient for industrial micropropagation. It is only a means to obtain a small amount of safe mother plants usable for studies like plant regeneration from leaf explants or mesophyll protoplasts. Indeed, the process, based on rooting plants, has a low multiplication coefficient (2-3 every 5 weeks), but it allows the maintenance of true-to-type plants. It is effective for all the Pelargonium species tested.

1. Under laminar flow hood, take rooted plantlet arising from seed or meristem.
2. Cut it into two parts: the apical one (1.5–2 cm with one unfolded leaf) and the basal one without roots and leaves (1 cm).
3. Introduce these explants in glass jars (4–5 explants per jar).
4. Use BM with MS (½×), 20 g/L sucrose, 2 g/L active charcoal, 3 g/L Phytagel, and 1 mg/L IAA (100 mL/jar) added before autoclaving (*see* Note 7).
5. Place jars in the growth chamber (24/22°C, 65–120 µmol/ m^2/s).
6. After 4–6 weeks, rooted plants can be fragmented again. Take young shoots from the initial basal part for rooting (steps 1–3). Cut again into two parts the initial apical part.
7. Repeat the treatments every 4–6 weeks until the number of mother plants is sufficient.
8. Maintain the required number of mother plants by apical part subculture every 6 weeks (Fig. 19.3a).
9. Use leaves if necessary as soon as in the 4th week.

Fig. 19.3. Micropropagation of *Pelargonium spp.* (**a**) *Pelargonium* x *hortorum* "Panaché Sud" 1 week after apex subculture on the rooting medium. (**b**) *Pelargonium* x *domesticum* "Autumn Haze" 1 week after apex subculture on the rooting medium. (**c**) *Pelargonium graveolens* "Grey Lady Plymouth" 6 weeks after apex subculture.

3.3. Explant Culture

The protocol described below is based on studies performed on PH-PaS; therefore, it is likely that some adjustments should be made with other PH or *Pelargonium* spp. It allows the regeneration of 7–9 shoots on each explant (*see* Note 8).

1. Use 4- to 5-week-old micropropagated plants.
2. Excise leaf discs (6 mm diameter) from the two youngest leaves.
3. Plate six to ten leaf discs, abaxial side down, per Petri dish (9–10 mm diameter) containing 20 mL of culture medium.
4. Use BM with 20 g/L sucrose, 0.5% PVP-10 as an antioxidant, 3.5 mM MES as pH buffer, and 2 g/L Gelrite. Add also 0.2 mg/L NAA, 0.5 or 1 mg/L BAP and 0.5 or 1 mg/L zeatin.
5. Seal Petri dishes with Parafilm; then, place cultures in the darkness or under low light intensity 12 μmol/s/m^2 at 24/22°C.
6. After 4 weeks of culture, transfer explants with regenerated shoots or buds (Fig. 19.4) into test tubes containing 10 mL of the same medium with diluted MS (½×), 0.2 mg/L NAA, 0.5 mg/L BAP, 0.5 mg/L zeatin, and 3 g/L Phytagel. At this step, growth regulators are added before autoclaving.
7. Maintain the test tubes in light (70 μmol/m^2/s) for shoot elongation.
8. Transfer elongated shoots (>1.5 cm) on the rooting medium used for micropropagation.
9. After 3–4 weeks, when plants are rooted, introduce them in the micropropagation cycle or acclimatise them in greenhouse if needed.
10. Acclimatise plants in small plastic pots (6 cm diameter).
11. Sterilise pot and small plastic greenhouse with sodium hypochlorite.

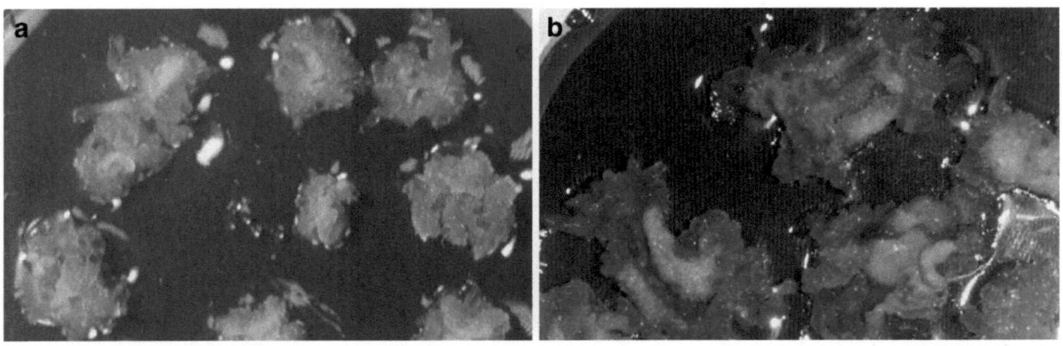

Fig. 19.4. Bud regeneration from leaf discs of in vitro grown plants of *Pelargonium* x *hortorum* "Panaché Sud." Leaf discs in Petri dish (9.5-cm diameter) on MS medium under low light 12μmol/s/m^2 by 12μmol/m^2/s; (**a**) 0.2 mg/L NAA, 0.5 mg/L BAP, and 0.5 mg/L Zeatin; (**b**) 0.2 mg/L NAA, mg/L BAP, and 1 mg/L Zeatin.

12. Fill pot with special substrate (*see* Subheading 19.2) and water with Cryptonol (fungicide).
13. Rinse with distilled water 24 h later, and plant in vitro-rooted plants in the pots.
14. Place the closed device in the greenhouse and water regularly.
15. Open partially after 2 weeks and totally after 4 weeks.

3.4. Protoplast Culture

The protocol described below was established with *P.* x *hortorum*: first on PH-A (13, 14) (34% colony formation, 12% regeneration) and was efficient in other cultivars including PH-DS (21%, 5%) (Fig. 19.5), PH-PuS (23%, 3%), and PH-PaS (13%, 7%). Then, it was improved using PH-PaS (16), leading to 30–40 % colony formation, 83% regeneration, and up to 7 shoots /callus. The optimised protocol is as follows (*see* Note 9):

1. Take the two youngest leaves from 4- to 5-week-old micropropagated plants under the laminar flow hood.
2. Put a leaf on a sterile paper, and cut it into thin strips (>1 mm large) with a scalpel, but take care to keep the upper epidermis quite intact to make easier the dipping of leaf in the Petri dish (9–10 cm diameter) containing 20 mL of the filter-sterilised enzyme solution.
3. Place the lower surface of the leaf on the enzyme solution, and repeat steps 2 and 3 until the entire enzyme surface is covered (Fig. 19.6a).
4. Put the Petri dishes for 6 h in light (70 µmol/m^2/s), in the growth chamber (24°C).
5. Use an enzyme solution containing 0.4% (w/v) cellulase Onozuka RS, 0.2% (w/v) pectinase (sigma), BM (macronutrients MS ½×), 0.5% PVP-10 as antioxidant, 3.5 mM MES as pH buffer and 0.5 M (91 g/L) mannitol as osmoticum.
6. After enzyme incubation, gently shake the mixture (60 rpm) for 15 min to isolate the protoplasts from leaf structures (Fig. 19.6b).

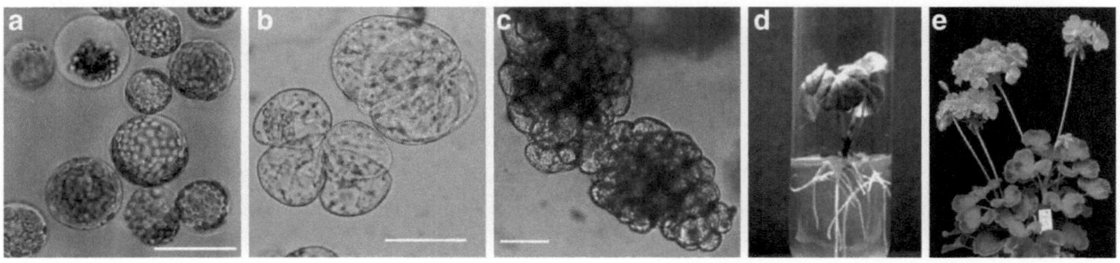

Fig. 19.5. Plant regeneration from mesophyll protoplasts of *P.* x *hortorum* "Deep Salmon" (**a**) freshly isolated protoplasts, (**b**) protoplast divisions after 7 days of culture, (**c**) protoplast-derived visible colonies after 21 days, (**d**) protoplast-derived plant after in vitro rooting, and (**e**) protoplast-derived plant flowering in the greenhouse. (bar = 50 µm).

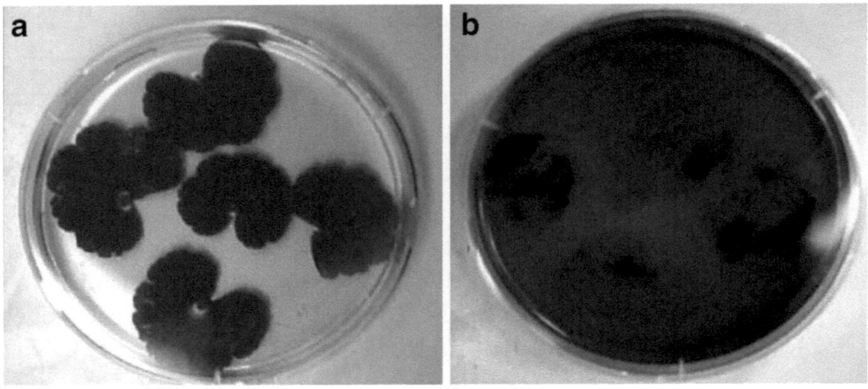

Fig. 19.6. Leaves of *Pelargonium* x *hortorum* "Alain" in the enzyme mixture. (**a**) at the beginning of the incubation process and (**b**) at the end, after shaking (15 min, 60 rpm).

7. Estimate the viability of protoplasts, considering as viable those where plastids are well distributed within a perfect spherical shape. Viability should be >80% to continue.
8. Dilute the mixture with an equal volume of CPW solution (22) containing 2.7 mg/L KH_2PO_4, 10.1 mg/L KNO_3, 24.6 mg/L $MgSO_4$, 16 μg/L KI, 2.5 μg/L $CuSO_4$, 14.8 mg/L $CaCl_2$ supplemented with 0.5 M mannitol.
9. Filter the mixture, directly over a sterile centrifugation tube (30 mL), through a stainless steel sieve (63 μm) to stop the largest leaf fragments.
10. Centrifuge the mixture (60 *g*, 10 min) to eliminate enzyme solution.
11. Suspend the pellet (around 2 mL) containing protoplasts and micro fragments, into 10 mL CPW solution supplemented with 0.5 M (175 g/L) sucrose CPW-S.
12. Centrifuge again, take the two upper mL containing only protoplasts, and suspend them again into 10 mL CPW-S.
13. Repeat the process once more.
14. Before the last centrifugation, depose 2 mL protoplast culture medium over the CPW-S containing the protoplasts.
15. Centrifuge and collect the protoplasts in the 2 mL culture medium.
16. Count the protoplast yield with a hematocytometer, under a microscope, and adjust the density in order to have a final concentration of 3×10^4 protoplasts per mL in the Petri dishes.
17. For culture, use BM (¼ MS macronutrients), 5 g/L sucrose, 0.5 % PVP-10, 3.5 mM MES, 0.5 M mannitol, 1.0 mg/L NAA, and 0.5 mg/L BAP.

18. Plate 3 mL of protoplast suspension per Petri dish (5 cm diameter) and maintain in darkness for 3 days prior transfer to the same growth conditions as the mother plants.
19. Every 10 days, add 1 mL of fresh modified culture medium (1/2 MS macronutrients, 10 g/L sucrose, and 0.42 M to 80 g/L mannitol).
20. After one month, transfer colonies into larger Petri dishes (9-cm diameter) containing 6 mL of modified culture medium (20 g/L sucrose and 0.37 M to 70 g/L mannitol, 1.0 mg/L NAA, 1.0 mg/L BAP, and 0.5 mg/L Zeatin).
21. After 1.5–2 months (Fig. 19.7a), isolate and put 35–50 microcalli (0.5–1 mm diameter) per Petri dish (95 mm diameter) in 20 mL of the same culture medium without mannitol, but solidified by 2 g/L Gelrite.
22. After 3–4 more weeks, calli reach a size of 5–7 mm diameter. Put 8–10 calli per Petri dish on BM free of mannitol, with MS macronutrients at full-strength, sucrose 30 g/L, MES, PVP, 2 g/L Gelrite, 0.2 mg/L IAA, 1.0 mg/L BAP, and 0.1 mg/L TDZ.
23. Maintain the culture under low light (12 μmol/s/m^2) in the growth chamber during 4 weeks (Fig. 19.7b).
24. For shoot elongation, subculture calli, with regenerated buds, for 3 more weeks, in test tubes containing 15 mL modified regeneration medium: MS macronutrients at half-strength, 20 g/L sucrose, 3 g/L Phytagel, 0.2 mg/L IAA, 1.0 mg/L BAP, and 0.1 mg/L TDZ. At this step, growth regulators are added before autoclaving (*see* Note 10).
25. Root individual shoots into similar medium, supplemented with 1.0 mg/L IAA added before autoclaving.
26. When plants are rooted, introduce them in the micropropagation cycle or acclimatise them in greenhouse if needed.

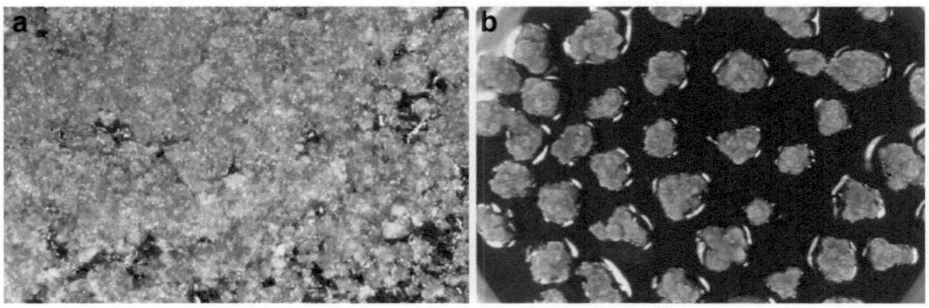

Fig. 19.7. Colony and callus formation from leaf protoplast of *Pelargonium* x *hortorum* "Panaché Sud." (**a**) protoplast-derived colonies after 1.5 month (bar: 5 mm); (**b**) callus formation from protoplast derived colonies after 4 weeks on medium free of mannitol (bar: 1 cm).

4. Notes

Among the methods reported, seed germination, meristem culture, and micropropagation maintain plant conformity. However, since cultivars of *Pelargonium* are often chimaera but also because callus is sometimes present, variations could appear from meristem, explant, and protoplast culture. These variations involve loss of variegation, colour modifications with reversion to initial type, and polyploidy (mainly $2n$ to $4n$). Therefore, plant conformity must be checked (phenotypic control, flow cytometry).

Although the protocols described in this paper work well in our conditions (greenhouses, growth chambers) and with our model genotypes, it is likely that they should be adapted due to genotypic or environmental effects on plants and explants. This adaptation should mainly deal not only with hormone ratio, concentration, and nature but also with in vitro environmental conditions. Some other possibilities will be mentioned below.

1. Although all the PH, PC, PG, and PD tested developed very well in test tubes (20 × 3 cm) closed with cellulose plugs, we found that glass jars with polyethylene cup were more suitable when leaves are needed for experiments (i.e. protoplast culture) (15).

2. If mother plants grown in the greenhouse are free of virus and other pathogens, it is possible to begin the propagation cycle directly after sterilization of the explants. Therefore, in the greenhouse, take apices (5–7 cm) from growing branches; plug the basal part with paraffin in order to limit hypochlorite penetration; dip quickly the explants in 70% ethanol, and replace ethanol with sodium hypochlorite (2.6% active chlorine) for 10 min; then, under the sterile laminar flow hood, rinse 4 times with sterile deionised water; cut off the basal part and introduce the explants for rooting in large test tubes (20 × 3 cm) containing 15 mL propagation medium. Finally, rooted plants could enter the micropropagation cycle. Note that the length of the hypochlorite step period could be increased depending on the cleanliness of the greenhouse and plants.

3. If seeds were pre-treated with a fungicide by the producer, disinfection period could be decreased to around 10 min. In contrast, if seeds are dirty, it is possible to double the active chlorine concentration.

4. As shown in Fig. 19.1, seeds germinate easily in darkness. If germination is poor in light, try to put the Petri dishes in darkness for some days. If the genotype studied is recalcitrant to regeneration from leaves, juvenile hypocotyls can be used as explants. In this case, the seeds must be kept in darkness to gain more explants.

5. If the plant contains naturally many phenols or if stressing greenhouse conditions enhanced it, it is possible to perform the first step on the same medium without BAP but supplemented with activated charcoal.
6. The medium containing 0.5 mg/L IAA and 0.1 mg/L BAP, resulting from a large screening and well adapted to *P.* x *hortorum* meristems, is also usable for *P.* x *domesticum*. Another one was adapted specifically to *P.* x *peltatum* (2), with 0.02 mg/L BAP and 0.1 mg/L GA3 40 mg/L Adenine instead of 0.5 mg/L IPA.
7. IAA in the medium helps rooting, and we always use it (Fig. 19.3b, c). However, some genotypes could need no auxin or a lower concentration. This should be tested. If IAA is added after autoclaving, the concentration could be lowered (i.e. from 1 to 0.5 mg/L), as it is a quite unstable hormone.
8. The protocol described plant regeneration from leaves of in vitro grown plant PH-PaS. We tested it also on PD-AH, PC-BJ, and PG-GLP. For both fragrant plants, the regeneration is direct from leaves and does not need an elongation phase (14). We never succeeded using leaves of greenhouse-grown plants of *P.* x *hortorum* may be due to their phenol content. In contrast, it was possible to use efficiently greenhouse leaves of *P.* x *peltatum* "Ville de Paris," where regeneration was up to 50 buds/explants after 6 weeks (Fig. 19.8). This work

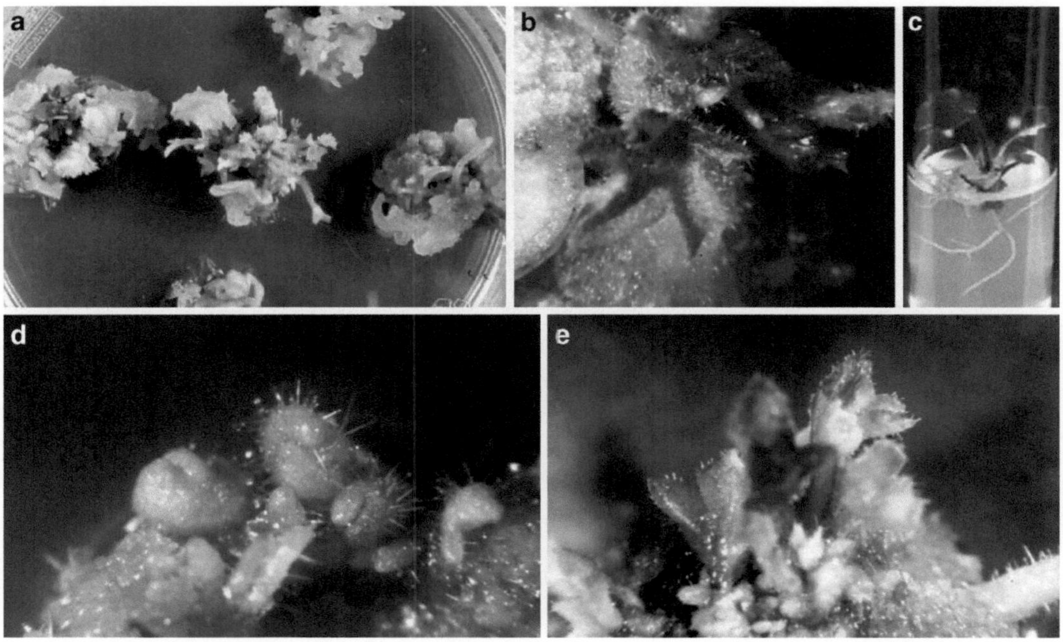

Fig. 19.8. Shoot regeneration from leaf of *Pelargonium* x *peltatum* grown in greenhouse. (**a**, **b**) "Ville de Paris" Lilac after 6 weeks on 0.2 mg/L IBA and 0.1 mg/L TDZ; (**c**) after rooting on 0.5 mg/L IAA; (**d**, **e**) "Ville de Paris" Rose after 2 weeks (**d**) and 5 weeks (**e**) on 0.2 mg/L IBA and 5 mg/L Zeatin.

exemplifies the genetic effect since "mutant Lilac" gives its optimum with 0.1 mg/L TDZ and 0.2 mg/L IBA, while "wild Rose" has its optimum with 5 mg/L Zeatin and 0.2 mg/L IBA. The protocol is as follows: take explants as for meristem culture; disinfect them according to note 2; then, follow the protocol for leaf explants (**steps 3–15**). Note that leaf discs are 0.8 mm diameter.

9. It is likely that the protocol could work quite easily with other cultivars of *Pelargonium* x *hortorum*. However, adaptation seems unavoidable. Main points to test are the osmoticum (nature and concentration), the enzymes (concentration), the length and temperature of the maceration period, and finally the ratio, concentration, and nature of growth regulators.
10. Addition of TDZ instead of Zeatin to the medium was important to improve regeneration (16). However, it was observed that many regenerated plants were tetraploid, while the mother plants were diploid. This fact must be kept in mind and cytokinin nature and concentration be revised if necessary (i.e. use the same hormones as for regeneration from leaf explants).

Acknowledgments

We would like to thank all the technical staff who worked on *Pelargonium* all along these years in the laboratory or in the greenhouse: Liliane Brisset, Denis Cesbron, Stéphanie Douai, Bernard Joire, Claire Marche, Anne Préveaux, Linda Voisine, and also the students. The French Ministry of Agriculture, INRA, BRG, Local Angers Council and Regional Council of Pays de la Loire supported this work.

References

1. Hassanein A, Dorion N (2006) Determining morphological and physiological parameters for selection of drought-tolerant geraniums (*Pelargonium* x *hortorum* L. H. Bailey). J Hort Sci Biotech 81:707–713
2. Gallard A, Panis B, Dorion N, Swennen R, Grapin A (2008) Cryopreservation of Pelargonium apices by droplet-vitrification. Cryolett 29:243–251
3. Kameya T (1975) Culture of protoplasts from chimeral plant tissue of nature. Jap J Genet 50:417–420
4. Bigot C (1983) Vegetative propagation in vitro from seedlings of F1 hybrids of *Pelargonium* x *hortorum* Bailey. Acta Hortic 131:61–70
5. Yarrow SA, Cocking EC, Power JB (1987) Plant regeneration from cultured cell-derived protoplasts of *Pelargonium* x *aridum* and *P.* x *peltatum*. Plant Cell Rep 6:102–104
6. Dunbar KB, Stephens CT (1989) Shoot regeneration of hybrid seed geranium (*Pelargonium* x *hortorum*) and regal geranium (*Pelargonium* x *domesticum*) from primary callus cultures. Plant Cell Tiss Org Cult 19:13–21

7. Dunbar KB, Stephens CT (1991) Plant regeneration from callus-derived protoplasts of *Pelargonium* x *domesticum*. Plant Cell Rep 10:417–420
8. Robichon MP, Renou JP, Jalouzot R (1995) Genetic transformation of *Pelargonium* x *hortorum*. Plant Cell Rep 15:63–67
9. Robichon MP, Renous JP, Jalouzot R (1997) Plant regeneration of ivy-leaved geranium through shoot organogenesis. Plant Cell Tiss Org Cult 49:209–212
10. Agarwal PK, Ranu RS (2000) Regeneration of plantlets from leaf and petiole explants of *Pelargonium* x *hortorum*. In vitro Cell Dev Biol Plant 36:392–397
11. Mithila J, Murch SJ, KrishnaRaj S, Saxena PK (2001) Recent advances in *Pelargonium* in vitro regeneration systems. Plant Cell Tiss Org Cult 67:1–9
12. Grapin A, Hupel C, Marche C, Dorion N (2001) First results of *Pelargonium* shoot apex cryopreservation. Acta Hortic 560:157–159
13. Nassour M, Dorion N (2002) Plant regeneration from protoplasts of micropropagated *Pelargonium* x *hortorum* 'Alain': effect of some environmental and medium factors on protoplast system efficiency. Plant Sci 163:169–176
14. Hassanein A, Dorion N (2005) Efficient plant regeneration system from leaf discs of zonal (*Pelargonium* x *hortorum*) and two scented (*P. capitatum* and *P. graveolens*) geraniums. Plant Cell Tiss Org Cult 83:231–240
15. Nassour M, Chassériaux G, Dorion N (2003) Optimization of protoplast-to-plant system for *Pelargonium* x *hortorum* 'Alain' and genetic stability of the regenerated plants. Plant Sci 165:121–128
16. Hassanein A, Dorion N (2006) High-efficiency colony formation and whole plant regeneration from mesophyll protoplasts of *Pelargonium* x *hortorum* 'Panaché Sud'. J Hort Sci Biotech 81:714–720
17. Hassanein A, Chevreau E, Dorion N (2005) Highly efficient transformation of zonal (*Pelargonium* x *hortorum*) and scented (*P. capitatum*) geraniums via *Agrobacterium tumefaciens* using leaf discs. Plant Sci 169:532–541
18. Hassanein A, Loridon K, Dorion N (2007) Comparison of two transformation method efficiency using *Pelargonium* x *hortorum* 'Panaché Sud' protoplasts and leaf discs as explants. Acta Hortic 743:127–132
19. Nassour M, Dorion N (2003) Somatic hybridisation between *Pelargonium* x *hortorum* and *P.* x *domesticum*, towards resistance to *Xanthomonas pelargonii*. Acta Hort 612:75–81
20. Murashige T, Skoog F (1962) A revised medium for rapid growth and bioassays with tobacco tissue culture. Physiol Plant 15:473–497
21. Morel G, Wetmore RH (1951) Fern callus tissue culture. Am J Bot 38:141–143
22. Frearson EM, Power JB, Cocking EC (1973) The isolation, culture and regeneration of *Petunia* leaf protoplasts. Dev Biol 33: 130–137

Chapter 20

Marigold Regeneration and Molecular Analysis of Carotenogenic Genes

Alma Angélica Del Villar-Martínez, Pablo Emilio Vanegas-Espinoza, and Octavio Paredes-López

Abstract

Marigold (*Tagetes erecta*) is an Asteraceous plant of industrial, ornamental and medicinal importance. *Tagetes erecta* species, popularly known as marigold, is grown as ornamental plant and is adapted to several agro climates. Inflorescences have been utilized as pigment source for food coloring, mainly of poultry skin and eggs. Lutein is the main pigment in marigold flowers. Some carotenoids are well known as provitamin A compounds. There are many reports on carotenoids and their effect on the prevention of certain ocular diseases, ischemic heart disease, strokes, photoprotection, immune response, aging and cancer. Marigold flowers are certainly a good source of carotenoids; they show very different pigmentation levels. This chapter describes the establishment of techniques for plant regeneration, gene expression, pigment extraction and pigment deposition in specific structures of marigold.

Key words: *Tagetes erecta*, Organogenesis, Asteraceae, Gene expression, Flower development, Plastids

1. Introduction

Carotenoids comprise a large group of secondary metabolites that are natural pigments present in most of higher plants. They are widely distributed in nature and participate in an important set of reactions in plants; moreover, some of them are the main dietary precursors of vitamin A (1). Genetic engineering provides good possibilities to obtain modified plants to improve pigment accumulation in high producing crops (2).

Marigold is an Asteraceous plant, and is being used in traditional Mexican medicine as antiparasitic, antispasmodic and against stomach and liver diseases (3) as antimutagenic and antioxidant (4).

S.M. Jain and S.J. Ochatt (eds.), *Protocols for In Vitro Propagation of Ornamental Plants*, Methods in Molecular Biology, vol. 589
DOI 10.1007/978-1-60327-114-1_20, © Humana Press, a part of Springer Science+Business Media, LLC 2010

Flowers are good source of carotenoids for food coloring, mainly of poultry skin and eggs; oleoresin is obtained from chopped flowers at processing industrial plants to be used for commercial purposes (5, 6). The uses of this plant underline the importance of establishing reliable plant regeneration, gene expression analysis and genetic manipulation systems aimed at increasing its carotenoids levels (7–9), by enhancing their concentration in the flowers and, at the same time, generating new cultivars with higher number of flowers per plant.

Marigold is also a good model to study carotenogenic process (6). *Tagetes erecta* flowers accumulate high levels of carotenoids, especially lutein being the main pigment in marigold flowers. Lutein isomer peaks have been identified analysing their retention time and absorbance maxima from spectra (10, 11).

There are a few reports on marigold in relation to plant tissue culture; the available studies use a wide range of explant sources, and different kind and combinations of plant growth regulators. However, there are no reports about the establishment of a marigold regeneration system or a cell culture for *in vitro* pigment production. The whole plant regeneration methodology has been carried out at our laboratory from marigold leaf explants using Murashige and Skoog (MS) medium (7, 12).

In higher plants, carotenoids are localized in plastids that, during flower development, are transformed from proplastids into chromoplasts (13). Ultrastructure of plastids from marigold flowers undergoes remarkable changes, including internal organization destruction throughout flower development (14).

Subcellular studies on *Tagetes erecta* callus cells have been done in our laboratory, whereby structures such as cell wall, vacuole, nucleus, nuclear membrane, intercellular spaces, proplastids, chloroplasts and chromoplasts were clearly identified. Organelle organization differences between cell types were evident in quantity and plastid appearance (unpublished).

Lutein is synthesized in plastids through non-mevalonate pathway (15, 16). Studies on carotenoid biosynthesis and regulation in different organs are necessary for further genetic modification (17). Participating genes in carotenoid biosynthetic pathway have been cloned from *T. erecta*; northern blot analysis shows differential expression of analysed genes, and *lcy-b* gene appears to be highly related to flower development and pigment accumulation (14, 18).

2. Materials

2.1. Plant Material

Seeds of a commercial line of marigold are used. *In vitro* germinated seeds are utilized as explant source for plant regeneration and greenhouse acclimation. Greenhouse-grown marigold plants are sampled during flower development at four stages. The plant tissues are harvested, frozen in liquid nitrogen and stored at −70°C until processed for RNA isolation and pigment analysis.

2.2. In vitro Plant Culture

Murashige and Skoog-prepared salt medium (Sigma-Aldrich) is used with or without added plant growth regulators. Medium is solidified adding Phytagel (Sigma-Aldrich) as gelling agent.

2.3. Blotting Material

Isolated and purified RNA's are transferred onto nylon Hybond+ membranes (Amersham Biosciences, Brazil); prehybridation and hybridation are carried out with Hybrisol^{R} (Oncor, USA) (50% formamide; 10% dextran sulfate; 1% SDS; and 100 μg/ml denatured salmon sperm DNA fragments). Films with radioactive signals are analysed using the Gel DOC 2000 and Quantity One Software V.4 (BIO-RAD, Italy).

2.4. Chromatographic Supplies

A YMC PACK C_{30} (YMC Inc., USA) reversed-phase column (5 μm; 25 cm × 4.6 mm i.d.) and lutein standard (Sigma-Aldrich, USA) are used for chromatographic analysis; using a HPLC 1,050 (Hewlett-Packard, USA).

2.5. Microscopy Requirements

Osmium tetroxide (OsO_4) and propylene oxide (C_3H_6O) are purchased from Sigma-Aldrich, USA. A MTX ultramicrotome (RMC, USA) is used in order to obtain thin sections, and observations are performed with a JEOL 1010 transmission electron microscope (JEOL Ltd., Japan).

3. Methods

3.1. Bud Induction from Leaf Explants, Plantlet Generation and Plant Adaptation to Greenhouse

1. Seeds are surface sterilized with absolute ethanol for 1 min, 70% ethanol for 5 min, 2% sodium hypochlorite for 15 min and 1% sodium hypochlorite for 15 min. As a surfactant, 0.1% Tween-20 is added to sodium hypochlorite solution.
2. A rinse with sterile deionized water is performed after each treatment. Finally, at least three rinses with sterile deionized water are carried out.
3. Surface sterilized seeds are germinated on hormone-free MS (12) medium containing 30 g/l sucrose and solidified with 3 g/l Phytagel.
4. All cultures are maintained at 25 ± 2°C under fluorescent light (50 μmol/m^{2}s) and 16 h light/8 h dark cycle.
5. Three-week-old plantlets are the explant source. Leaf portions about 0.25 cm^2 are placed on MS media containing 17.1 μM IAA and 13.3 μM BA.

6. After thirteen days, shoots are transferred to hormone-free MS medium for plant elongation and rooting. (*See* Note 1).
7. The well-developed plantlets are then transferred to pots containing sterile soil and protected with plastic covers, modified on the second and fifth day to decrease the internal humidity.
8. The pots are maintained in a growth chamber for 1 week. A solution of half-strength MS salts and sucrose is added on the second and fifth days. After the eighth day, the plants are moved to a greenhouse to reach their complete development. An example is shown in Fig. 20.1.

3.2. Northern Blot Analysis

1. Total RNA (20 µg) samples are loaded under denaturing conditions on a 1.2%, w/v agarose gel containing formaldehyde.
2. The RNA's are transferred onto Hybond+ membranes, with hybridizations performed under high stringency conditions.
3. The membranes are incubated for 2 h at 42°C in 20 ml prehybridization solution Hybrisol^{R}.
4. RNA on membranes is probed with the full-length cDNA for each carotenogenic gene.
5. Denatured and ^{32}P-labeled probe is added to hybridization solution.
6. Each hybridization is carried out in Hybrisol^{R} at 42°C for 16–18 h using 1–2 ng/ml radiolabeled probe.
7. After hybridization, membranes are transferred as quickly as possible to 50–100 ml washing solution for three washing periods decreasing successively the SSPE concentration: SSPE 4×/SDS 0.1%, SSPE 2×/SDS 0.1%, SSPE 1×/SDS 0.1% (all solutions are in w/v relationship), at room temperature during 20 min each one.

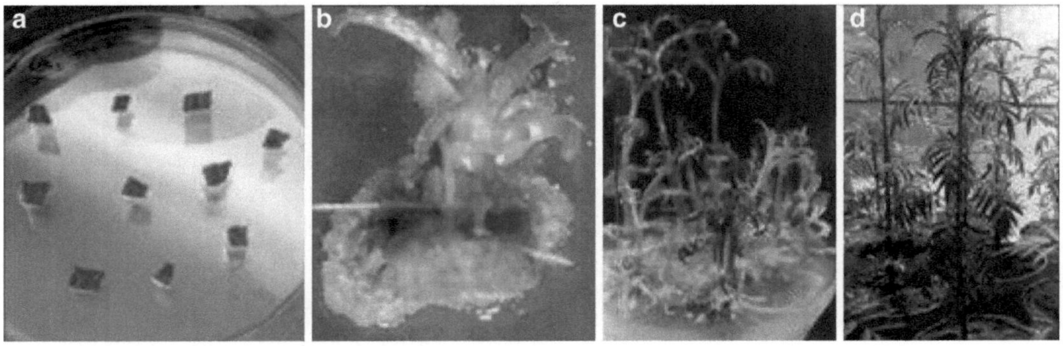

Fig. 20.1. Regeneration process from leaf explant to greenhouse-adapted plants. (**a**) leaf explants in regeneration medium; (**b**) emerging bud; (**c**) plantlet development; and (**d**) greenhouse-adapted plants.

8. The membranes are removed from washing solution and then placed into exposure cassette with an X-ray film for a suitable time, typically 5 h.
9. Films with signals are scanned using the Gel DOC 2000 and Quantity One Software V.4. The transcript levels are determined. An example of the results obtained is shown in Fig. 20.2. (*See* Note 2).

3.3. Analysis of Carotenoid Pigment Content in Marigold Flowers by High Performance Liquid Chromatography

1. Carotenoids are extracted from flowers at four developmental stages.
2. Fresh tissue samples are extracted with HEAT (hexane:absolute ethanol:acetone:toluene, 10:6:7:7 v/v/v/v) and 40% methanolic KOH solution.
3. Separations are carried out with an YMC PACK C_{30} reversed-phase column (5 µm; 25 cm × 4.6 mm i.d.).
4. A solvent gradient is programmed into an HPLC analyser model 1050 as follows: solvent A (methanol) is delivered isocratically from 0 to 6 min; then, a 4 min linear gradient to 5% solvent B (isopropyl alcohol) is applied, followed by 5 min isocratically with 5% solvent B; then, a 10 min linear gradient to 25% solvent B is used followed by a 5 min linear gradient to 50% solvent B, a 10 min linear gradient to 75% solvent B, 25% solvent C (hexane), and finishing with 10 min isocratically with the final mixture.

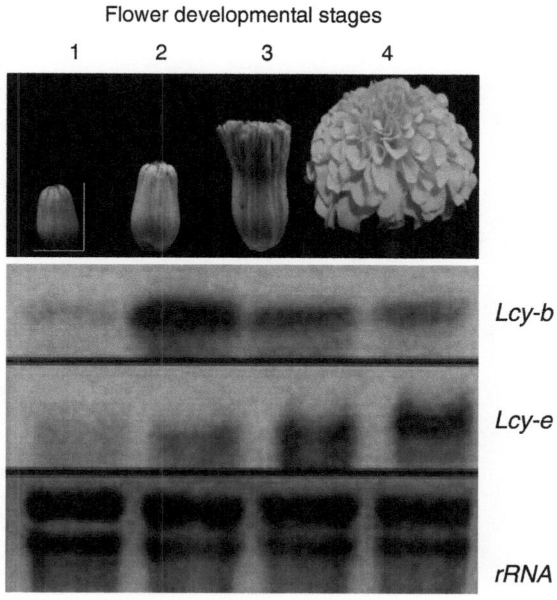

Fig. 20.2. mRNA analysis in developing flowers of marigold. Differential expression is observed at different flower developing stages. (1) closed bud; (2) semi-open bud; (3) open flower; and (4) fully open flower.

5. The column is re-equilibrated between samples for 15 min with solvent A.
6. Pigments are monitored at 450 nm, and their spectra are analysed using a diode array detector (10).
7. Lutein isomer peaks are identified by analysing their retention time and absorbance maxima from spectra as described (10) using a lutein standard. Figure 20.3 shows lutein content from different flowering stages extractions.

3.4. Transmission Electron Microscopy

1. Small pieces of marigold petals are cut and fixed for 12 h in 4% formaldehyde (from aqueous alkaline hydrolysis at 60°C of paraformaldehyde) and 2% glutaraldehyde in 0.1 M cacodylate buffer, pH 7.4 at 4°C.
2. After fixation, samples are washed overnight in 0.25 M sucrose in the same buffer. Post-fixation is carried out in 1% OsO_4 in cacodylate buffer for 12 h at 4°C. (*See* Note 3).
3. Tissue samples are dehydrated through graded (10% steps) ethanol series from 10% to absolute (over OsO_4); propylene oxide is the intermediate solvent for infiltration.
4. The first infiltration step is carried out with complete resin diluted with propylene oxide (1:1, w/v) for 24 h in a closed jar with silica gel with moisture indicator.
5. A second infiltration step is carried out with complete undiluted resin in a rotatory shaker for 2–4 h at room temperature.

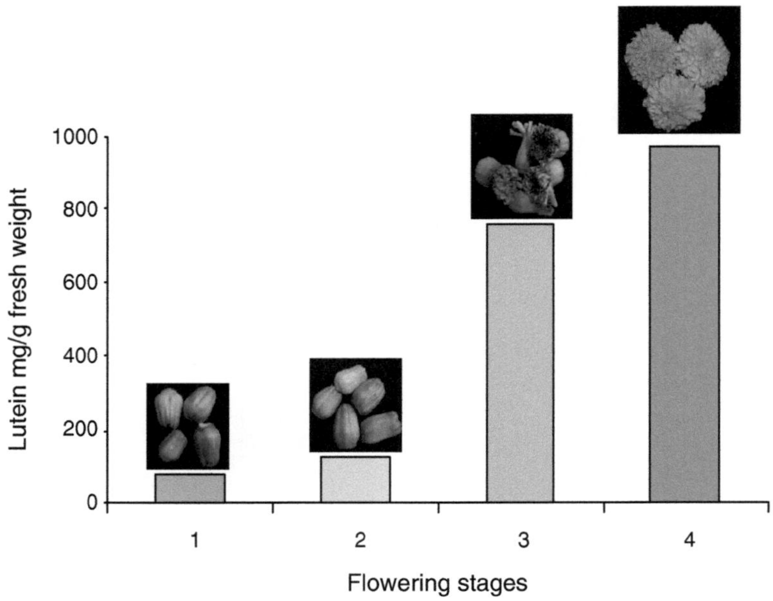

Fig. 20.3. Lutein content in different flowering stages of marigold. (1) closed bud; (2) semi-open bud; (3) open flower; and (4) fully open flower.

Fig. 20.4. Transmission electron micrographs of callus and marigold flowers. (1) green callus; (2) yellow callus; (3) closed bud; and (4) fully opened flower. Key: *cw* cell wall; *g* grana; *im* inner membrane; *li* lipid vesicles; *om* outer membrane; *r* ribosomes; *v* vacuole.

6. After the second infiltration step, embedded samples are transferred to flat embedding molds filled with fresh undiluted resin and are polymerised in a convection oven at 60°C for 36 h.
7. To select samples for TEM observation, 1 μm thick sections are obtained with a glass knife and stained with toluidine blue. From the selected areas, thin sections (70–90 nm) are cut with a diamond knife in an MTX ultramicrotome. (*See* Note 4).
8. Thin sections are attached to a 300-mesh copper grid, contrasted with lead citrate and uranyl acetate, and observed at 80 kV in a JEOL 1010 transmission electron microscope. An example of results from flower and callus samples is shown in Fig. 20.4. (*See* Note 5).

4. Notes

1. Plant regeneration efficiency may differ depending on the genotype.
2. There are some considerations about gene expression method – gene architecture, specificity, sensitivity, sample requirements, and costs, among others – because of the technique complexity.

3. Sample preparation for transmission electron microscopy and probe labeling use dangerous substances as radioactive elements (^{32}P) and poisonous reagents (OsO_4); they must be used under restricted security conditions.
4. A method of plant regeneration, a system of cell culture from marigold leaf, the gene expression in marigold of carotenoid-biosynthesis related genes, an ultrastructural method for analysis of structures related to sites of pigment accumulation, and a method for genetic transformation by particle bombardment (8) are now available.
5. Hence, with these basic techniques, marigold may be manipulated by genetic engineering in order to produce new kind of carotenoids or any other specific metabolite. Currently, tools exist to improve the potential traits of marigold in order to obtain flowers with modifications in level and type of containing pigment, thus enhancing its commercial potentiality as an ornamental plant.

Acknowledgments

We are grateful to Consejo Nacional de Ciencia y Tecnología (CONACYT), Instituto Polittgécnico Nacional (IPN), and to CONCYTEG-Gto., México for financial support.

References

1. Diretto G, Welsch R, Tavazza R, Mourgues F, Pizzichini D, Beyer P, Giulano G (2007) Silencing of beta-carotene hydroxylase increases total carotenoid and beta-carotene levels in potato tubers. BMC Plant Biol 7:11–18
2. Yonekura-Sakakibara K, Saito K (2006) Review: genetically modified plants for the promotion of human health. Biotechnol Lett 28:1983–1991
3. Guzmán-Maldonado SH, Paredes-López O (1998) Functional products of plants indigenous to Latin America: Amaranth, quinoa, common beans, and botanicals. In: Mazza G (ed) Functional foods. Biochemical and processing aspects. CRC Press, Technomic Publishing Lancaster, Pennsylvania, USA
4. Delgado-Vargas F, Paredes-López O (2003) Natural colorants for food and nutraceutical uses. CRC Press. Boca Ratón, Florida, USA
5. Delgado-Vargas F, Paredes-López O, Ávila-González E (1998) Effects of sunlight illumination of marigold flower meals in egg yolk pigmentation. J Agric Food Chem 46: 689–706
6. Delgado-Vargas F, Jiménez AR, Paredes-López O (2000) Natural pigments: carotenoids, anthocyanins and betalains – characteristics, biosynthesis, processing and stability. Crit Rev Food Sci Nutr 40:173–289
7. Vanegas PE, Cruz-Hernández A, Valverde ME, Paredes-López O (2002) Plant regeneration via organogenesis in marigold. Plant Cell Tissue Organ Cult 69:279–283
8. Vanegas PE, Valdez-Morales M, Valverde ME, Cruz-Hernández A, Paredes-López O (2006) Particle bombardment, a method to gene transfer in marigold. Plant Cell Tissue Organ Cult 84:359–363
9. Rout GR, Mohapatra A, Jain Mohan S (2006) Tissue culture of ornamental pot plant: a critical review on present scenario and future prospects. Biotechnol Adv 24:531–560
10. Delgado-Vargas F, Paredes-López O (1996) Correlation of HPLC and AOAC methods to

assess the all-trans-lutein content in marigold flowers. J Sci Food Agric 72:283–290

11. Delgado-Vargas F, Paredes-López O (1997) Effects of enzymatic treatments of marigold flowers on lutein isomeric profiles. J Agric Food Chem 45:1097–1102
12. Murashige T, Skoog F (1962) A revised medium for rapid growth and bioassays with tobacco tissue cultures. Physiol Plant 15:472–497
13. Wurbs D, Ruf S, Bock R (2007) Contained metabolic engineering in tomatoes by expression of carotenoids biosynthesis genes from the plastid genome. Plant J 49:276–288
14. Del Villar-Martínez AA, García-Saucedo PA, Cárabez-Trejo A, Cruz-Hernández A, Paredes-López O (2005) Carotenogenic gene expression and ultrastructural changes during marigold flowering. J Plant Physiol 162:1046–1056
15. Dubey VS, Bhala R, Luthra R (2003) An overview of the non-mevalonate pathway for terpenoid biosynthesis in plants. J Biosci 28:637–646
16. Del Villar-Martínez AA, Ma R, Paredes-López O (1999) Molecular characterization of cDNAs encoding 3-hydroxy-3methylglutaryl coenzyme A reductase in marigold (*Tagetes erecta*). J Plant Physiol 155:205–211
17. Sandmann G, Römer S, Fraser PD (2006) Understanding carotenoid metabolism as a necessity for genetic engineering of crop plants. Metab Eng 8:291–302
18. Moehs CP, Tian L, Osteryoung KW, DellaPenna D (2001) Analysis of carotenoid biosynthetic gene expression during marigold petal development. Plant Mol Biol 45: 281–293

Chapter 21

Somatic Embryogenesis in In Vitro Culture of *Leucojum vernum* L.

Agata Ptak

Abstract

Procedures for somatic embryogenesis (SE) in in vitro culture of spring snowflake have been developed from different types of explants like scales and leaves isolated from bulbs, ovaries and fruits. Various plant growth regulators were tested including a cytokinin – benzyladenine (BA) and various concentrations of the exogenous auxins 3,6-dichloro-2-methoxybenzoic acid (Dicamba), 2,4-dichlorophenoxyacetic acid (2,4-D) or 4-amino-3,5,6-trichloropicolinic acid (Picloram). Fruit explants, cultured on medium containing Picloram and BA, ensured the highest percentage of callusing and such calli were most efficient in inducing somatic embryos. The addition of abscisic acid (ABA) in combination with polyethylene glycol (PEG) stimulated somatic embryo maturation. Torpedo-stage embryos developed into plants in the presence of BA and 1-naphthaleneacetic acid (NAA). The formation and growth of adventitious bulbs required that the plantlets be chilled at 5°C in the dark for 6 weeks. After chilling, the bulbs grew well in darkness 25°C. High sucrose concentration in the medium was necessary for obtaining large bulbs.

Key words: *Leucojum vernum* L., Embryogenic callus, Somatic embryos, Bulbs

1. Introduction

Leucojum vernum L. (spring snowflake), a monocotyledonous species belonging to the *Amaryllidaceae*, is a garden flower cultivated in Poland and throughout the world (1). The spring snowflake is also gladly planted in parks. It is one of the earliest spring flowering plants.

Leucojum vernum grows widely in the Karpaty region and Ukraine. In Poland two subspecies occur: *vernum* and *carpaticum*. The spring snowflake is under strict species protection (2). The plant has often been transplanted from natural stands into

S.M. Jain and S.J. Ochatt (eds.), *Protocols for In Vitro Propagation of Ornamental Plants*, Methods in Molecular Biology, vol. 589
DOI 10.1007/978-1-60327-114-1_21, © Humana Press, a part of Springer Science+Business Media, LLC 2010

gardens. By growing this plant in gardens the devastation of natural stands might be prevented.

Leucojum vernum is also a medicinal plant, rich in alkaloids (lycorine, homolycorine, haemanthamine) with antiretroviral activities, while galanthamine and lycorine anticholinesterase activity has been found as well (3, 4).

Natural vegetative propagation of *Leucojum vernum* is slow. *Leucojum* can be propagated by seeds and vegetatively using offsets, bulb chipping, or twin scaling (1). In vitro cultures could be an alternative way of propagating this plant. The most convenient way of regenerating the plant is through somatic embryogenesis, already shown to strongly improve the multiplication rate of in vitro bulbous plants. Considering the *Amaryllidaceae* family, somatic embryogenesis was described, for example, in *Narcissus sp.* (5–9), *Galanthus sp.* (10), *Nerine sp.* (11), *Hippeastrum* x *hybridum* (12) and *Leucojum aestivum* (13). No plantlets were regenerated through somatic embryogenesis in *Leucojum vernum*. In the available literature no protocol regarding micropropagation of that plant can be found either.

The process of producing spring snowflake plants through somatic embryogenesis consists of several steps, which include the initiation of embryogenic callus, the development and maturation of embryos, their conversion into plants and induction of bulbs. Many factors influence somatic embryogenesis in *Leucojum vernum* in vitro culture. This paper deals with the effects of explant type, exogenous growth regulators, sucrose concentration and temperature treatment on different steps of somatic embryogenesis of spring snowflake.

2. Materials

1. Clean, healthy bulbs (obtained from the commerce) (*see* Note 1).
2. Ovaries (obtained from The Botanical Garden of the Jagiellonian University in Krakow) (*see* Note 2, Fig. 21.1a, b).
3. Young fruits (obtained from The Botanical Garden of the Jagiellonian University in Krakow) (*see* Note 3).
4. 70% (v/v) EtOH, sterile water.
5. Any commercially available bleach solution containing approx. 5% of available chlorine. Domestos bleach solution is used in this study.
6. 90×25 mm Petri dishes and Erlenmeyer flasks (250 ml) containing 40 and 50 ml, respectively, of Murashige and Skoog (MS) (14) medium supplemented with 0.8% agar (purified agar from Difco) and different growth regulators (from Sigma-Aldrich), the sucrose concentration being in accordance with the experiment (*see* Table 21.1, Note 4).

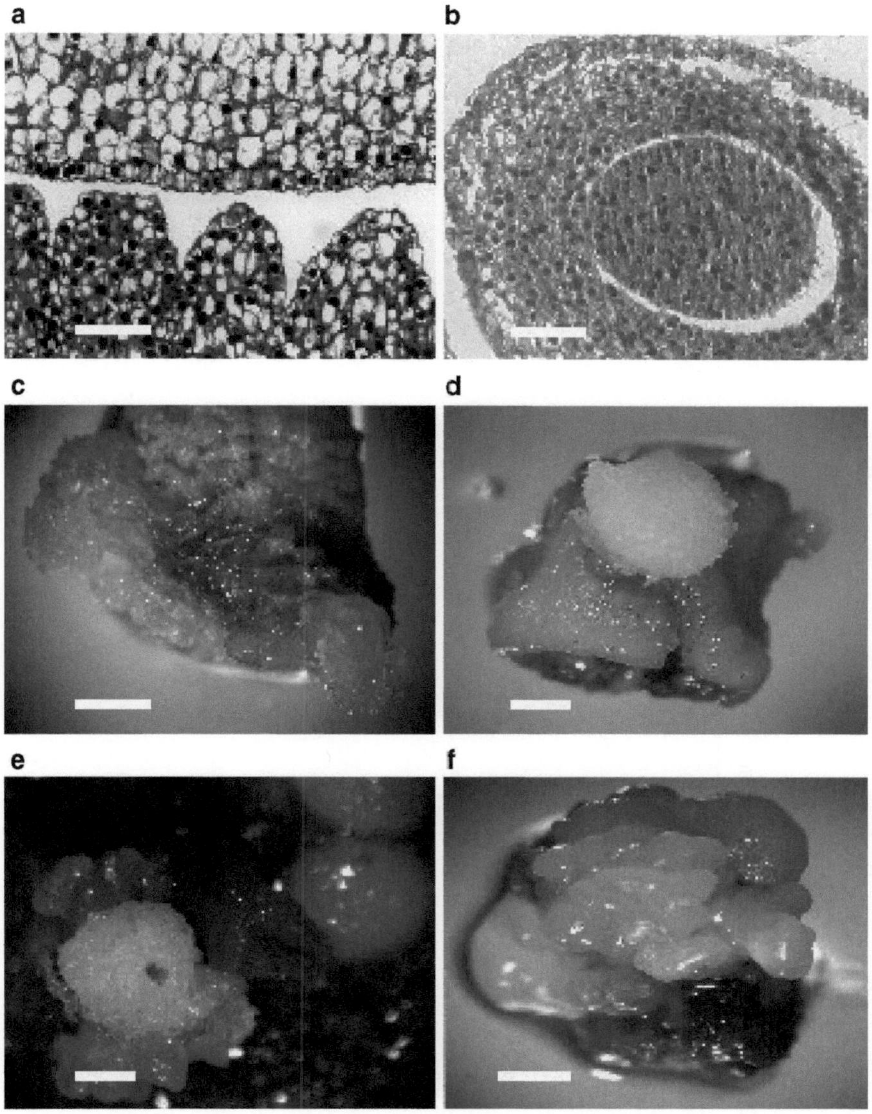

Fig. 21.1. Induction of embryogenic callus in spring snowflake: (**a**) longitudinal section of ovary initial explants (*bar*= 200 μm); (**b**) cross section of ovary initial explants (*bar*= 300 μm), (**c**) scale culture (*bar*= 1 mm), (**d**) leaf culture (*bar*= 1 mm), (**e**) ovary culture (*bar*= 1 mm), (**f**) fruit culture (*bar*= 1 mm).

7. Partec CA II instrument (Munster, Germany).
8. Rotary shaker.

3. Methods

3. Callus Initiation from Scale Explants

1. Remove roots and dry tunic and wash the bulbs with water.
2. Remove the outer scale and 1/3 of the upper part of bulbs and halve them longitudinally.

Table 21.1
Constituents of *Leucojum vernum* in vitro culture media (sucrose and PEG in g/l, all other compounds in μM)

	Component									
	Sucrose	2,4-D	Dicamba	Picloram	NAA	BA	ZEA	ABA	PEG	pH
Callus initiation	30	25	–	–	–	0.5	–	–	–	5.5
	30	50	–	–	–	0.5	–	–	–	5.5
	30	–	25	–	–	0.5	–	–	–	5.5
	30	–	50	–	–	0.5	–	–	–	5.5
	30	–	–	25	–	0.5	–	–	–	5.5
	30	–	–	50	–	0.5	–	–	–	5.5
Callus multiplication/ embryo induction	30	–	25	–	–	0.5	–	–	–	5.5
	30	–	–	25	–	0.5	–	–	–	5.5
Embryo maturation	30	–	–	–	1.0	–	5.0	–	–	5.8
	30	–	–	–	–	–	–	15	–	5.8
	30	–	–	–	–	–	–	–	45	5.8
	30	–	–	–	–	–	–	15	45	5.8
Embryo germination	30	–	–	–	0.5	5.0	–	–	–	5.8
Bulb formation	30	–	–	–	0.5	5.0	–	–	–	5.8
	60	–	–	–	0.5	5.0	–	–	–	5.8
	90	–	–	–	0.5	5.0	–	–	–	5.8

3. Rinse the bulbs for 30 s with 70% (v/v) EtOH and sterilize for 15 min in 15% (v/v) Domestos on a rotary shaker. Then rinse three times with sterile water.
4. Isolate scales. Cut explants of about 4×4 mm (Fig. 21.2a) and place them on initiation medium (5 explants/dish) (*see* Table 21.1) at 25±2°C in darkness.
5. Subculture every 4 weeks onto fresh initial medium. Eliminate brown explants and explants without callus (*see* Note 5, Table 21.2, Fig. 21.1c).

3.2. Callus Initiation from Leaf Explants

1. Remove roots, dry tunic and scales. Separate leaves from the bulbs and wash them with water.
2. Rinse the leaves for 30 s with 70% (v/v) EtOH. Sterilize for 15 min in 15% (v/v) Domestos. Rinse three times with sterile water.
3. Cut explants of about 2×2 mm (Fig. 21.2b). Culture explants on initiation medium (5/dish) (Table 21.1) at 25±2°C in darkness.

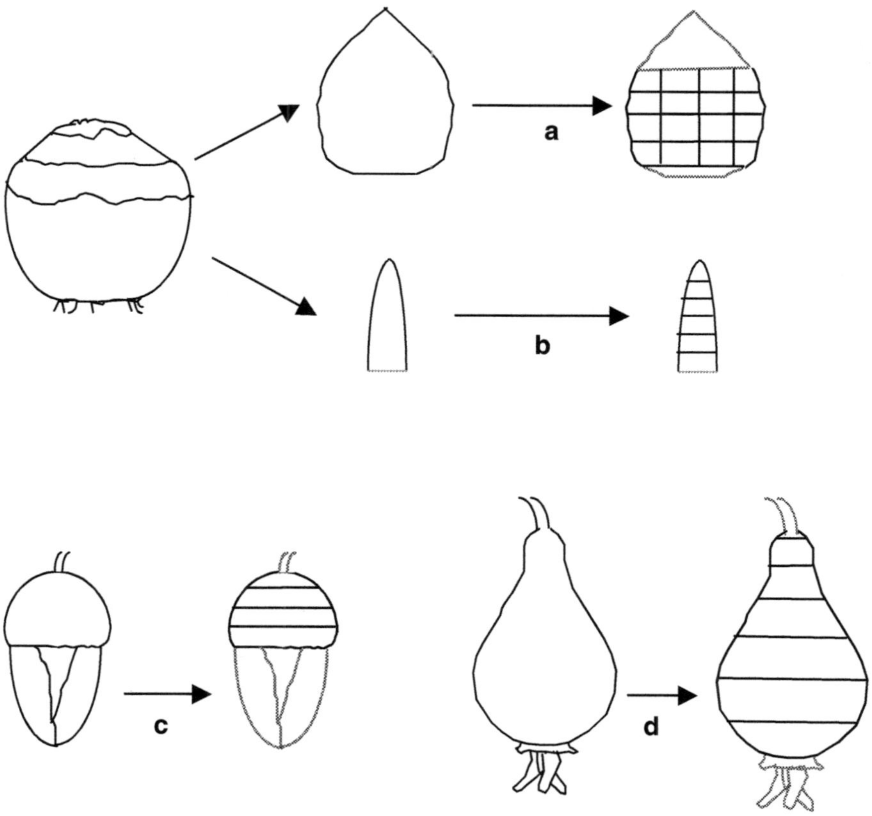

Fig. 21.2. Preparation of *Leucojum vernum* explants: (**a**) scale, (**b**) leaf, (**c**) ovary, (**d**) fruit.

Table 21.2
Effects of growth regulators and explant type on the percentage of explants forming embryogenic callus

Growth regulators [μM]	Explant type scale	leaf	ovary	fruit
2,4-D 25+BA 0.5	0.0[a] a	0.0 a	10.0 a	0.0 a
2,4-D 50+BA 0.5	0.0 a	0.0 a	17.5 ab	0.0 a
Dicamba 25+BA 0.5	10.5 ab	12.0 ab	15.0 ab	20.0 ab
Dicamba 50+BA 0.5	30.0 ab	10.0 ab	23.0 ab	23.8 ab
Picloram 25+BA 0.5	31.0 ab	8.5 ab	32.5 ab	40.0 b
Picloram 50+BA 0.5	31.1 ab	10.0 ab	41.5 b	81.3 c

[a]Means followed by different letters within columns are significantly different at $\alpha \leq 0.05$, Duncan's test

4. Subculture every 4 weeks onto fresh initial medium. Eliminate brown explants and explants without callus (*see* Note 5, Table 21.2, Fig. 21.1d).

3.3. Callus Initiation from Ovary Explants

1. Wash the ovary with water. Rinse for 30 s with 70% (v/v) EtOH, sterilize for 15 min in 15% (v/v) Domestos, and rinse three times with sterile water.
2. Cut the sterilized ovary into thin slices (about 1–2 mm) (Fig. 21.2c).
3. Culture explants on initiation medium (5/dish) (Table 21.1), at 25 ± 2°C in darkness.
4. Subculture every 4 weeks onto fresh initial medium. Eliminate brown explants and explants without callus (*see* Note 5, Table 21.2, Fig. 21.1e).

3.4. Callus Initiation from Fruit Explants

1. Wash the fruits with water, rinse for 30 s with 70% (v/v) EtOH. Then, sterilize for 15 min in 15% (v/v) Domestos and rinse three times with sterile water.
2. Cut the sterilized fruits into thin slices (about 1–2 mm) (Fig. 21.2d) and place 5 explants/dish on initiation medium (Table 21.1), at 25 ± 2°C in darkness.
3. Subculture every 4 weeks onto fresh initial medium. Eliminate brown explants and explants without callus (*see* Note 5, Table 21.2, Fig. 21.1f).

3.5. Multiplication of Callus and Induction of Somatic Embryos

1. Separate the embryogenic callus, obtained in media enriched with Dicamba and Picloram (50 μM) in combination with BA (0.5 μM), from primary explants (*see* Note 6, Fig. 21.3a).

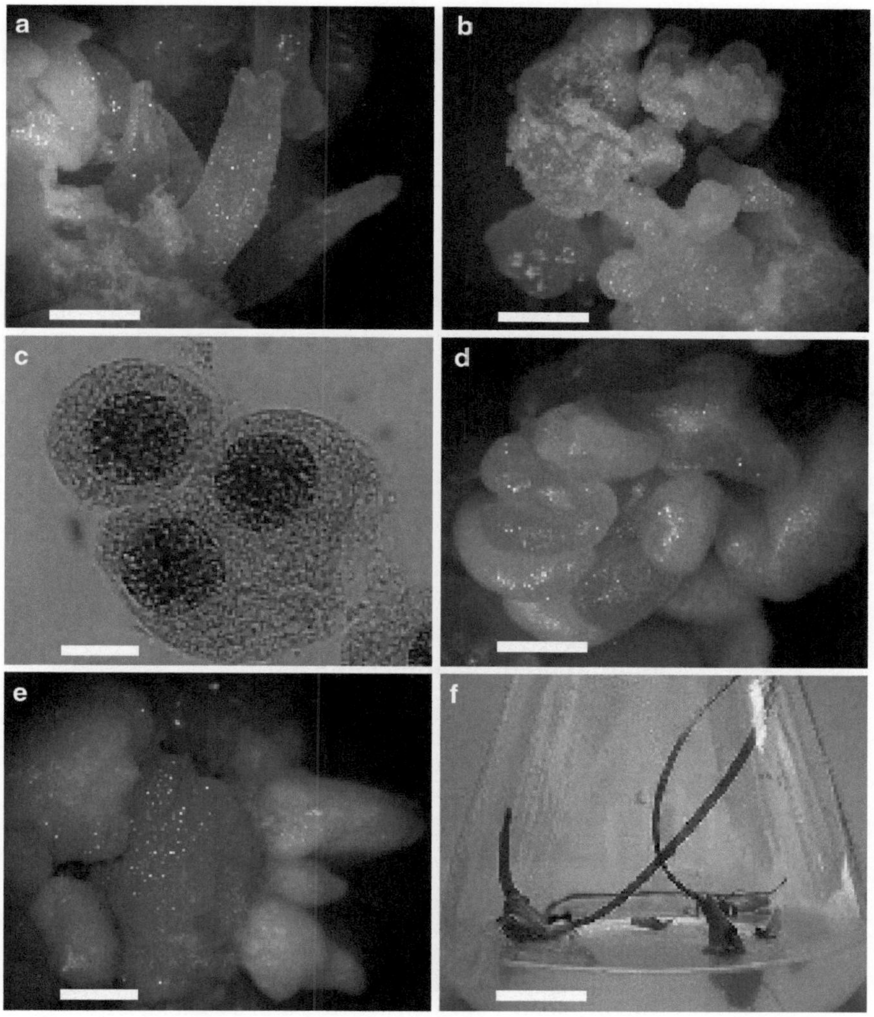

Fig. 21.3. Somatic embryogenesis and bulb formation in spring snowflake: (**a**) root induction (*bar*= 1 mm), (**b**) embryogenic callus multiplication (*bar*= 1 mm), (**c**) isolated embryogenic cells (*bar*= 25 μm), (**d**) somatic embryos at the globular stage (*bar*= 1 mm), (**e**) somatic embryos at the torpedo stage (*bar*= 1 mm), (**f**) regenerated bulbs with leaves and roots (*bar*= 10 mm).

2. Select nodular, yellow callus, and excise it with a scalpel (*see* Note 7) (Fig. 21.3b, 3c).
3. Culture the callus on multiplication medium (1 g/dish) (Table 21.1) at 25 ± 2°C in darkness.
4. Subculture every 4 weeks onto fresh medium (*see* Notes 8 and 9, Table 21.3, Fig. 21.3d).

3.6. Maturation of Somatic Embryos

1. Put 500 mg of somatic embryos at the globular stage onto maturation medium (Table 21.1).
2. The somatic embryos can be cultured at 25 ± 2°C in darkness for 4 weeks (*see* Note 10, Table 21.4, Fig. 21.3e).

Table 21.3
Effects of growth regulators and explant type on the callus multiplication rate and somatic embryo formation

	Multiplication rate		No. of embryos per gram of callus	
Explant type	Dicamba 25+BA 0.5	Picloram 25+BA 0.5	Dicamba 25+BA 0.5	Picloram 25+BA 0.5
Scale	3.5[a] a	3.6 a	2.0 a	2.1 a
Leaf	3.6 a	3.7 a	3.1 a	10.0 b
Ovary	3.6 a	4.2 ab	3.1 a	10.4 b
Fruit	3.8 a	4.5 b	10.0 b	19.1 c

[a]Means followed by different letters within columns are significantly different at $\alpha \leq 0.05$, Duncan's test

Table 21.4
Maturation of somatic embryos

Medium compounds	No. of mature embryos per gram of globular embryos
NAA 1+ZEA 5	25[a] a
ABA 15	38 c
PEG 45	26 b
ABA 15+PEG 45	45 c

[a]Means followed by different letters within columns are significantly different at $\alpha \leq 0.05$, Duncan's test

3.7. Germination of Somatic Embryos

1. Transfer individual somatic embryos at the cotyledonary stage to Petri dishes containing germination medium (5 somatic embryos/dish) (Table 21.1).
2. Maintain the culture at 5°C in the dark for about 2–3 weeks (*see* Note 11).
3. Maintain the culture at 25 ± 2°C under light (30 μmol m²/s⁻) for about 8 weeks.
4. Subculture every 4 weeks onto fresh medium.

3.8. Induction and Growth of Bulbs

1. Analyze the in vitro plantlets by flow cytometry (*see* Note 12, Fig. 21.4) (15).
2. Culture the regenerated plantlets on bulb growth medium at 5°C in the dark for about 6 weeks (three plantlets/Erlenmeyer flasks) (*see* Notes 13 and 14, Table 21.1).

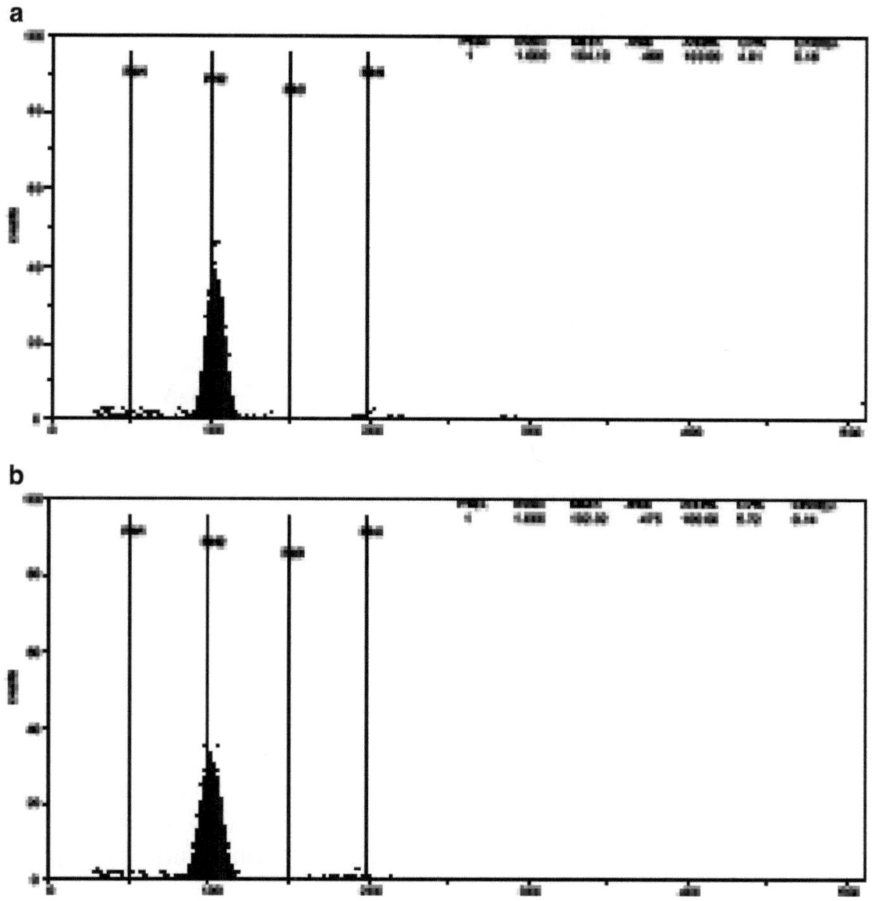

Fig. 21.4. Histograms of the leaf nuclear DNA content in spring snowflake: (**a**) control – mother plants, (**b**) plant regenerated in in vitro culture.

3. Transfer the plantlets onto bulb growth medium and culture them at 25 ± 2°C in the dark for about 8 weeks (*see* Note 15, Fig. 21.3f).

4. Notes

1. The bulbs were stored in darkness at 5°C for 12 weeks (from September). Although bulbs stored for more than 3 months could be used for callus initiation, fungal contamination was observed and the number of explants showing callus development decreased.
2. The ovaries were at the immature stage (before the embryo sac formation). A histological study was carried out using the paraffin method (16). The ovaries were separated from closed green flower buds, about 10 mm long.

3. The young fruits were at the early stage of seed formation.
4. All media components were autoclaved. Only ABA was added by filter sterilization.
5. After about 6 weeks of culture, embryogenic (nodular, yellow) callus developed. Callus formation varied depending both on growth regulators used and on the initial explant.
6. The callus derived from explants cultured on medium containing 2, 4-D induced root formation. Somatic embryos developed only sporadically.
7. Embryogenic callus can be accompanied by the development of colorless callus incapable of somatic embryogenesis. The embryogenic callus, irrespective of its origin, was composed of meristematic cells containing centrally located nuclei (squash preparations of callus were made in acetocarmine).
8. The highest rates of callus multiplication (4.5) and somatic embryo formation (19.1/1 g callus) were observed with fruits used as primary explants, on the medium supplemented with 25 μM Picloram and 0.5 μM BA.
9. Calli, irrespective of their origin, maintained their capacity to induce somatic embryos for approximately 1 year. After that time calli can turn brown and wet.
10. After 4 weeks of culture, somatic cotyledonary-stage embryos can be observed. Embryos are considered mature when they posses fully grown cotyledons and are opaque white-yellowish.
11. Sometimes somatic embryos can be dormant and they need a cold treatment.
12. Flow cytometric analysis of the regenerants was carried out. The analysis of the ploidy level of spring snowflake did not reveal any changes.
13. Before the formation of adventitious bulbs, the plantlets must be chilled at 5°C in the dark to break dormancy.
14. Larger bulbs can be obtained by subculturing the plantlets into the medium with high (9%) sucrose concentration.
15. Bulbs with roots and leaves can be planted in soil.

References

1. Le Nard M, de Hertogh A (eds) (1993) The physiology of flower bulbs. Elsevier, Amsterdam, London, New York, Tokyo
2. Piekos-Mirkowa H, Mirek Z (2003) Atlas roślin chronionych. Multico, Warszawa (in polish)
3. Szlávik L, Gyuris Á, Minárovits J, Forgo P, Molnár J, Hohmann J (2004) Alkaloids from *Leucojum vernum* and antiretroviral activity of *Amaryllidaceae* alkaloids. Planta Med 70(9):871–873
4. Lilienfeld S (2002) Galantamine – a novel cholinergic drug with a unique dual mode of action for the treatment of patints with Alzheimer's disease. CNS Drug Rev 8(2):159–176

5. Sellés M, Viladomat F, Bastida J, Codina C (1999) Callus induction, somatic embryogenesis and organogenesis in *Narcissus confusus*: correlation between the state of differentiation and the content of galanthamine and related alkaloids. Plant Cell Rep 18:646–651
6. Chen L, Zhu X, Gu L, Wu J (2005) Efficient callus induction and plant regeneration from anther of Chinese narcissus (*Narcissus tazetta* L. var. *chinensis* Roem). Plant Cell Rep 24(7):401–407
7. Sage D, Lynn J, Hammatt N (2000) Somatic embryogenesis in *Narcissus pseudonarcissus* cvs. Golden Harvest and St. Keverne. Plant Sci 150(2):209–216
8. Anbaris S, Tohidfar M, Hosseini R, Haddad R (2007) Somatic embryogenesis induction in *Narcissus papyraceus* cv. Shirazi. Biotechnology 6(4):527–533
9. Malik M, Bach A (2005) Induction of somatic embryogenesis in *Narcissus* L. 'Carlton' ovary culture. Book of abstracts, COST 843. Final Conference/COST 843 and COST 851 Joint Meeting. Stara Lesna, Slovakia, pp 197–181
10. Bach A, Pawłowska B, Warchoł M (2005) Phenolic compounds as a biological marker in snowdrop (*Galanthus sp.*) organogenesis and somatic embryogenesis. Book of abstracts, COST 843. Final Conference/COST 843 and COST 851 Joint Meeting. Stara Lesna, Slovakia, pp 117–119
11. Lilien-Kipnis H, Azizbekova N, Ziv M (1994) Scaled-up proliferation and regeneration of *Nerine* in liquid cultures. Part II. Ontogeny of somatic embryos and bulblet regeneration. Plant Cell Tissue Organ Cult 39(2):117–123
12. Huang CL, Chang KC, Okubo H (2005) In vitro morphogenesis from ovaries of *Hippeastrum* x *hybridum*. J Fac Agr Kyushu Univ 50(1):19–25
13. Ptak A, Cierniak O (2003) Regeneration of summer snowflake (*Leucojum aestivum*) in in vitro cultures. Biotechnologia 4(63):16–22
14. Murashige T, Skoog F (1962) A revised medium for rapid growth and bioassays with tobacco tissue culture. Physiol Plant 15:473–497
15. Doležel J (1991) Flow cytometric analysis of nuclear DNA content in higher plants. Phytochem Anal 2:143–154
16. Ptak A, Bach A, (2007) Somatic embryogenesis in tulip (*Tulipa gesneriana* L.) flower stem cultures. In Vitro Cell Dev Biol Plant 43:35–39

Chapter 22

In Vitro Propagation of Australian Native Ornamental Plant, *Scaevola*

Chui Eng Wong and Prem L. Bhalla

Abstract

In this chapter, we describe a robust method for the micropropagation of Australian fan flower, *Scaevola*, a native plant increasingly being used in the ornamental horticulture industry. Shoot segments from different species of *Scaevola* can be successfully multiplied following this protocol. Multiple shoots can be obtained in hormone-free Murashige and Skoog medium. The regenerated shoot is rooted on hormone-free medium within 4–6 weeks. In vitro grown plantlets readily adapt to glasshouse conditions.

Key words: Australian fan flower, Micropropagation, In vitro culture, Scaevola, Australian native plants

1. Introduction

Australia has a treasure trove of native plants and many of these have great horticulture significance. The macadamia nut (*Macadamia tetraphylla*) that has its origin in southern Queensland and north-eastern New South Wales is the only edible native horticultural crop extensively commercialised (1). While this represents an orchard crop, there are other native plants that are also widely used in the Australian ornamental horticulture industry. One example is a vigorous woody climber of the genus *Pandorea*, from the family of *Bignoniaceae* that is much sought after for its large flowers and long flowering period (2).

Another increasingly used species of Australian native in ornamental horticulture industry is *Scaevola* from the family of *Goodeniaceae*. This is a remarkable floriferous plant with fan-shaped flowers (Fig. 22.1). The species *S. aemula*, commonly known as fan flower, has the typical colouration of purple, blue or

S.M. Jain and S.J. Ochatt (eds.), *Protocols for In Vitro Propagation of Ornamental Plants*, Methods in Molecular Biology, vol. 589, DOI 10.1007/978-1-60327-114-1_22, © Humana Press, a part of Springer Science+Business Media, LLC 2010

Fig. 22.1. Flowering *Scaevola aemula* (**a**) and *Scaevola albida* (**b**). Close up of fan shaped flower of *Scaevola albida* (**c**).

mauve and the blooms last from spring until the end of summer (Fig. 22.1a). Popular commercial cultivars include *S. albida* variety marketed as "Mauve Clusters" (Fig. 22.1b, c), *S. striata* sold as "Pink Perfection" or "Misty Blue" and *S. phlebopetala*. *Scaevola* is extremely versatile and is widely used as a ground cover in Australia and in hanging baskets, window boxes and garden beds in Europe and America (3, 4).

The ornamental plant industry relies on the introduction of new products with attractive attributes. *Scaevola* has significant genetic potential for the generation of new cultivars for the local as well as international ornamental market. The breeding potential of *Scaevola*, however, is limited due to the poor germination of its seeds (5). The use of micropropagation would thus facilitate the introduction of new commercial cultivars, as it allows rapid in vitro multiplication.

There are a number of ways to micropropagate *Scaevola*. Indirect plant regeneration via somatic embryogenesis or from callus is possible with *S. aemula* R. BR (6, 7) or *S. phlebopetala* (8), respectively. A simple direct shoot regeneration method is described here ((9), Fig. 22.2) that enables the consistent production of a large number of in vitro cultured plants in a short period of time and is robust for different *Scaevola* genotypes. Following this protocol, cultured shoots start to proliferate on micropropagation medium within 1–2 weeks and well-developed branches develop within 4 weeks. These branches can be excised and cultured again allowing the regeneration of at least 25 plantlets from five explants within ~6 weeks. These cultured shoots can serve as a source of continuous supply of plantlets as they can be repeatedly cut and subcultured onto fresh medium. Direct shoot regeneration from different explants such as leaf, node, stem of *Scaevola aemula* R. BR has also been reported (10). There are also methods available to test the genetic variation of the propagated plants (3, 11, 12). The development of these techniques facilitates not only the commercial propagation of cultivars of interest but also the improvement of these plants using genetic engineering.

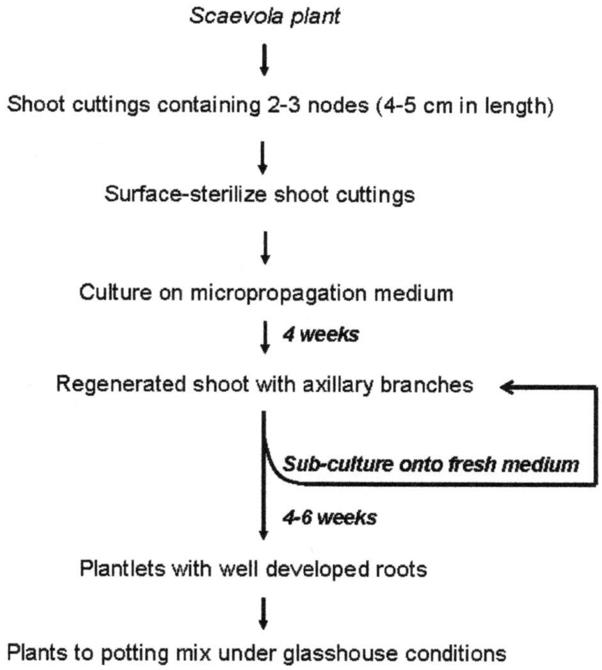

Fig. 22.2. Outline of in vitro micropropagation of *Scaevola*.

2. Materials

2.1. Plant Material

Commercial mature flowering plants of *Scaevola* obtained from nursery were used. The plants were maintained under normal glasshouse conditions. We have successfully used the protocol described here to micropropagate *Scaevola aemula*, *S. albida*, *S. striata*, *S.glandulifera*, *S. hookeri* and *S. ramonissima* (9).

Shoot cuttings (approximately 3–4 cm) were used as starting material. Old leaves and developing flowers or buds were pruned before surface-sterilisation. Surface-sterilisation was carried out in 50 ml sterile plastic tubes (Greiner Bio-One, North Carolina, USA). The shoot cuttings were cultured in sterile, transparent plastic disposable cups.

2.2. Media

Micropropagation medium is essentially Murashige and Skoog medium (13), but with macronutrients at half strength as detailed below:

Macronutrients: 800 mg/L ammonium nitrate, 950 mg/L potassium nitrate, 123 mg/L magnesium sulphate 7 hydrated, 85 mg/L of KH_2PO_4, 7 mg/L l-cysteine, 7.5 mg/L FeSO4 $7H_2O$, 14 mg/L EDTA Na_2 $2H_2O$, 27 mg/L myoinositol.

Minor salt stock: Use 1 ml in total volume of 1 L of the medium. (Stock: Dissolve 620 mg boric acid, 2,230 mg $MnSO_4$, 860 mg $ZnSO_4.7H_2O$, 25 mg Na_2MoO_4 $2H_2O$, 2.5 mg copper sulphate, 2.5 mg cobalt chloride in 100 ml distilled water. Store at –20°C.) (*see* Note 1).

Vitamin stock: Use 1 ml in total volume of 1 L of the medium. (Stock: Dissolve 100 mg Nicotinic acid, 1.0 g Thiamine hydrochloride, 100 mg pyridoxine hydrochloride, 10 g myo-inositol in 100 ml distilled water. Can be stored at –20°C up to 3–4 months) (*see* Note 1).

Calcium chloride stock: Use 2.9 ml in total volume of 1 L of the medium. (Stock: Dissolve 15 g of Calcium chloride dihydrate in 100 ml distilled water. Can be stored at 4°C up to 3–4 months).

Potassium iodide stock: Use 1 ml in total volume of 1 L of the medium. (Stock: Dissolve 75 mg of potassium iodide in 100 ml distilled water; store in a dark glass bottle at 4°C. Make fresh every 1–2 months).

Add sucrose (20 g/L). Adjust pH of the medium to pH 6.0 with 1 N NaOH and add 6 g/L agar (Sigma-Aldrich, New South Wales, Australia); autoclave at 121°C for 20 min.

2.3. Other Supplies

1. 100 ml of sterile distilled water in bottles.
2. Plastic disposable transparent drinking cups with compatible lids (240 ml capacity; Dalton Packaging, Victoria, Australia) and sterilisation by gamma rays (Steritech, Victoria, Australia).
3. Micropore surgical tape (3 M; Victoria, Australia).
4. Cutting mix: 10 L pearlite, 4 L sieved peat, 36 L of 6 mm pine bark and 37.5 g dolomite.
5. Potting mix: 320 L fine pine bark, 40 L fine sand, 500 g osmocote, 400 g dolomite and 500 g gypsum.

3. Methods

3.1. Shoot Cutting Preparation

1. Obtain shoot cuttings, 4–5 cm long with 2–3 nodes from *Scaevola* plants (*see* Note 2). Prune old leaves, developing flowers or buds.
2. Surface sterilise five shoot cuttings per tube in 1.6% sodium hypochlorite (Sodium hypochorite, 12.5% (wt/vol) (Chem-Supply, Cat. no. ST044)) (*see* Notes 3 and 4) with detergent (0.5 ml of 20% SDS in 50 ml) in 50 ml sterile plastic tubes (Greiner Bio-One, North Carolina, USA) by continuous shaking for 20 min.

The following steps should be performed in sterile conditions in a laminar flow hood:

Rinse shoot cuttings with sterile distilled water at least three times.

3.2. Shoot and Root Initiation

1. Cut the ends (~0.4 cm) of sterilised shoot cuttings before placing them (five per cup) vertically into 40 ml micropropagation medium in sterile disposable drinking cups (Fig. 22.3a).
2. Seal the cup closed with its compatible lid using Micropore tape and incubate them under a 16-h light regime (50 μE/ $m^{-2} s^{-1}$) at a constant temperature of 25°C ± 2°C.
3. When shoots (axillary branches) develop from axillary buds (approximately 4 weeks) (Fig. 22.3b), cut the regenerated shoots into segments with 1–2 nodes of 2–3 cm long.
4. Subculture the cut segments by placing them vertically onto fresh medium as described above. Incubate the shoot cuttings under the same growing conditions as above.
5. If desired, newly developed axillary branches can be cut again and subcultured after 4 weeks. Otherwise, rooted plantlets (*see* Note 5) are ready to be transferred to glasshouse conditions (Fig. 22.3b, c) (*see* Note 6).

3.3. Acclimation and Transfer of Plantlets to Soil

1. Rinse roots of the regenerated plantlets in warm water to remove tissue culture medium and transfer into pots containing the cutting mix (*see* Notes 7 and 8). For this, fill pots with the cutting mix and water the pots well before plantlet transfer. Place the regenerated shoot into the cutting mix by making a hole, depth depending upon the root length, in the soil. Compact the soil around the shoot and water lightly.

Fig. 22.3. In vitro microprogation stages. (**a**) Shoot cuttings on the medium at the shoot initiation stage. (**b**, **c**) Regenerated plants with roots.

2. Keep the plantlets in a mist chamber before transferring to potting mix, as described above for cutting mix, and maintain under glasshouse conditions, 25 ± 3°C day and 16 ± 3°C night.

4. Notes

1. MS minor salts and vitamin stocks are stored frozen and thus ensure all solutions are completely thawed before taking any aliquots. All stock solution of vitamins prepared can be stored at 4°C up to 4 months.
2. Vigorous growing healthy plant is critical. We have routinely grown the plants obtained from nursery or from the wild in potting mix and maintained under glasshouse conditions.
3. Sodium hypochlorite should be stored in dark container below 25°C up to 4 months. Buy a fresh stock of sodium hypochlorite every 2–3 months as sodium hypochlorite degrades over time.
4. A maximum of five shoot cuttings per tube can be used without compromising the sterility of the explants.
5. 92–100% of the regenerated shoots are able to root on the micropropagation medium within 4–6 weeks (9). Micropropagation medium containing IBA (1 mg/L) can be used for rooting of the regenerated shoots. Root initiation usually occurs within one week on IBA containing medium. However, no difference in percent of regenerated shoots with roots with IBA is found after 4–6 weeks (9).
6. Plantlets that are optimum for the transfer to glasshouse conditions develop a minimum of 4–5 roots, 2–3 cm long and with 4–5 leaves per plantlet.
7. Do not let shoots wilt during the transfer. Use well-cleaned disinfected or new pots; cutting and potting mix should be autoclaved to kill bacteria and fungi.
8. Regenerated plantlets may be transferred directly into potting mix.

Acknowledgments

We would like to thank Australian Research Council for financially supporting research in our laboratory.

References

1. Mulwa RMS, Bhalla PL (2000) In vitro shoot multiplication of *Macadamia tetraphylla* L. Johnson. J Hortic Sci Biotechnol 75:1–5
2. Kancherla SL, Bhalla PL (2003) Plant regeneration of the Australian native ornamental genus, Pandorea. J Hortic Sci Biotechnol 78:148–153
3. Swoboda I, Bhalla PL (1997) RAPD analysis of genetic variation in the Australian fan flower, Scaevola. Genome 40:600–606
4. von Hentig W-U, Ehlers D (1991) Mutterpflanzenhaltung und stecklingsvermehrung bei *Scaevola aemula*. Dtsch Gartenbau 6:359–361
5. Wrigley RJ, Fagg M (1996) Australian Native Plants. Reed Books, Australia
6. Wang YH, Bhalla PL (2004) Somatic embryogenesis from leaf explants of Australian fan flower, *Scaevola aemula* R. Br. Plant Cell Rep 22:408–414
7. Wang YH, Bhalla PL (2006) Plant regeneration from cell suspensions initiated from leaf- and root-derived calli of the Australian ornamental plant *Scaevola aemula* R.BR. Propagation Ornamental Plants 6:55–60
8. Bhalla PL, Xu HL (1999) Plant regeneration from callus of Australian fan flower, Scaevola. J Plant Physiol 154:374–378
9. Bhalla PL, Sweeney K (1998) Micropropagation of Scaevola – Australian native of ornamental horticulture. Aust J Exp Agric 38:399–401
10. Bhalla PL, Sweeney K (1999) Direct in vitro regeneration of the Australian fan flower, *Scaevola aemula* R. Br. Sci Hortic 79:65–74
11. Mulwa RMS, Bhalla PL (2007) Assessment of clonal stability of in vitro regenerated shoots of Macadamia tetraphylla by RAPD analysis. Aust J Agric Res 58:253–257
12. Jain A, Apparanda C, Bhalla PL (1999) Evaluation of genetic diversity and genome fingerprinting of Pandorea (Bignoniaceae) by RAPD and inter-SSR PCR. Genome 42:714–719
13. Murashige T, Skoog F (1962) A revised medium for rapid growth and bioassays with tobacco tissue cultures. Physiol Plant 15:473–497

Chapter 23

Micropropagation of Tulip: Production of Virus-Free Stock Plants

Małgorzata Podwyszyńska and Dariusz Sochacki

Abstract

We describe here a new tulip micropropagation method based on the cyclic shoot multiplication in presence of the thidiazuron (TDZ), which enables the production of virus-free stock plants, speeds up breeding, and provides new genotypes for the market. In our novel protocol, cyclic shoot multiplication can be performed for 2–3 years by using TDZ instead of other cytokinins, as 6-benzylaminopurine (BAP) and N^6-(-isopentyl)adenine (2iP). It makes possible to produce 500–2,000 microbulbs from one healthy plant. There are six main stages of tulip micropropagation. Stage 0 is the selection of true-to-type and virus-free plants, confirmed by ELISA. Fragments of flower stems isolated from bulbs are used as initial explants. Shoot multiplication is based on the regeneration of adventitious shoots, which are sub-cultured every 8 weeks. In the Stage 3, the specially prepared shoots are induced by low temperature treatment to form bulbs which finally develop on a sucrose-rich medium at 20°C. Bulbs are then dried for 6 weeks and rooted in vivo. The number of multiplication subcultures should be limited to 5–10 cycles in order to lower the risk of mutation. Virus indexing should be repeated 3–4 times, at the initial stage and then during shoot multiplication. Genetic stability of micropropagated shoots can be confirmed using molecular markers.

Key words: *Tulipa*, In vitro propagation, Thidiazuron, Virus-free plants, Virus indexing

1. Introduction

Among the top ten flowers sold worldwide, the tulip takes the third place. The world tulip bulb and flower production as well as its marketing is localized in a few countries. The Netherlands dominates the production of tulip bulbs and accounts for 87% of the worldwide area, with 10,800 hectares (1). Japan, France, and Poland follow among the other 14 tulip producing countries. Commercially, the propagation of tulips is based on the vegetative

S.M. Jain and S.J. Ochatt (eds.), *Protocols for In Vitro Propagation of Ornamental Plants*, Methods in Molecular Biology, vol. 589, DOI 10.1007/978-1-60327-114-1_23, © Humana Press, a part of Springer Science+Business Media, LLC 2010

production of daughter bulbs by the buds located in the axils of the mother bulb scales. Such a type of propagation can cause an accumulation of viruses and their subsequent spreading within daughter bulbs. Viral diseases can drastically affect the yield and quality of tulip bulbs and flowers. Of the 22 viruses reported to infect tulip five are the most common, widespread and causing serious diseases: *Tulip breaking potyvirus* (TBV), *Lily symptomless carlavirus* (LSV), *Cucumber mosaic cucumovirus* (CMV), *Tobacco necrosis necrovirus* (TNV), and *Tobacco rattle tobravirus* (TRV) (2). When propagating or cultivating tulips, it is essential to start with healthy virus-free bulbs (3). Micropropagation of selected virus-free plants and elimination of viruses during in vitro culture are the best methods to ensure that the breeding and planting stocks are healthy. The new tulip micropropagation method is based on a cyclic proliferation of adventitious shoots in the presence of thidiazuron (TDZ) (4, 5). Because of the cyclic shoot multiplication, this technique enables the production of virus-free stock plants, speeds up breeding, and provides new genotypes for the market, which is very important given the very low natural propagation rate of tulips. Each bulb producing only two or three bulbs a year, 20 or more years can pass between an initial crossing and the commercial release of a new cultivar.

The new protocol was generally elaborated on the basis of previously reported methods (6–8), where direct shoot organogenesis was obtained in the presence of NAA combined with the common cytokinins BAP or 2iP. Independently of the method, the final product of the tulip in vitro propagation is a microbulb. In systems where shoots regenerate directly on initial explants, bulb development is induced in shoots during the second as well as the last stage of the micropropagation. Such methods allow to obtain about 200 microbulbs from one donor bulb. In our novel protocol, the cyclic shoot multiplication can be performed for 2–3 years due to the application of TDZ instead of BAP or 2iP, whereby it is possible to produce 500–2,000 of microbulbs from one healthy plant. Micropropagation via adventitious shoot regeneration in a long-term culture system involves a higher risk of mutation than the propagation through axillary bud proliferation (9), but no methods based on the micropropagation of lateral shoots have been developed in tulips so far. Therefore, the genetic stability of plants produced was verified on the basis of morphological and molecular markers (10, 11). The results showed that the novel method can increase the risk of mutation, especially when shoots are micropropagated for more than 4 years. Limiting the period of in vitro propagation to 3 years and monitoring of the plant material with DNA markers can further reduce the risk of mutated plant production.

2. Material

2.1. Donor Plant Preparation

Pots with perlite soaked with aqueous solution of fungicide, e.g., 0.2% captan.

2.2. Initial Phase of Culture

1. Plant material: cooled bulbs of *Tulipa* sp. obtained from selected true-to-type and virus-free plants (*see* Subheadings 23.3.1 and 23.3.7).
2. Surface disinfection in aqueous solutions of 3% Chloramine T and 0.1% $HgCl_2$ with two drops of Tween 20 per 100 mL (*see* Note 1). Using $HgCl_2$ requires protective gloves, as $HgCl_2$ is toxic for skin and by inhalation, harmful for aquatic organisms. This chemical reacts with aluminum tools.
3. Sterile deionised water.
4. Initial medium (Table 23.1) dispensed into 50–100 mL Erlenmeyer flasks or test tubes 2.5 cm in diameter.

Table 23.1
Media for tulip micropropagation based on modified mineral composition of Murashige and Skoog (12) medium and organic compounds of Nishiuchi (6)

	Concentration (mg/L)				
	Micropropagation stages				
Compounds	Culture initiation	Shoot multiplication	Last subculture prior to cooling	Liquid medium added in eighth week of the last subculture prior to cooling	Bulb induction and formation
Macroelements					
KNO_3	1,900	1,900	1,900	1,900	1,900
KH_2PO_4	170	170	170	170	170
$MgSO_4 \cdot 7H_2O$	370	370	370	370	370
NH_4NO_3	1,650	1,650	1,650	1,650	1,650
$CaCl_2 \cdot 2H_2O$	440	440	440	440	440
Microelements					
$MnSO_4 \cdot 4H_2O$	223	22.3	22.3	22.3	22.3
H_3BO_3	6.2	6.2	6.2	6.2	6.2
$CuSO_4 \cdot 5H_2O$	0.025	0.025	0.025	0.025	0.025

(continued)

Table 23.1 (continued)

Compounds	Concentration (mg/L)				
	Micropropagation stages				
	Culture initiation	Shoot multiplication	Last subculture prior to cooling	Liquid medium added in eighth week of the last subculture prior to cooling	Bulb induction and formation
$CoCl_2{\cdot}6H_2O$	0.025	0.025	0.025	0.025	0.025
$ZnSO_4{\cdot}7H_2O$	8.6	8.6	8.6	8.6	8.6
KI	0.83	0.83	0.83	0.83	0.83
$Na_2MoO_4{\cdot}2H_2O$	0.25	0.25	0.25	0.25	0.25
FeEDDHA	96	96	96	96	96
Organic compounds					
Adenine	40	40	40	–	–
Nicotinic acid	5	5	5	1	1
Pyridoxine HCl	1	1	1	1	1
Inositol	100	100	100	100	100
Thiamine	10	10	10	1	1
Glycine	20	20	20	2	2
Glutamine	200	200	200	–	–
Casein hydrolysate	1,000	1,000	1,000	–	–
Sucrose	30,000	30,000	30,000	30,000	70,000
Activated charcoal	–	–	–	–	1,000
Agar	6,000	6,000	6,000	–	6,000
Growth regulators					
TDZ	0.5–2	0.1–0.2	–	–	–
2iP	–	–	1–5	–	–
NAA	1	0.1	0.1	–	–
Paclobutrazol	–	–	–	1	–
pH	5.6				

2.3. Shoot Multiplication and Bulb Formation

1. Medium for a given stage (Table 23.1) disposed into the standard plant culture vessels, e.g., 300 mL food jars covered with plastic lids or Vitro Vent Containers (Duchefa Biochemie BV, The Netherlands).
2. Solution of 0.1% methyl jasmonate (MeJA) (Duchefa Biochemie BV, The Netherlands) dissolved in 10% ethanol (v/v) (*see* Note 2).

2.4. Rooting of Bulbs and Further Growing of Plants

1. Plastic boxes (60×40 cm with the inner depth of at least 8.5 cm) with openwork bottom.
2. Rooting substrate composed of peat with addition of perlite, sand, or vermiculate (4:1 or 3:1 v/v), supplemented with complete fertilizer at 1 g/L, pH 6–7, soaked with fungicides against soil born diseases and gray mold (*Botrytis cinerea*).

2.5. Virus Indexing by ELISA

1. Antibodies and reagent sets for detection of the viruses: TBV, LSV, TNV (Wageningen UR/Applied Plant Research, Flowerbulbs, Nursery & Fruit, Lisse, The Netherlands), CMV and TRV (Loewe Biochemica GmbH, Germany), for group of potyviruses (Agdia Inc, USA).
2. ELISA plates: Nunc-Immuno plates Maxisorp F 96.
3. pNPP substrate tablets (Sigma).
4. Buffers: coating buffer, pNPP substrate buffer, conjugate buffer, PBS-T wash buffer (phosphate-buffered saline, pH 7.4, containing 0.1% Tween 20), indirect extraction buffer, ECl buffer (Loewe Biochemica GmbH, Agdia Inc., USA). Also buffer packs for DAS-ELISA and indirect-ELISA are offered by above mentioned companies and many others.

3. Methods

3.1. Stage 0. Selection and Preparation of Donor Plants

1. Selection of visibly healthy and true-to-type plants should be done when they are in the flowering phase, in spring. The plants should be indexed by ELISA against the most common tulip viruses: TBV, LSV, CMV, TRV, TNV, and potyviruses (*see* Subheading 23.3.7). The virus-free plants should be labeled. In June, the bulbs obtained from these plants are harvested, stored at standard conditions in a well-ventilated storage room at 17–20°C until October, and then dry cooled at 5°C until forcing in January-February (*see* Note 3).
2. Selected healthy bulbs should be planted into pots with perlite soaked with aqueous solution of captan and forced in darkness at 9°C for 5–14 days in order to obtain a 2–5 cm long leaf tip. The recommended bulb size is 10 cm diameter.

3.2. Stage 1. Initiation of in Vitro Shoot Culture

1. Rooted bulbs with a "nose" (leaf tip) at 2–5 cm are washed with tap water. One or two external scales are removed and the bulbs are surface disinfected with 3% Chloramine T for 20 min, and then rinsed with sterile water. Flower stems 2–3 cm long, covered with one internal bulb scale are isolated from the forced bulbs, disinfected with 0.1% $HgCl_2$ for 10 min (on a shaker), and rinsed twice with sterile water. Then internal bulb scale and leaves are removed carefully not to damage the stem epidermis (*see* Note 4).
2. Leaves which serve as samples for virus indexing should be put into plastic bags and kept in a refrigerator until testing.
3. The 12–20 initial explants, 1–2 mm thick stem slices, are cut from the stem, and the 2–4 mm long basal part is removed (Fig. 23.1). Explants are placed on initial medium (Table 23.1) in 100 mL flasks or in test tubes and are incubated in darkness

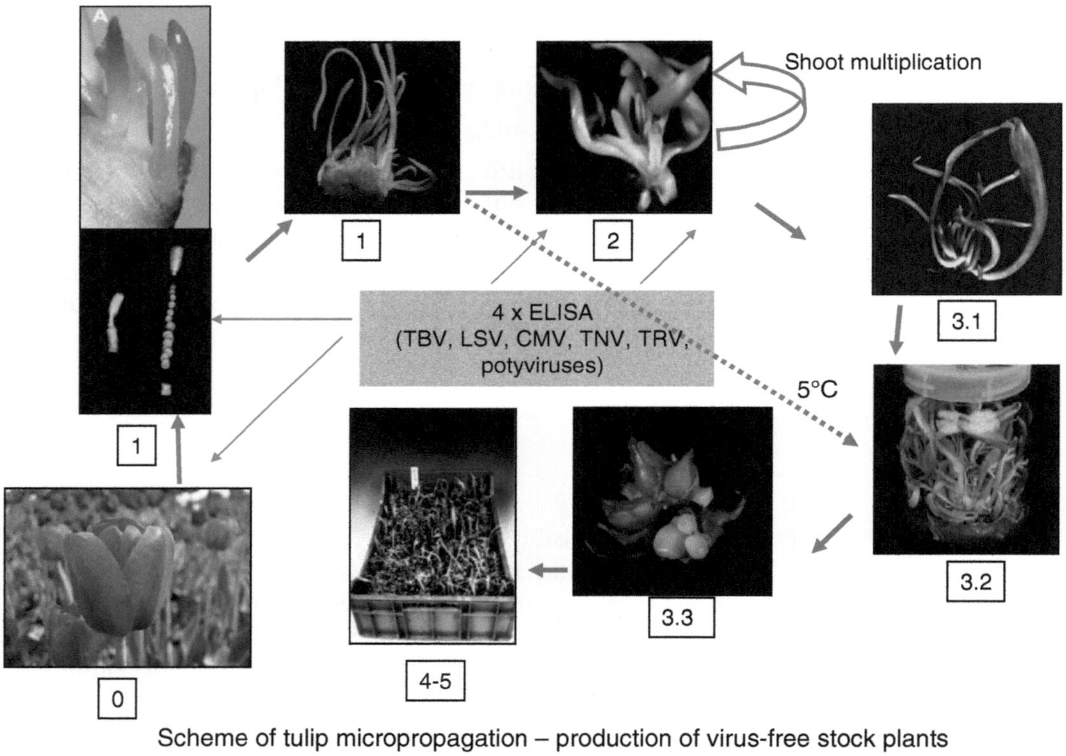

Fig. 23.1. Micropropagation stages: (**0**) Selection of donor plants; (**1**) Initiation of in vitro shoot culture, initial explants – flower stem slices 1–2 mm thick and direct regeneration of leaf-like structures on flower stem explant 10 weeks after exposure to induction medium (Reproduced from ref. 4 with permission from Acta Soc. Bot. Pol.); (**2**) Multiplication stage, stabilised shoot cultures during multiplication stage (Reproduced from ref. 20 with permission from Acta Hort); (**3.1**) Shoots with elongated leaves just before low temperature treatment (Reproduced from ref. 20 with permission from Acta Hort); (**3.2**) Development of bulbs, shoots forming bulbs, 2 weeks after cooling at 5°C; (**3.3**) Drying of bulbs (Reproduced from ref. 20 with permission from Acta Hort). (**4-5**) Growing of plants. Plants grown in an insect-proof tunnel during the first season.

for 2–3 months at 15–20°C. Each explant should be labeled with a code carrying the information about the donor plant and explant. These codes for shoot clones derived from a single explant will serve to localize precisely the detected mutations or virus infections and eliminate any undesirable plant material.

4. Subsequently, all explants regenerating leaf-like structures are transferred onto the multiplication medium supplemented with TDZ and NAA, both at 0.1–2 mg/L (Table 23.1) and cultured at 20°C under 16-h photoperiod from warm, white fluorescent lamps (Philips TLD 36 W/95; 40 μmol/m^2/s). The regenerating structures develop gradually into characteristic forms (leaf-like structures) with cotyledonary leaves growing upwards and stolons growing downwards into the culture medium. After 2 months culture in light, the drying cotyledonary leaves are cut off and their bases with the stolons and adventitious shoots developing on their tips are placed upside down in a fresh medium for proper development of the shoots (*see* Note 5). The percentage of explants regenerating leaf-like structures depends on genotype and range from 70 to 100%. A single explant can produce 2.5–20 leaf-like structures.

3.3. Stage 2. Multiplication of Shoots

1. Developing adventitious shoots form clusters that are divided into smaller ones and subcultured on fresh multiplication medium (Table 23.1) every 8 weeks, under the same conditions as for Stage 1. The multiplication rate, i.e., the number of smaller clumps obtained from one initial clump, varies from 1.2 to 3, depending on cultivar (*see* Note 6). In order to lower the risk of mutation occurrence, the number of subcultures should be limited to 5–10 cycles, i.e., 1–2 years of shoot multiplication at 8 week subculture period. Genetic stability can be confirmed using molecular markers (*see* Note 7).
2. Two or three series of virus indexing should be done after 8–10 months of culture, i.e., by the second or third multiplication subculture, and then after 14–20 months of cyclic multiplication. The samples should be collected from each shoot clump or one shoot clump per container, depending on the amount of plant material (*see* Subheading 23.3.7).

3.4. Stage 3. Development of Bulbs

Bulb formation is the essential stage in tulip in vitro cultures because only bulbs show high rooting ability and good field performance. In standard micropropagation protocols, bulbs develop from the shoots formed on the initial explants isolated from a floral stem (7, 13) or bulb scale (14). In the novel system, the bulbing process is induced in shoots multiplied in vitro for 1–2 years. Independently of the method, the induction of bulb formation requires cold treatment (5°C) for 10–12 weeks, and transplantation in a medium without cytokinin and containing a high sucrose (6–7%) concentration. Generally, the tulip bulb development proceeds in vitro similarly as in natural conditions (in vivo) (15). Bulb formation in vitro is controlled with a complex of external (temperature, sucrose level, exogenous growth regulators) and internal (hormonal and carbohydrate status) factors, which have to change and occur in a specific sequence.

Stage 3 of tulip micropropagation, bulb development, includes three phases and takes 9–10 months (Fig. 23.1). The first phase, preparing shoots for bulb formation is performed through modifying and prolonging the last multiplication subculture. This phase is associated with the decrease of endogenous cytokinin and increase of endogenous abscisic acid (ABA) in shoots as well as the development of their dormancy, which is manifested by the inhibition of shoot multiplication. This phase proceeds at high (summer) temperature of 20°C. The second phase is the induction of bulb development by cooling shoots; it is associated with the release of shoot dormancy under low (winter) temperature treatment of 4–5°C for 12–14 weeks, with the increase of endogenous cytokinin and decrease of endogenous ABA. In the third phase, shoots form bulbs on the sucrose-rich medium at high (Preparation) temperature of 20°C.

1. Shoot Preparation for the process of bulb formation involves prolonging of up to 14 weeks the last multiplication subculture prior to cooling (*see* **Note 8**). TDZ in the solid medium is replaced with 2iP (Table 23.1). Then, a liquid medium with a growth retardant, either ancymidol or paclobutrazol (Duchefa, The Netherlands) (Table 23.1), is added at the eighth week of this subculture. The volume of liquid medium should be about 70% of the solid medium volume, and shoots are grown on such double-phase medium for a further 6 weeks.

2. For the induction of bulb formation, shoots in clusters (not divided) are transplanted in sucrose-rich medium without growth regulators and cooled at 5°C, in darkness for 12–14 weeks (Table 23.1). The volume of the medium should be about twice higher than that used for shoot multiplication.

3. For subsequent bulb formation, the shoots are cultured in the same sucrose rich-medium for 2–3 months at 20°C in light, under a 16 h photoperd from warm-white fluorescent lamps as above but at 55 µmol/m2/s. Usually, 6–8 weeks after cooling, the green bulbs developing at shoot bases are well visible. MeJA treatment is recommended at this time in order to accelerate the bulbing process and to improve bulb quality. MeJA solution should be added into the culture vessels with the bulbing shoots under sterile conditions (on a clean table), (2.5–5 mL of 0.1% MeJA solution per 100 mL of a culture medium).

During 2 weeks, leaves dry gradually and bulbs cover themselves with a brown skin, i.e., tunic, the bulb formation is finished. If the bulbs are not well formed and leaves are still partly green, the treatment with MeJA can be repeated 4 weeks after the first treatment (*see* Note 9).

3.5. Stage 4. Drying of Bulbs

After being taken out of the containers, the bulbs are rinsed with tap water, dried and stored at 20°C for at least 6 weeks. Such a storage period is necessary for the root primordium development, which is a prerequisite for the subsequent rooting of bulbs in soil (15).

3.6. Stage 5. Rooting of Bulbs and Subsequent Growing of Plants

Microbulbs can be rooted in two seasons: (1) during autumn, outdoors, starting from October and (2) during winter, indoors, at 9°C, starting from January. In both, the bulbs are planted in plastic boxes filled with a growing substrate.

1. Autumn rooting. Boxes with bulbs are transferred into an insect-proof tunnel in late September up to late October. Bulbs are protected against frost with a bark chippings layer (5 cm). If frost below −10°C is forecast, trays are covered with unwoven fabric and straw. In early spring (March), the fabric and straw are removed, and the plants are fertilized twice alternately, with a complete fertilizer and calcium nitrate. Subsequently, plants are watered as needed and fertigation is applied weekly with calcium nitrate and a complete fertilizer during April and May. The plants are protected against *Botrytis tulipae* and *B. cinerea* by a weekly spray with chemicals according to the recommendation of the advisory service in the area.

2. Winter rooting. In January, the bulbs in plastic boxes are placed in a growing room at 9°C for 3 months. In April, the boxes with plants are transferred outdoors into an insect-proof tunnel. The plants are fertigated and watered as mentioned above.

3. Harvesting of bulbs. The bulbs are harvested in early summer. After lifting, the bulbs are dried, cleaned and stored at 20°C until planting for the second growing cycle in late September or October. Before planting, the bulbs should be disinfected with fungicides against *Fusarium* and *Rhizoctonia* according to the recommendation of the local advisory service. Also the soil in the insect-proof tunnel should be disinfected every year with soil disinfectant. Bulb reproduction after the first growing cycle depends on the initial weight of microbulbs, genotype and winter temperature. It ranges from 40 to 50% for small microbulbs (weighing less than 100 mg), 60–80% for microbulbs weighing100–300 mg and 100% for bulbs larger than 300 mg.

3.7. Virus Indexing by ELISA

1. Plants are sampled as follows: (1) at the first indexing term, fresh leaves of tulip donor plants are sampled in April/May, at the beginning of flowering; (2) at the second indexing term, leaves of sprouts are taken from forced bulbs of donor plants; (3) at the third, fourth and possibly fifth indexing term, the leaves are sampled from in vitro cultures after 8, 10 and 14–20 months of shoot culturing in vitro.
2. Leaves should be triturated in phosphate-buffered saline, pH 7.4 (for DAS-ELISA) or in an indirect sample extraction buffer (for indirect-ELISA), using a roller press. All other stages of ELISA should be performed according to published procedures or by following the instructions accompanying the proprietary reagents.
3. Tests are performed with DAS-ELISA (16) using commercial kits with specific antibodies for the detection of TBV, LSV, TNV (Wageningen UR/Applied Plant Research, Flowerbulbs, Nursery & Fruit, Lisse, The Netherlands), CMV and TRV (Loewe Biochemica GmbH, Germany). Indirect ELISA with monoclonal antibodies for the group of potyviruses (17) are performed according to the protocol provided with the commercial kit (Agdia Inc, USA), with minor modifications (18).
4. The results are measured with a Multiscan Plus Reader (Labsystems, Finland), at 405 nm. The samples are counted (recorded) as infected when the value of absorbancy (A_{405}) is at least twice higher than the A_{405} for negative control (19). In addition, the samples whose A_{405} exceeds the negative control by 50–99%, are recorded as "virus-suspected" (±) (*see* **Note 10**).
5. The virus-infected shoot-clones should be removed generally, except for the genotypes totally infected. In such cultivars, the most vigorous shoots are selected and micropropagated for further indexing (*see* Note 11).

4. Notes

1. Other alternative disinfection methods can be used, e.g., the solution of 2–3% NaOCl for 10 min in the first step of surface sterilization, instead of Chloramine T treatment.
2. MeJA solution should be prepared in sterile conditions just before application. It is recommended that MeJA should be dissolved first in ethanol, then added to sterile water and stirred incessantly.
3. In tulips, among different initial explants such as axillary buds, fragments of bulb scale or floral stem, the highest regeneration

capacity was found in short floral stems isolated in November or December from the bulbs stored at 17°C rather than at higher temperatures of 20–23°C. On the other hand, a high regeneration potential was also noted in the flower stems isolated from bulbs subjected to the normal cold treatment used for forcing (4, 7). In the presented micropropagation method, the initial explants are isolated from the floral stem excised from bulbs cooled for 12 weeks at 5°C. Tulips belong to a group of bulbous plants in which the flower stems require low temperature to grow sufficiently in length. Thus, after a 9–12-week cooling period, the flower stems, developing within the bulb, reach 2–3 cm and can be cut into 15–20 initial explants.

4. This is very important not to damage the stem epidermis because a primary regeneration takes place in the epidermal and subepidermal tissue (4).
5. The primary regeneration of the so-called leaf-like structures can be regarded as a four-step process: (1) development of the growing cotyledonary leaf; (2) formation of a probable root primordium at the base of cotyledonary leaf, as an axis, opposite to the initial explant (these two steps proceed during the first 2 months in darkness); (3) development of a very short, or a longer, stolon growing downwards to the medium; (4) development of shoot meristems on the stolon tips (stolons and shoot buds develop in third and fourth month of initial culture). In this micropropagation method, the further development of primary root has never been observed. The cyclic shoot multiplication obtained in this method has a nature of adventitious shoot formation, first on the stolon tip and then on the meristematic tissue originating from the primary stolon. The remaining of such meristematic tissue at the basis, on which the clusters of buds and shoots develop, has been the prerequisite for further success in tulip micropropagation. Removal of such basal tissue and separation of individual shoots markedly decreased the formation of individual shoots.
6. The tulip model cultivars 'Blue Parrot,' 'Prominence' and 'Giewont,' used for this study on tulip micropropagation, had a satisfactory multiplication rate of 1.8–2.5. However, among thirty other genotypes propagated in vitro according to the novel protocol, 50% of them exhibited an unsatisfactory efficiency of shoot multiplication. Such cultivars showed an inhibition of regeneration after the fourth to sixth multiplication cycle, that was associated with leaf-drying and formation of irregular swollen structures at the shoot base, attributed to the dormancy development. In some of those cultivars, the multiplication rate could be increased by cooling shoots (on multiplication medium) at 5°C for 10 weeks.

7. This micropropagation method was verified by the evaluation of genetic stability of the plants produced (10, 11). Reliable information on somaclonal variation introduced during in vitro propagation can be obtained when more than 30% of the plants have flowered. This can occur as early as in the third-fifth growing season, depending on the genotype. Genetic changes are particularly dangerous if they occur at an early stage of micropropagation, as this can involve a large production of mutated plants, because phenotypic evaluation is only possible several years later. Therefore, somaclonal variation of plants in our study was analyzed based on the phenotypic observation and using molecular markers. The results showed that the frequency of somaclonal variation depended on the genotype and micropropagation time. Generally, in the model cultivars 'Blue Parrot,' 'Prominence,' and 'Giewont' whose plants derived from 1 to 3-year-old cultures, phenotypic variation was not detected or was observed in less than 10%. However, within the plants derived from the 4-year-old cultures, all plants of "Blue Parrot" and half of the plants of "Prominence" were changed. Most of the off-type plants showed minor changes, such as lily-like flowers, a changed color and size of flower, although some plants did have strongly malformed flowers or leaves, leaf variegation was also observed. All phenotypic changes appeared to be stable and transmitted through vegetative propagation. Most of these changes were confirmed by ISSR and AFLP genetic analyzes (11).
8. Shoots at eighth week of multiplication subculture are unable to respond to the factor inducing bulb formation, i.e., low temperature due to the inappropriate physiological phase. By prolonging the last subculture prior to cooling, replacing TDZ with the fast metabolized cytokinin 2iP, and treating with inhibitors of gibberellin synthesis, the shoots develop dormancy and become sensitive to the low temperature treatment (20).
9. Some tulip genotypes propagated in vitro reveal a low capacity for bulb formation, e.g., the model cultivar "Blue Parrot." Not only bulb formation efficiency in such cultivars is below 50%, but also bulbs take 5 months to develop. Bulbs are irregular, not covered with tunic, and liable to desiccation. MeJA treatment improves both the bulb quality and efficiency of bulb formation by 50% (21).
10. It is recommended that only plant material for which the A_{405} value is close to negative control be subjected to further micropropagation. Shoot clones recorded as "virus-suspected" should be tested for the given virus again. The succeeding testing of individual propagules shows both virus-positive

and virus-negative reactions. This indicates the nonuniform distribution of virus in a rapidly proliferating culture. Thus, it is possible to obtain virus free plants (22, 23). It should also be taken into account that testing with ELISA could be insufficiently sensitive to detect the given viruses when present at low concentration.

11. Systematic testing against viruses and elimination of virus-positive plant material lead to the production of virus-free plants.

Acknowledgments

This work was supported by grants No. 5 PO6C 01410 and PBZ-MIN-007/P04/3003 (Poland)

References

1. Buschman JCM (2005) Globalisation – flower – flower bulbs – bulb flowers. Acta Hortic 673:27–34
2. Mowat WP (1995) Tulip. In: Lebenstein G, Lawson RH, Brunt AA (eds) Virus and virus-like diseases of bulb and flower crop. John Wiley and Sons, Chichester, UK, pp 353–383
3. Anonymous (2002) Classification scheme for tulip. Bull OEPP 32:115–121
4. Podwyszyńska M, Marasek A (2003) Effect of thidiazuron and paclobutrazol on regeneration potential of flower stalk explants in vitro and subsequent shoot multiplication. Acta Soc Bot Pol 72:181–190
5. Sochacki D, Podwyszyńska M (2006) Detection of *Tulip breaking virus* (TBV) in tulips by ELISA technique. Adv Agric Sci Probl Issues 510:593–599 (in Polish with English abstract and captions)
6. Nishiuchi Y (1979) Studies on vegetative propagation of tulip. 2. Formation and development of adventitious buds in the excised bulb scale cultivated in vitro. J Jpn Soc Hortic Sci 44:99–105
7. Le Nard M, Ducommun C, Weber G, Dorion N, Bigot C (1987) Observations sur la multiplication in vitro de la tulipe (*Tulipa gesneriana* L.) à partir de hampes florales prélevées chez des bulbes en cours de conservation. Agronomie 7:321–329
8. Taeb AG, Alderson PG (1990) Shoot production and bulbing of tulip in vitro related to ethylene. J Hortic Sci 65:199–204
9. Jain SM, De Klerk G-J (1998) Somaclonal variation in breeding and propagation of ornamental crops. Plant Tissue Cult Biotechnol 4:63–75
10. Podwyszyńska M (2005) Samoclonal variation in micropropagated tulips based on phenotype observation. J Fruit Ornam Plant Res 13:109–122
11. Podwyszyńska M, Niedoba K, Korbin M, Marasek A (2006) Somaclonal variation in micropropagated tulips determined by phenotype and DNA markers. Acta Hortic 714:211–219
12. Murashige T, Skoog F (1962) A revised medium for rapid assays with tobacco tissue cultures. Physiol Plant 15:473–497
13. Rice RD, Alderson PG, Wright NA (1983) Induction of bulbing of tulip shoots in vitro. Sci Hortic 20:377–390
14. Nishiuchi Y (1980) Studies on vegetative propagation of tulip. 4. Regeneration of bulblets in bulb scale segments cultured in vitro. J Jpn Soc Hortic Sci 49:235–239

15. Le Nard M, De Hertogh AA (1993) *Tulipa*. In: De Hertogh AA, Le Nard M (eds) The physiology of flower bulb. Elsevier Science Publishers, The Netherlands, pp 617–682
16. Clark MF, Adams AN (1977) Characteristics of the microplate method of enzyme-linked immunosorbent assay for the detection of plant viruses. J Gen Virol 34:475–483
17. Jordan R, Hammond J (1986) Analysis of antigenic specificity of monoclonal antibodies to several potyviruses. Phytopathology 76:1091
18. Sochacki D, Orlikowska T (2005) The obtaining of *Narcissus* plants free from potyviruses via adventitious shoot regeneration in vitro from infected bulbs. Sci Hortic 103:219–225
19. Clark MF, Lister RM, Bar-Joseph M (1988) ELISA techniques. In: Weissbaum A, Weissbaum H (eds) Methods for plant molecular biology. Academic, New York, p 527
20. Podwyszyńska M (2006) Improvement of bulb formation in micropropagated tulips by treatment with NAA and paclobutrazol or ancymidol. Acta Hortic 725:679–684
21. Podwyszyńska M (2006) Effect of ethylene, auxin and methyl jasmonate on bulb formation in vitro in tulip shoot cultures. Adv Agric Sci Probl Issues 510:461–469 (in Polish with English abstract and captions)
22. Lilien-Kipnis H, Stein A, Levy S (1992) The course of bean yellow mosaic virus elimination from *Gladiolus* by meristem tip culture. Acta Hortic 325:709–714
23. Podwyszyńska M, Sochacki D, Kaminska M (2005) Application of tulip in vitro culture for propagation of virus-free plants. Adv Agric Sci Probl Issues 504:667–674 (in Polish with English abstract and captions)

Chapter 24

In Vitro Propagation of Ornamental Myrtus (*Myrtus communis*)

Barbara Ruffoni, Carlo Mascarello, and Marco Savona

Abstract

The Myrtle (*Myrtus communis* L.) is an evergreen shrub typical of the Mediterranean area; it is an interesting plant with multipurpose use. The ornamental use takes into account the production of green cut branches for indoor decoration and production of pot plants for gardening. In this species, there is a great variability in the natural germplasm around the Mediterranean coasts for type and size of fruit, plant architecture, leaf size and internode length. Selected genotypes have been successfully sterilized and cultured in vitro. The shoots were multiplied on MS (16) salts and vitamins, with 0.5 mg/L BA and 0.2 mg/L IAA. Clones showed variation of multiplication rate and rooting percentage. IAA or IBA at 0.5 mg/L increased the rooting percentage and noticed differences in root number and length. The sucrose concentration can affect rooting, such as light intensity during the in vitro rooting phase can modulate biomass production and chlorophyll content. The combination of these factors enhanced the frequency rate of acclimatization.

Key words: Rooting, Acclimatization, PPFD, Chlorophylls

1. Introduction

The evergreen shrub myrtle (*Myrtus communis* L., Myrtaceae family) is well known for its multipurpose value: it is an important medicinal plant for the scent and the essential oils extracted from the leaves whose active principles are used in the chemical and medicinal industry. It is used for the industry related to the production of liqueurs (alcoholic infusion of the mature fruits), food additives and handycrafts. The ornamental trade has had increasing importance in the Italian and European markets in which green cut branches with flowers or colored fruits (berries) were fruitfully commercialized, for the pot plant production and

S.M. Jain and S.J. Ochatt (eds.), *Protocols for In Vitro Propagation of Ornamental Plants*, Methods in Molecular Biology, vol. 589
DOI 10.1007/978-1-60327-114-1_24, © Humana Press, a part of Springer Science+Business Media, LLC 2010

for gardening. Myrtle is also important for the reforestation of coastal zones injured by fires.

In intensive cultivation for the ornamental market, the plants were first propagated by seed or by woody cuttings. Recently, studies on morphological, agronomical, chemical and molecular characterization of different genotypes have been published (1–3). Furthermore this species, for its peculiar characters, can act as a model plant for in vitro culture of woody or semi-woody shrubs.

Micropropagation trials were first reported by Uhring (4) who referred difficulties in the rooting phase. Later several in vitro multiplication protocols were reported (5–11) based on the use of Benzyladenine at several concentration with eventual addition of NAA, IAA, GA_3 or GSH. Experiments with native genotypes from Iran, Portugal, Italy and Spain were reported (5, 12, 13).

It is well known that the quality of plantlets cultured in vitro is the bottleneck affecting the acclimatization stage of micropropagated plants of many species, rooting and acclimatization are the points to be focused in order to obtain protocols suitable for massive commercial propagation systems.

The presence of photosynthetic pigments as chlorophylls a and b has been evaluated in *in vitro* plantlets in *Doritenopsis* in relationship to relative humidity and temperature (14) in order to determine photosynthetic efficiency. Data are presented on the chlorophyll content related to acclimatization performances to have physiological precocious markers for the prediction of ex vitro performances (15).

2. Materials

2.1. Media Preparation for In Vitro Culture

The culture media should be prepared using commercial salt combination in order to standardize the micropropagation:

1. Dissolve Murashige and Skoog (MS) (16) DUCHEFA® cod M 0221.0010 powder containing a micro and macro elements complex (43.0 g) in 9 L of deionized water to prepare a 10× stock solution; while stirring the water, add the powder and stir until complete dissolution; bring the medium to the final volume (10 L) filling water; use 1 L of this solution (4302.09 mg) for each L of culture medium.
2. Dissolve MS vitamins (16) DUCHEFA® cod M 0409.0250 powder containing 25.80 g mixed vitamins, to prepare 250 mL 1,000× stock solution in deionized water and stir until completely dissolved, eventually warm the solution up to 30°C. Use 1 mL vitamin stock solution (1.032 g) for each L of culture medium.
3. Add any desired stable supplement (sucrose, ascorbic acid and citric acid) weighting the powder and dissolve by stirring.

4. Growth regulator stocks: To prepare a 1 mg/mL stock solution: accurately weight 100 mg of the plant growth regulator 6-Benzylaminopurin (BA), Indole-3-butyric acid (IBA), Indole-3-acetic acid (IAA) and add in a volumetric flask 20–30 mL ethanol 99.8%. Gently stir up to a complete solution; once completely dissolved, bring slowly the volume to 100 mL with distilled water at room temperature. Store the stock solution as recommended (for BA, IBA, 4°C; for IAA in dark). Add the PGR as requested for each propagation phase.
5. After adding all the components, while stirring, adjust the pH of the medium using NaOH 1 N or HCl 1 M.
6. Add 8 g/L of any technical agar, heat until clarity of the solution stirring the medium on an electric plate or heat in a microwave.
7. Dispense the medium into the culture vessels preparing 16 vessels of 500 mL capacity for each L of medium; 62.5 mL medium for each vessel
8. Sterilize the medium in autoclave at 121°C, 1 atm (1.01325 bar) for 20 min and allow the medium to cool and solidify prior plant inoculation.

2.2. Sterilization Agents

1. Ethanol 70% prepared with 70 mL ethanol 99.8% and 29 mL distilled water.
2. Aqueous NaOCl solution prepared by mixing one part of the commercial bleach hypochlorite (5% free chlorine) and three parts distilled water (1.25% free chlorine) plus two drops Tween 20®as surfactant.

2.3. Chlorophyll Determination

1. Weigh 100 mg fresh leaves.
2. Grind and homogenized with mortar and pestle.
3. Extract the chlorophyll in darkness with 80% acetone.
4. Measure and read the filtered extract solution at 645 and 663 nm using a UV-VIS spectrophotometer.
5. Determine the chlorophyll a and b according to Lichtenthaler and Buschmann (17).

3. Methods

3.1. In Vitro Establishment, Multiplication and Rooting

1. Excise shoot apexes and internodes of young branches of *Myrtus* ornamental genotypes from selected adult plants.
2. Rinse the material first in a detergent solution with few drops of liquid dish soap, then sterilize with 70% ethanol for 30 s and with the NaOCl solution, 1.2% of active chlorine, for 20 min and rinse twice.

3. Culture the aseptic explants on the medium for *in vitro* establishment (Table 24.1). Containing the antioxidant agents, 100 g/L ascorbic acid and 10 g/L citric acid that were added in the first two subcultures to avoid the oxidation of the exudates from the basal part of the explants.
4. Cut the explants dividing apex and internode, cut and discard the explant base, transfer both explant types to multiplication medium (Table 24.1) and subculture every 28 days. Set the growth chamber conditions for the multiplication phase at 23°C ± 1 with a photoperiod of 16 h of light at a PPFD of 30 μmol/m^2/s. Calculate the multiplication rate as SEM (shoot *per* explant *per* month) at the end of each subculture.
5. Separate the shoots only longitudinally and then put onto 0.5 mg/L IBA or IAA 0.5 mg/L (7) for root induction (*see* Note 1) or 0.5 mg/L IAA+0.1 mg/L BA for rooting and biomass formation (15) (*see* Note 2), light intensity of PPFD 25 μmol/m^2/s or less (15 μmol/m^2/s PPFD) and at 16 h photoperiod (*see* Note 3) containing 15 or 30 g/L sucrose (*see* Note 4).

3.2. Acclimatization

1. After 40 days, evaluate plant fresh weight, cluster height, N. of leaves per shoot, N. of shoot per cluster in order to calculate a biomass index (*see* Note 5).
2. Before the transfer to the acclimatization greenhouse, combine the chlorophyll (a + b) content with the Biomass Index,

Table 24.1
Media composition for in vitro propagation of *Myrtus* ornamental genotypes

	In vitro establishment	Multiplication	Rooting	Rooting and biomass increase
MS (1962) salts	1 L	1 L	1 L	1 L
MS (1962) vitamins	1 mL	1 mL	1 mL	1 mL
Ascorbic acid	100 mg/L	–	–	–
Citric acid	10 mg/L	–	–	–
BA	0.5	0.5	–	0.1
IAA	0.2	0.2	0.5	0.5
IBA (alternative to IAA)	–	–	*0.5*	–
Sucrose	30 g/L	30 g/L	15 or 30 g/L	30 g/L
Technical agar	8 g/L	8 g/L	8 g/L	8 g/L
pH	5.6	5.7	5.7	5.7
PPFD	30 μmol/m^2/s	30 μmol/m^2/s	30 μmol/m^2/s	25 μmol/m^2/s

Table 24.2
Evaluation of the acclimatization and rooting percentages and the chlorophyll content (mg/g of fresh weight) related to the presence of different PGR combinations in the medium and to the PPFD supplied during culture; averages of 3 genotypes. Within each column, values with different letters are different at *p*=0.05 (SKN test). From ref. 15

	Acclimatization (%)			Rooting(%)			Chlorophyll(mg/g f.w.)		
PPFD µmol/m²/s	100	50	25	100	50	25	100	50	25
MS0	0	9.4	0	1.03	5.16	12.4	0.76 ab	2.71 c	1.81 b
BA (0.1 mg/L)	61.5	100	75	0	0	0	1.69 b	0.89 a	1.55 b
IAA (0.5 mg/L)	11.8	25	0	8.30	21.8	32.1	0.28 a	1.33 b	0.39 a
IAA (0.5 mg/L +BA 0.1 mg/L)	88.9	75	50	1.03	3.09	0	1.57 b	1.43 b	1.27 b

with this value it is possible to understand the possible acclimatization performance also for the non-rooted clusters (*see* Note 6). For *Myrtus* ornamental genotypes, the acclimatization performances are better related to the biomass total fresh weight and the total chlorophyll content per cluster rather than the rooting percentage shown at the end of the in vitro culture (Table 24.2).

3. Transfer the rooted and non-rooted shoots to a 40% peat: 60% perlite substrate (by volume) and place for 60 days in the greenhouse with a relative humidity of 80% supplied by a fog or mist system (10 s every 30 min) for the first 30 days then decrease the humidity up to 60%. The mean temperature can vary between 20 and 25°C and lighting can be supplied by natural light with 50% shade provided by polyester-aluminum net.
4. If a fog greenhouse is not available, cover the plants with a paper foil tunnel placed under mist for the first 15 days (see Note 7).

4. Notes

1. The rooting aptitude varied among the genotypes (Table 24.3); the particular miniaturized architecture can affect the poor rooting efficiency either in case of cuttings or in the case of in vitro rooting. The best rooting percentages were achieved in the presence of 0.5 or 1 mg/L IBA, or 0.5 mg/L IAA (Fig. 24.1). The two auxins induced a different root morphology: with IBA, microshoots produced short, big and fleshy roots, with IAA, the microshoots produced long and thin roots.

Table 24.3
Evaluation of micropropagation-related features during subculture with the Rooting and biomass increase medium: multiplication rate (SEM = shoot *per* explant *per* month), rooting percentage, acclimatization percentage and chlorophyll content (mg/g of fresh weight) related to the genotype, average of all treatments. Within each column, values with different letters are different at $p=0.05$ (SKN test). From ref. 15

	Multiplication rate (SEM)	Rooting (%)	Acclimatization (%)	Chlorophyll (mg/g f.w.)
Clone I	5.52 c	1.71 a	85.0 c	1.50
PA2	1.84 a	7.93 b	59.2 b	1.75
CG5	2.30 b	18.64 c	39.8 a	1.90

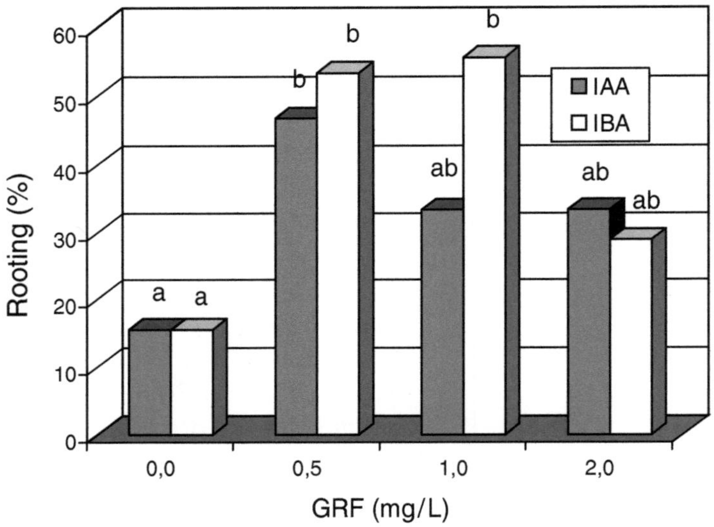

Fig. 24.1. Rooting percentages in the presence of different concentrations of IAA or IBA. From ref. 13.

2. In the case of medium containing a small amount of BA, shoot proliferation occurred during rooting phase (Fig. 24.2).
3. Testing several treatments with different PPFD, the multiplication rate appeared dependent on the clone (Table 24.2); the height of the explants decreased significantly at the increase of PPFD (Fig. 24.3) and at the highest light intensity, the multiplication rate value was significantly lower than at the other conditions. But in any case, the rooting percentage calculated for all of three genotypes is clearly related to the presence of the auxin IAA (Fig. 24.4) and, also in this case, indirectly related to the PPFD levels (Fig. 24.5) (15).

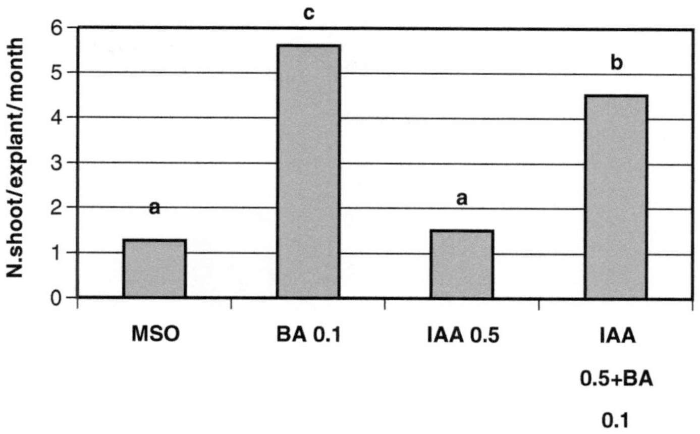

Fig. 24.2. Multiplication rate in the presence of auxin or cytokinin or with the combination of them. From ref. 15.

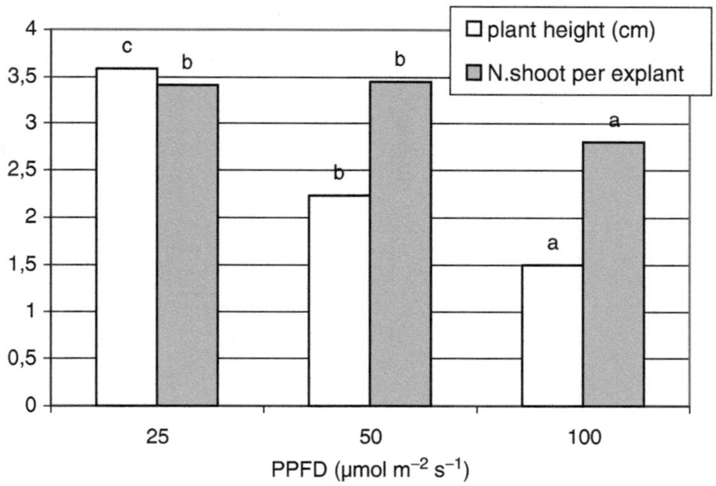

Fig. 24.3. Height of the explants and multiplication rate during culture in the presence of Rooting and biomass induction medium. From ref. 15.

4. In another experiment evaluating the interaction between sucrose and light supply during root induction (13), it was observed that root differentiation at 15 or 30 g/L sucrose was significantly higher than at 5 g/L only at low light level (15 μmol/m^2/s PPFD) (Fig. 24.6). The root length was directly related to the sucrose concentration and slightly improved by a high photon flux but an interaction between sucrose level and light intensity can be evidenced in the number of developed roots (Fig. 24.7).
5. The total biomass index per cluster (Number of leaves/shoot *per* Number of shoots/cluster) is related to the presence of a small amount of BA in the rooting medium and also to the

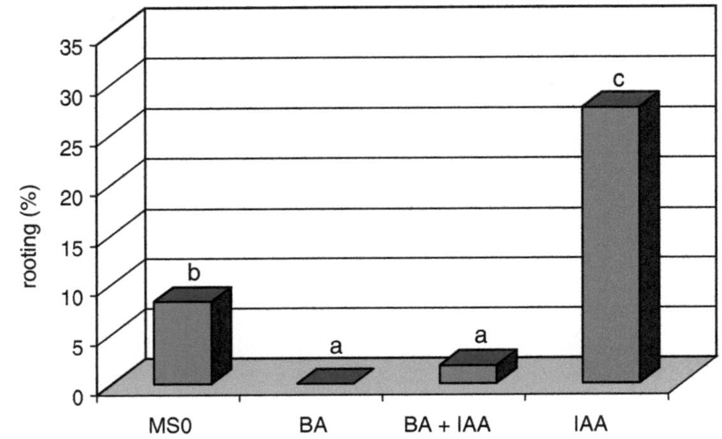

Fig. 24.4. Rooting percentage related to the presence of BA in the medium. From ref. 15.

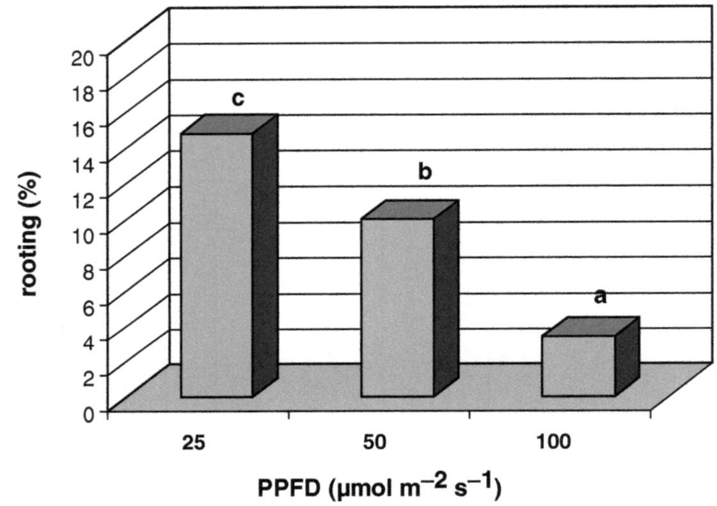

Fig. 24.5. Rooting percentage related to the PPFD supplied during culture. From ref. 15.

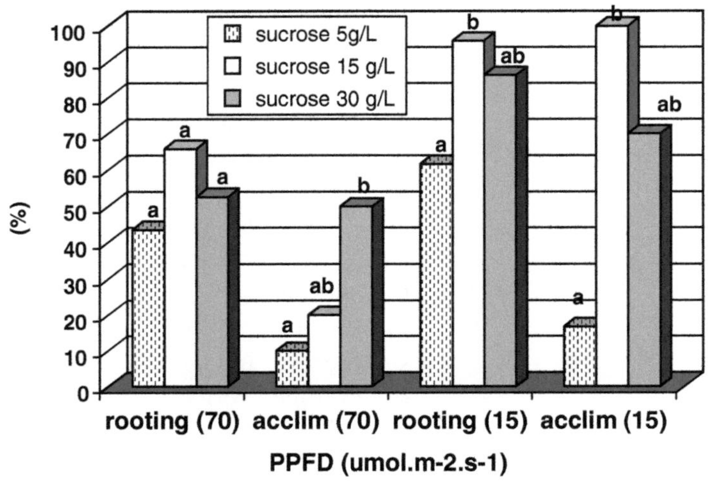

Fig. 24.6. Rooting percentage and acclimatization performances of *Myrtus* genotypes related to sucrose concentration and light intensity supply during culture. From ref. 13.

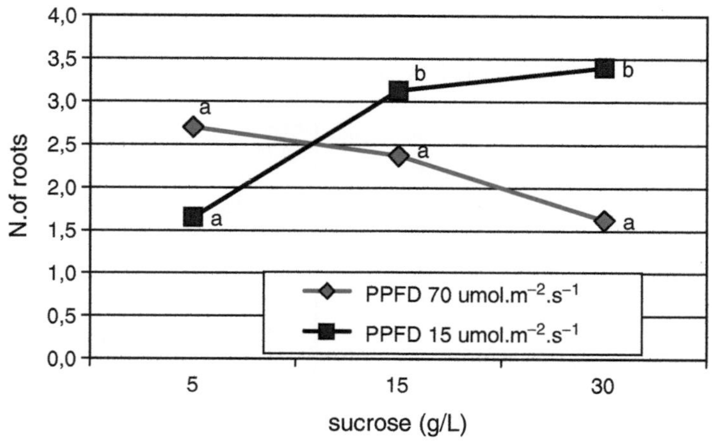

Fig. 24.7. Root number developed in relationship to light intensity and sucrose concentration. From ref. 13.

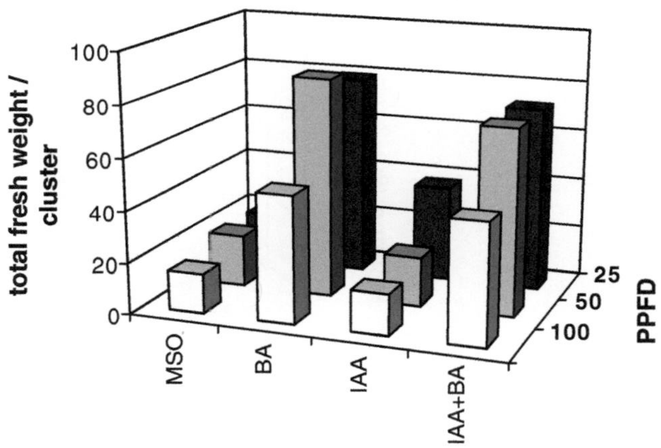

Fig. 24.8. Total biomass fresh weight per cluster (N. of leaves/shoot *per* N. of shoots/cluster) related to the PGR in the media and to the PPFD applied during the subculture. From ref. 15.

applied PPFD; high fresh weights were produced under the light, 50 or 25 μmol/m²/s PPFD (Fig. 24.8).

6. The chlorophylls content varied in absolute values among the genotype tested (Table 24.2), but the same tendency can be observed regarding the light intensity; in effect the lowest chlorophyll amount can be detected in the presence of the highest PPFD, so it is confirmed that low PPFD are able to concentrate the chlorophylls in the cells. There is a an indirect relationship between chlorophylls and light intensity (Fig. 24.9). An interesting effect can be also described for the PGR on chlorophylls content, the highest value detected is in the absence of PGR and the lowest values appeared with only auxins in the media; the addition of Benzyladenine alone or

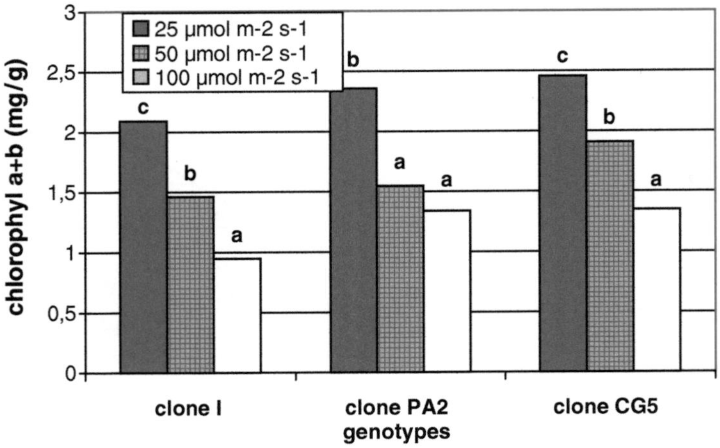

Fig. 24.9. Chlorophills content related to genotype and light intensity. From ref. 15.

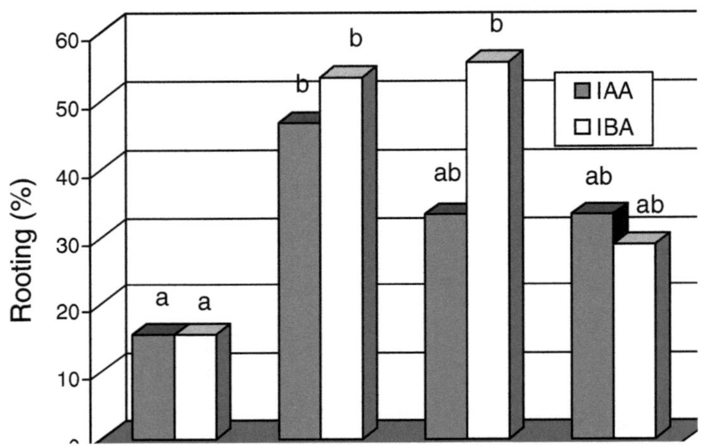

Fig. 24.10. Chlorophylls content related to the presence of BA in the medium. From ref. 15

with IAA increased significantly the pigment presence, and this is particular evident for clone I (Fig. 24.10).

7. The paper tunnel under mist permits to avoid the continuous hydration-dehydration stress during mist treatments and again to avoid the necessity to increase the number of water (mist) supplies that, for this plant, are deleterious and that produce basal rot. After several trials with different materials, it has been demonstrated that the paper cover helped very fruitfully to maintain the microclima on the vegetative part of the plants with a slow variation of the relative humidity between mist supplies (15).

8. Genotypic differences were seen in shoot multiplication and rooting (15).

9. Several ornamental clones (PA2, DCCOD1 and AR6 (7)), characterized by long internodes and large leaves propagated easy while some other clones generally more interesting for their compact shape require more efforts to establish efficient propagation protocols.
10. The acclimatization phase is a complex mix of different variables including the genotypic factor (13).
11. Light intensity supplied during the last phase of in vitro culture can influence the further acclimatization phase (13).
12. It is not really important to transfer *ex vitro* plants already rooted (Figs. 24.11 and 24.12) rather maintain a constant high relative humidity around the aerial part of the shoots to form new roots either by fogging or covering with transparent plastic cups for the first 15 days. A completely acclimatized plant is shown in Fig. 24.13 (15).
13. The total chlorophyll content can be used as marker to evaluate the plant potentiality to survive and growth (15).

Acknowledgments

The results reported in this paper comes from research financed by the Italian Ministry of Agriculture, Fund "Programma di sviluppo per il Mezzogiorno d'Italia: ricerca ed innovazione tecnologica" (Project 205/7303/05 FRUMED "Valorizzazione delle produzioni florovivaistiche del meridione"); publication N.57.

Fig. 24.11. Rooting phase of ornamental Myrtus.

Fig. 24.12. Rooted explants ready for transfer to acclimatization greenhouse.

Fig. 24.13. Acclimatized plant of ornamental *Myrtus* with miniaturized architecture, 60 days after the transfer to the greenhouse.

References

1. Cervelli C, Ruffoni B, Castello S, Savona M (2001) Variability of vegetative traits with ornamental value in spontaneous *Myrtus communis* L. genotypes. Proc Convegno Biodiversità 2: 135–144
2. Mulas M, Perinu B, Francescani AHD, Peana I (2001) Variability on fruit quality of different mirtle genotypes. Atti Convegno Biodiversità Bari 1:103–108
3. Bruna S, Portis E, Cervelli C, De Benedetti L, Schiva T, Mercuri A (2007) AFLP-based genetic relationships in the Mediterranean myrtle (*Myrtus communis* L.). Sci Hortic 113(4):370–375
4. Uhring J (1983) *In vitro* propagation of *Sedum* and *Myrtus* cultivars. HortScience 18(4):453–454
5. Khosh-Khui M, Shekafandeh A, Azarakhsh H (1984) Micropropagation of myrtle. Sci Hortic 22:133–137
6. Nobre J (1994) *In vitro* shoot proliferation of *Myrtus communis* L., from field grown plants. Sci Hortic 58:253–258
7. Ruffoni B, Gazzano A, Costantino C (1994) *Myrtus communis* micropropagation: indolacetic acid effect on rooting. Italus Hortus 1: 8–12
8. Lucchesini M, Massai R, Gucci R (1998) Radicazione in vitro di Mirto (*Myrtus communis* L.) in contenitori permeabili agli scambi gassosi. Atti IV Giornate Scientifiche SOI, Sanremo 1–3 aprile 1:43–44
9. Lucchesini M, Mensuali-Sodi A, Serra G (2001) Propagazione *in vitro* e studio del microambiente in specie mediterranee. Flortecnica 4:66–68
10. Morini S, Freudiani F, D'Onofrio C (2002) Indagini sulla micropropagazione del mirto. Italus Hortus 9(2):41–48
11. Rigodi MP, Cani MR, Chessa G, Zurru R (2004) Prime valutazioni sulla risposta alla coltura in vitro di cloni di *Myrtus communis* L. Italus Hortus 11(4):315–317
12. Parra R, Amo-Marco JB (1996) Effect of plant growth regulators and basal media on *in vitro* shoot proliferation and rooting of *Myrtus communis* L. Biol Plant 38:161–168
13. Ruffoni B, Airò M, Fascella G, Mascarello C, Zizzo G, Cervelli C (2003) Propagation and acclimatization of ornamental *Myrtle* genotypes. Acta Hort 616:255–258
14. Jeon MW, Ali MB, Hahn EJ, Paek KY (2004) Photosynthetic pigments, morphology and leaf gas exchange durino ex vitro acclimatization of micropropagated CAM Doritaenopsis plantlets under relative humidity and air temperature. Environ Exp Bot 55:183–194
15. Mascarello C, Melis R, Mantovani E, Ruffoni B (2009) Evaluation of the role of light intensity and hormonal treatment during *in vitro* rooting of *Myrtus* micropropagated plants related to acclimatization performance. A&MP 2007 SYMP. Acta Hort 812.379–3
16. Murashige T, Skoog F (1962) A revised medium for rapid growth and bioassays with tobacco tissue culture. Physiol Plant 15:478–487
17. Lichtenthaler HK, Buschmann C (2001) Chlorophylls and carotenoids – Measurement and characterisation by UV-VIS. In: Current Protocols in Food Analytical Chemistry *(CPFA)*, Suppl 1. John Wiley, New York, pp F4.3.1–F4.3.8

Chapter 25

Sunflower Propagation

M.L. Mayor, G. Nestares, T. Vega, R. Zorzoli, and L.A. Picardi

Abstract

Sunflower (*Helianthus annuus* L.) has been cultivated both as an oilseed and as an ornamental plant. Several protocols have been described for the micropropagation, direct plant regeneration by organogenesis being acceptable for this plant species. Besides a strong genotype dependency, the type and ontogenic stage of explants, environmental conditions of the culture, and media composition affect sunflower organogenesis. Several problems have hindered the ability to regenerate normal shoots; the most common being hyperhydricity and precocious flowering. This chapter describes a protocol for direct shoot regeneration from cotyledons developed and established in our laboratory, as well as the improvement regenerated shoot quality.

Key words: *Helianthus annuus* L., Micropropagation, Direct organogenesis, Cotyledonary explant, Shoot induction, Hyperhydricity, Precocious flowering, Acclimatization

1. Introduction

Sunflower (*Helianthus annuus* L.) is cultivated mainly as a staple crop for edible oil production, as well as a source of protein and biorenewable energy. However, one of the early uses of this species was as an ornamental plant due to the colorful and variety of its flowers. Furthermore, over the last 10 years, the public interest in ornamental sunflowers increased, allowing an expansion of the market by introducing new varieties for cut flower, pot plant, or gardening purposes.

Sunflower is described as a recalcitrant species to in vitro propagation. Since early 1980s, a variety of techniques to regenerate this species have been described, including the most successful ones: somatic embryogenesis and organogenesis (1–3). Studies

S.M. Jain and S.J. Ochatt (eds.), *Protocols for In Vitro Propagation of Ornamental Plants*, Methods in Molecular Biology, vol. 589
DOI 10.1007/978-1-60327-114-1_25, © Humana Press, a part of Springer Science+Business Media, LLC 2010

on direct regeneration of a wide range of genotypes showed that shoot organogenesis from mature cotyledons occurred without an intermediate stage of cell dedifferentiation (callus formation) (4–7). Plant regeneration by organogenesis is highly variable and dependent on several factors such as genotype, type and ontogenic stage of explants, environmental conditions of the culture, and media composition (1). The genotypic dependence for in vitro manipulation has been reported in the literature, e.g., recalcitrant performance of restorer inbred lines with low organogenic response to in vitro tissue culture (8, 9).

Organogenic regeneration has been obtained from leaves, shoot tips or embryonic axes, immature embryos, hypocotyls, and cotyledons; the last three types being the most commonly used explants in tissue culture of this species (3). Cotyledon explants are frequently used for organogenic regeneration because of the year-round availability, ease of culture initiation, and applicability to a number of genotypes (6). Moreover, this type of explant produces a relatively high frequency of regenerated fertile plants either from ungerminated seeds or from young plantlets. Culture conditions, such as temperature, day length, concentration and type of growth regulators, and gelling agent also have an impact on the quality of the regenerated shoots.

Several problems have hindered the ability to regenerate normal shoots in sunflower micropropagation. In vitro quality of regenerated shoots has been mainly affected by premature flowering and hyperhydricity (1, 2, 10). The appearance of premature flower heads on regenerated shoots creates difficulties in the development and subsequent establishment of a root system; and generally impedes the recovery of viable seeds. Hyperhydric shoots, previously known as vitreous, have a water-soaked, translucent, deformed and (or) fleshy appearance and are very difficult to acclimatize and transfer to soil. Difficulties in rooting and acclimatization have also been mentioned in several protocols (11). Efforts have been made to enhance the quality of regenerated shoots. Silver nitrate added to the culture media diminished hyperhydricity of primordia and shoots (10). Reduced temperature and shorter day length were useful in controlling precocious flowering under in vitro conditions (12). Rooting could be improved by transferring the regenerated shoots to a medium containing only auxins, specifically naphthalene-acetic acid (NAA), plus activated charcoal which avoids callus formation on the base of the shoots (6).

A suitable protocol for direct shoot regeneration from cotyledons has been developed and established in this laboratory and will be described herein. Modifications to the basic protocol for the improvement in the quality of the regenerated shoot are described.

2. Materials

2.1. Stock Solutions

2.1.1. Murashige and Skoog (MS) Basal Salts Solutions (13)

Micronutrient Solution (10X)

1. Dissolve the following amount of salts in 100 mL ddH_2O (double distilled water).
2. Boric acid (H_3BO_3) – 620 mg.
3. Cobalt chloride ($CoCl_2 \cdot 6H_2O$) – 2.5 mg.
4. Cupric sulfate ($CuSO_4 \cdot 5H_2O$) – 2.5 mg.
5. Manganese sulfate ($MnSO_4 \cdot 4H_2O$) – 2,230 mg.
6. Sodium molybdate ($Na_2MoO_4 \cdot 2H_2O$) – 2.5 mg.
7. Zinc sulfate ($ZnSO_4 \cdot 7H_2O$) – 860 mg.

Iron Chelated Solution

1. Dissolve 27.8 g ferrous sulfate ($FeSO_4 \cdot 7H_2O$) in 350 mL ddH_2O. Heating may be required to facilitate dissolution of the salt.
2. Dissolve 37.3 g of Na_2-EDTA in 350 mL ddH_2O. Heating may be required.
3. To obtain the final solution, pour EDTA solution into ferrous sulfate solution.
4. Use ddH_2O to bring the solution to a final volume of 1 L.

The order described above is required to yield a crystal clear yellow solution.

Vitamin Solution

Dissolve the following amount of vitamins in 100 mL ddH_2O:

1. Glycine – 200 mg.
2. Tiamine·HCl – 10 mg.
3. Pyridoxine·HCl – 50 mg.
4. Nicotinic Acid – 50 mg.
5. Myo-inositol 10 g.

MS Basal Solution

1. Weigh the following macronutrients:
   - Ammonium nitrate (NH_4NO_3) – 16.5 g.
   - Potassium nitrate (KNO_3) – 19.0 g.
   - Potassium phosphate monobasic (KH_2PO_4) – 1.70 g.
   - Magnesium sulfate·$7H_2O$ ($MgSO_4 \cdot 7H_2O$) – 3.70 g.
   - Calcium chloride·$2H_2O$ ($CaCl_2 \cdot H_2O$) – 4.40 g.
2. Stir 300 mL ddH_2O in an Erlenmeyer and add macronutrients in the above order. Before adding each macronutrient, make sure that the added salt is completely dissolved.
3. Add 10 mL Potassium Iodide (KI, 830 mg/L, stored in brown glass bottles at 4°C), 10 mL micronutrient Solution, 10 mL Iron chelated solution, and 10 mL vitamin Solution.

4. Add ddH_2O to bring the solution to a final volume 1,000 mL.
5. Aliquot in 10 flasks each of them with 100 mL solution and store at −20°C.

2.1.2. Growth Regulators Stock Solutions

2.1.2.1. Preparation of 1 mg/mL Indole Acetic Acid (IAA) and Napthalene Acetic Acid (NAA) Stock Solution

1. Weigh 100 mg IAA (Sigma®) or 100 mg NAA (Sigma®) and pour into a 100 mL volumetric flask or other glass container.
2. Dissolve the growth regulator by adding 2–5 mL 96° Ethanol.
3. Once completely dissolved, bring the volume with warm ddH_2O (40°C). Stir the solution until ethanol evaporates. Store at 4°C.

2.1.2.2. Preparation of 1 mg/mL Kinetin (KIN) Stock Solution

1. Dissolve 100 mg KIN (Sigma®) in 2–5 mL 1 N hydrochloric acid (HCl).
2. Add 80 mL ddH_2O (60°C). Once at room temperature, adjust volume to 100 mL with ddH_2O. Store at 4°C.

2.2. Culture Media

1. Add 50 mL 10× MS basic salts solution from Subheading 25.2.1.1 in 700 mL ddH_2O.
2. Pour 10 g sucrose and stir until completely dissolved.
3. Bring the medium to the final volume of 1 L by adding ddH_2O.
4. Adjust the solution to pH 6.0 using 0.1 N sodium hydroxide (NaOH).
5. Add 8 g Agar-Agar (Plant Tissue Culture Grade, Sigma®), heat at 100°C until the solution have a clear appearance.
6. Aliquot in "*baby-food*" flasks (20–25 mL) or in test tubes (10 mL), seal with aluminum foil and autoclave for 20 min at 121°C.

2.2.2. Induction Medium

1. Dilute 100 mL 10× MS basic salts solution from Subheading 25.2.1.1 in 700 mL ddH_2O.
2. Pour 30 g sucrose and stir until completely dissolved.
3. Add growth regulators: 1 mL IAA stock solution (from 2.1.2.1) and 2 mL KIN stock solution (from 2.1.2.2).
4. Bring the medium to a final volume 1 L with ddH_2O.
5. Once the final volume is achieved, adjust medium pH to 6 using 0.1 N NaOH.
6. Add 9 g Agar-Agar (Sigma, Plant Tissue Culture Grade). Heat until the solution is crystal clear (*see* Note 1).
7. Aliquot in Erlenmeyer flasks, seal with aluminum foil and autoclave (121°C, 20 min).
8. Once sterilized pour in sterile Petri dishes.

2.2.3. Shoot Development Medium

1. To prepare this medium use the same protocol as Subheading 25.2.2.2 but omit step 3.
2. Before sterilization aliquot in test tubes (10 mL), seal with aluminum foil and autoclave for 20 min at 121°C.

2.2.4. Rooting Medium

1. To prepare this medium, use protocol Subheading 25.2.2.2 but, in step 3 replace 1 mL IAA by 1 mL NAA. Do not use KIN.
2. Before sterilization aliquot in test tubes (10 mL), seal with aluminum foil and autoclave for 20 min at 121°C.

3. Methods

3.1. Germination of Seeds and Type of Explant to Be Used

1. Submerge achenes into distilled water for 15 min to facilitate the removal of pericarps.
2. Sterilize dehulled seeds by soaking in 96° ethanol for 5 s, followed by submersion in 50% (v/v) commercial bleach for 15 min. The addition of a wetting agent such as Tween 20 is recommended to reduce surface tension and to better surface contact. (*see* Note 2).
3. Under sterile conditions (sterile flux camera) decant sterilization solution and rinse the seeds three times with sterile distilled water (*see* Note 3).
4. Place seeds on tubes containing germination medium and incubate at 25 ± 2°C for 2 days in darkness and then for another 2 days at a 12 h photoperiod, 50 µmol/m^2/s.
5. Cotyledons are excised from the seedlings. The entire embryo is removed and discarded.
6. Cuts are made at 1 mm of the distal end of each cotyledon, and the resulting central sections are divided in three parts obtaining (from the cotyledonary node area to the distal area): proximal (P), middle (M), and distal (D) regions. Each of them could be further subdivided longitudinally into two segments (*see* Note 4) (Fig. 25.1).

3.2. Primordia and Shoot Induction

1. Proximal cotyledonary regions tend to have the highest regeneration potential (14). Hence, these regions are used as explants and placed in petri dishes with abaxial side in contact with the induction culture medium (*see* Note 5).
2. Petri dishes are sealed with plastic wrap (PVDC film) and incubated for 21 days at controlled conditions: 25 ± 2°C; 12-h photoperiod; 50 µmol/m^2/s.
3. After the incubation period, three types of response can be obtained: organogenesis (primordia, shoots and/or roots),

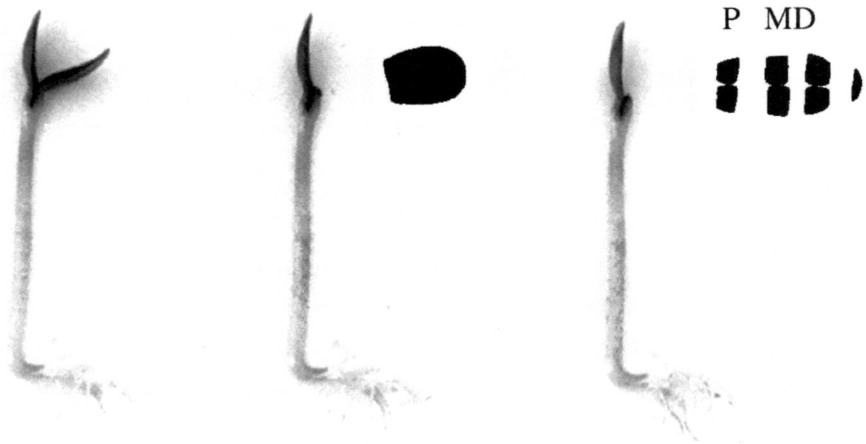

Fig. 25.1. Explant production for shoot induction initiation. From left to right: 2-day old plantlet with cotyledons, detached cotyledon, and consequent subdivision in explants. Cuts are shown transversally and along the main axis, yielding two explants each region. Proximal (P), middle (M) and distal (D) cotyledonary regions are indicated. Figure modified from ref. 14.

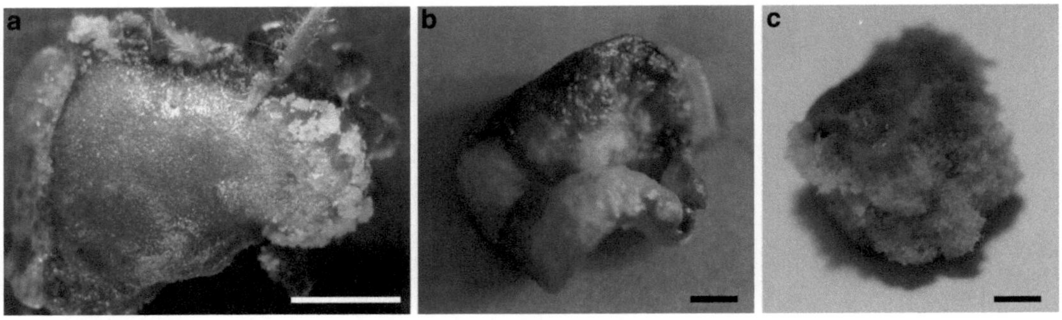

Fig. 25.2. Type of responses obtained after 21 days of incubating cotyledonary explants in the induction medium: (**a**) primordia development: depending on its size and appearance, primordia can be transferred to shoot development medium; (**b**) hypertrophy: primordia and (or) shoots are not developed; (**c**) callus formation: friable callus formation.

hypertrophy, and callus (Fig. 25.2). Hypertrophic explants present an increased cellular volume (swelling) and also may show necrosis from the borders towards the center of the explant. No shoots or primordia can be obtained from such tissues.

4. Shoots and primordia could be either normal or abnormal. The most common abnormal types of shoots are hyperhydric and flowered (Fig. 25.3). The addition of silver nitrate and modification of the day length and temperature could reduce hyperhydricity as well as precocious flowering. The corresponding protocols are described in Notes 6 and 7.

3.3. Shoot Development

1. Once the incubation period finished, petri dishes with explants are open and examined under flux chamber. Those explants that developed small shoots and primordia (less than 5 mm) are transferred to the shoot development media (Fig. 25.4a, b, c).

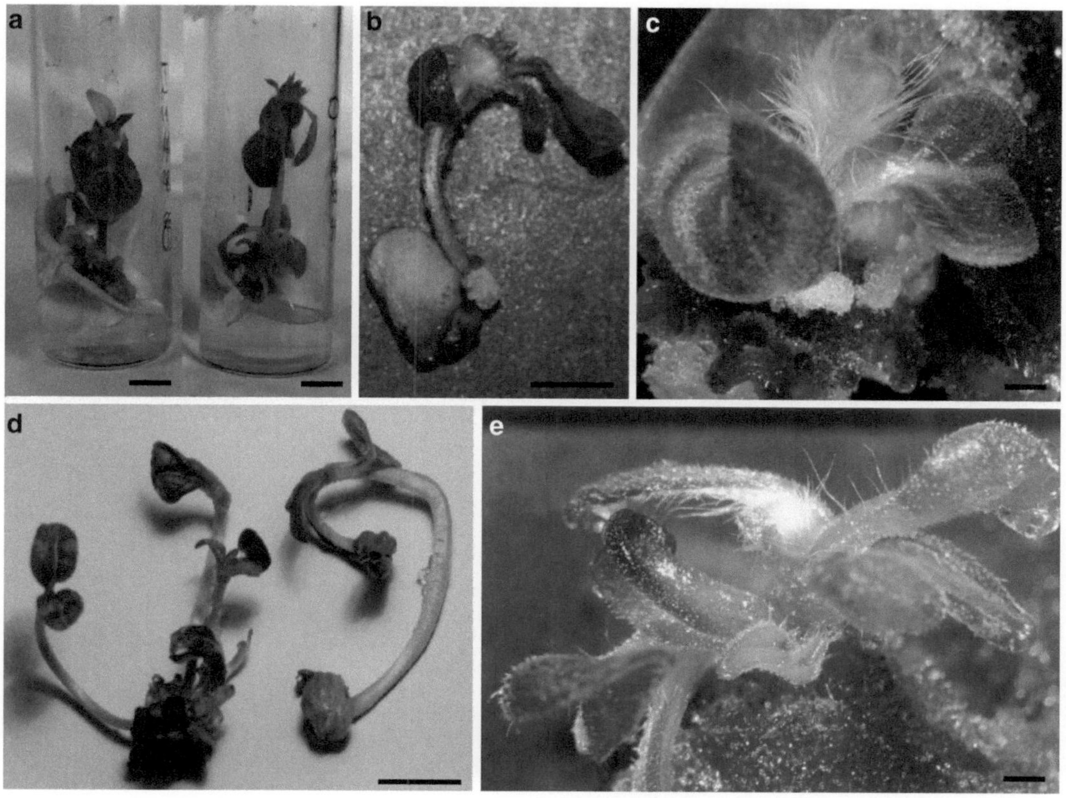

Fig. 25.3. Induced shoots with abnormal development: (**a–c**) precocious flowering: from left to right sunflower heads in elongated shoots, some of them also developed ray flowers. (**d**, **e**) hyperhydricity: note the glassy and deformed appearance of shoots. Extreme cases of hyperhydricity developed a grassy appearance of otherwise normal shoots.

2. Allow the transferred explants to elongate in the shoot developing medium for a maximum period of 2 weeks (Fig. 25.4d, e, f). This medium does not contain growth regulators to reduce the appearance of hyperhydric shoots and (or) roots.
3. Commonly, an explant can yield multiple shoots and (or) primordia. This type of explants can be partitioned into sub-explants and transferred to several test tubes containing the shoot developing medium.

3.4. Rooting

Normal shoots with no roots developed can be transfer to test tubes containing rooting media described in Subheading 25.2.2.4 to allow development of the rooting system in previous acclimatization (Fig. 25.5a, b).

3.5. Acclimatization and Soil Transfer

1. Remove PVC from tubes and leave them 1–2 days in the growing chamber at the same temperature and day length (Fig. 25.5c).
2. Remove regenerated plantlets and wash agar debris. Plantlets are left in test tubes with distilled water for 3–4 days, under the same conditions described before (Fig. 25.5d).

Fig. 25.4. Primordia and shoot's induction and development: (**a**–**c**) shoots and (or) primordia regenerated on induction culture medium; (**d**–**f**) shoots elongated on shoot development medium.

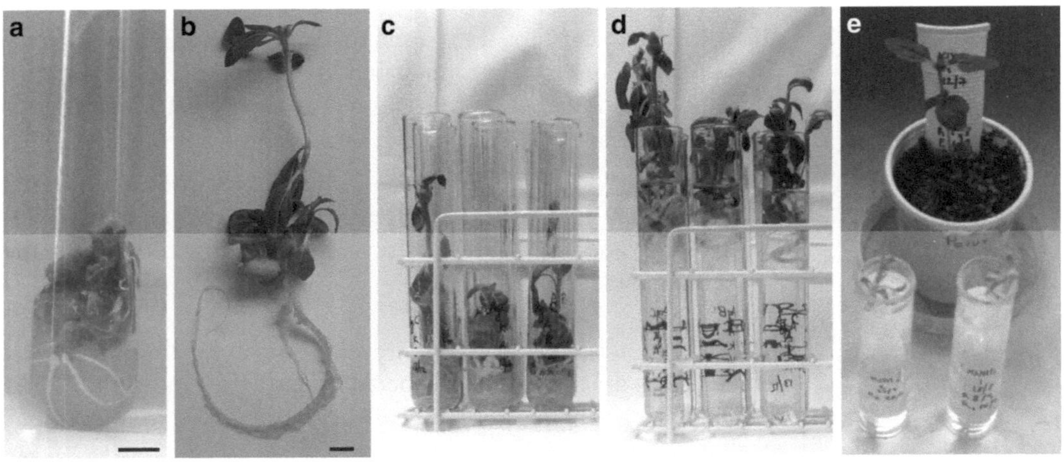

Fig. 25.5. Rooting, acclimatization, and hardening of micropropagated shoots and plantlets. (**a**) Shoots placed on rooting media ready for the next step in the acclimatization process. (**b**) Normal shoot with elongated rooting system, previous hardening. (**c**) Normal shoots in test tubes with rooting media; plastic cover has been removed as part of the acclimatization process. (**d**) Normal shoots in test tubes with ddH_2O. Agar-agar debris was removed. (**e**) Normal plantlet transferred to soil.

3. Transfer plantlets to small pots containing sterile soil and perlite (2:1 ratio) (Fig. 25.5e). Cover the pots with a plastic bag and incubate for a period of 7–10 days in the growth chamber. This will reduce the stress during this period. The plastic bag keeps the moisture generating a suitable environment for the plantlet to survive and develop. By day 7, remove the plastic bag and let the plants adjust to the new environment until day 10 (Note 8).
4. Transfer plant to pots with regular soil and move them to the greenhouse at 22–25°C, 12 h day length.

4. Notes

1. The type and concentration of gelling agent is very important for in vitro regeneration in this species. Agar-Agar has proved to be better than other gelling agents in promoting shoot induction as well as in controlling hyperhydricity. An increase in agar concentration can help in reducing hyperhydricity yet it could affect the regeneration ratio.
2. Achenes can be also sterilized using the same protocol as described for the dehulled seed. In some inbred lines and hybrids, sterilization of achenes was useful to reduce contamination from the pericarp.
3. Cotyledons from mature seed can be used as explants. In this situation, germination of seeds can be avoided, and the seeds are ready to be sterilized.
4. Cotyledons from mature seed can be excised to be placed in induction media. Proximal regions can be obtained as described in Subheading 25.3.1, step 6. It is important to remove the seed teguments before planting since they could have a detrimental effect on regeneration.
5. Even though both faces of the cotyledon are efficient in regenerating shoots, it is recommended first to screen the regeneration ability of the genotypes at both cotyledonary faces. In some genotypes, middle regions also have some regeneration potential and could also be cultivated to regenerate shoots.
6. Silver nitrate added to the culture media can reduce hyperhydricity in genotypes with high regeneration ability. Protocol Subheading 25.2.2.2 can be modified by adding 5 μM silver nitrate after step 3. Although this amount of silver nitrate is very conservative, it balances between a number of induced shoots and percentage of hyperhydric shoots. If the genotype has high regeneration ability, the amount of silver nitrate can be increased to 7.0 μM. Despite decreasing the number of

regenerated shoots, an increase in the level of silver nitrate raise the amount of normal shoots obtained. The use of silver nitrate during shoot primordia induction could also have a positive effect on later plant development at the shoot elongation step.

7. The average rate of flowering plants is genotype-dependent. Reducing the temperature (20°C) and shortening day length (8–16 h day/night) reduced in vitro flowering.
8. Better results could be obtained when incubation is carried out in growth chamber with moisture control.

References

1. Alibert G, Aslane-Chanabé C, Burrus M (1994) Sunflower tissue and cell culture and their use in biotechnology. Plant Physiol Biochem 32:31–34
2. Bidney DL, Scelonge CJ (1997) Sunflower biotechnology. In: Sunflower technology and production. Ch. 8. ASA-CSSA-SSSA, Madison, Wisconsin, USA
3. Fambrini M, Fisichella M, Pugliesi C (2001) Enhanced morphogenetic potencial from *in vitro* regenerated plants of genus *Helianthus*: An overview. Recent Res Dev Plant Biol 1:35–54
4. Ceriani MF, Hopp HE, Hahne G, Escandón AS (1992) Cotyledons: an explant for routine regeneration of sunflower plants. Plant Cell Physiol 33:57–164
5. Nestares G, Zorzoli R, Mroginski L, Picardi L (1996) Plant regeneration from cotyledons derived from mature sunflower seeds. Helia 19:107–112
6. Baker CM, Muñoz-Fernandez N, Carter CD (1999) Improved shoot development and rooting from mature cotyledons of sunflower. Plant Cell Tiss Org Cult 58:39–49
7. Vega T, Nestares G, Pratta G, Zorzoli R, Gattuso S, Picardi L (2007) Biochemical and histological changes associated to *in vitro* responses in sunflower cotyledonary explants. In Vitro Cell Dev Biol Plant 43:415–422
8. Nestares G, Zorzoli R, Mroginski L, Picardi L (2002) Heritability of *in vitro* plant regeneration capacity in sunflower. Plant Breeding 121:366–368
9. Mayor ML, Nestares G, Zorzoli R, Picardi LA (2006) Analysis for combining ability in sunflower organogenesis-related traits. Aust J Agric Res 57:1123–1129
10. Mayor ML, Nestares G, Zorzoli R, Picardi LA (2003) Reduction of hyperhydricity in sunflower tissue culture. Plant Cell Tissue Organ Cult 72:99–103
11. Hazarika BN (2006) Morpho-physiological disorders in *in vitro* culture of plants. Sci Hort 108:105–120
12. Ivanov P, Enchiva J, Ivanova I (1998) A protocol to avoid precocious flowering of sunflower plantlets *in vitro*. Plant Breed 117: 582–584
13. Murashige T, Skoog F (1962) A revised medium for rapid growth and bioassays with tobacco tissue cultures. Physiol Plant 15:473–497
14. Vega T, Nestares G, Zorzoli R, Picardi L (2006) Responsive regions in sunflower cotyledons. Acta Physiol Plant 28:427–431

Chapter 26

Clonal Propagation of *Cyclamen persicum* Via Somatic Embryogenesis

Traud Winkelmann

Abstract

Cyclamen (*Cyclamen persicum*) is an economically important ornamental pot plant with local use as cut flower as well. Traditionally, it is propagated via seeds, but interest is given in vegetative propagation of parental lines as well as superior single plants. Somatic embryogenesis is an efficient in vitro propagation method for many cyclamen cultivars. Starting from ovules of unpollinated flowers, callus is induced and propagated in a medium containing 2,4-dichlorophenoxyacetic acid (2,4-D) and 6-(γ,γ-dimethylallylamino) purine (2iP). Transfer to hormone-free medium results in the differentiation of somatic embryos, which afterwards germinate on the same medium. These first culture stages take about 6–7 months and are carried out in complete darkness. Two to four months after the transfer to light, plantlets develop which can be acclimatized in the greenhouse. The regenerated plants are characterized by low percentages of somaclonal variation. This protocol has proven useful not only for clonal propagation, but also for artificial seed preparation, cryopreservation, genetic transformation and protoplast regeneration.

Key words: In vitro, Multiplication, Ornamental, Regeneration, Somaclonal variation

1. Introduction

In Europe and Japan, cyclamen is well known as an ornamental pot plant mainly for indoor use, and the annual worldwide production is estimated at 150–200 million plants. Conventionally, cyclamen are propagated via seeds, which are produced by manual pollination and, therefore, are expensive (up to 0.20 € per seed). Since cyclamen are suffering from inbreeding depression, the production and propagation of parental lines of F_1 hybrids dominating the market are difficult. Thus, breeders are interested in the vegetative propagation of their breeding lines. Moreover, if an efficient, reasonably priced vegetative multiplication became reality, breeders could produce clonal varieties just by multiplying selected elite plants.

S.M. Jain and S.J. Ochatt (eds.), *Protocols for In Vitro Propagation of Ornamental Plants*, Methods in Molecular Biology, vol. 589
DOI 10.1007/978-1-60327-114-1_26, © Humana Press, a part of Springer Science+Business Media, LLC 2010

Cyclamen cannot be propagated from cuttings or tuber division, thus in vitro techniques have to be applied for vegetative propagation. In this paper, a protocol for multiplication via somatic embryogenesis will be detailed. The elaborate literature overview is included in previous publications (1–3), while this chapter contains the step-wise description of the complete regeneration pathway. In addition, the embryogenic cells are a well-suited material for cryopreservation (4) and a source for protoplast isolation (5, 6) and currently their use for genetic transformation by *Agrobacterium tumefaciens* is being investigated. First steps towards the production of artificial seeds have been done (7–10).

The process of somatic embryogenesis can be divided into distinct stages with different requirements. First competent cells have to be identified from which embryogenic cells can be initiated. In the case of cyclamen ovules from closed, unpollinated flower buds are the optimal explant type, because (a) they are available on adult plants, which have been tested for their breeding or economic value, (b) they can be harvested without the destruction of the plant, and (c) and most important, they very rarely give rise to contaminations in tissue culture. This first induction phase, which needs the presence of the strong synthetic auxin 2,4-D (2,4-dichlorophenoxyacetic acid), results in cyclamen in the formation or embryogenic callus, and therefore, this regeneration pathway has to be regarded as indirect somatic embryogenesis. As long as the embryogenic cells are cultivated on 2,4-D containing medium, they divide and proliferate, while the formation of somatic embryos is inhibited. If mass propagation is aspired, this can only be achieved by the propagation of embryogenic cells, either as callus cultures on solid medium or as suspensions in liquid medium ((11), *see* Note 1).

The second step in somatic embryogenesis involves the realization phase, in which somatic embryos differentiate and develop through characteristic stages, i.e., globular, heart, torpedo and cotyledon stage. In cyclamen, one of the two cotyledons is suppressed very early, so that zygotic as well as somatic embryos develop only one cotyledon. Cyclamen require darkness for germination, and thus all steps in somatic embryogenesis should be carried out in darkness as well. In the last step, mature somatic embryos germinate into small plantlets, which can be adapted to light and transferred to soil.

2. Materials

2.1. Surface sterilization

1. Ethanol (70%).
2. Sodium hypochlorite (NaOCl, 2.4–2.8% active chlorine): Dilution of commercially available NaOCl solution (12–14%

active chlorine) with tap water: 1:5, plus one droplet of Tween 20 (detergent) per 250 ml. Care should be taken with this bleaching and caustic solution, which should be prepared freshly on demand.

3. Autoclaved deionized water in sterile beakers.

2.2. Culture Media and Sieves

1. Callus induction medium for culture of ovules and for callus culture and maintenance (Table 26.1). Store at room temperature in darkness for 2 months max.

Table 26.1
Composition of culture media, all based on Murashige and Skoog (MS) medium *(14)*

Substance	Callus induction medium	Differentiation medium
Macroelements (1/2 MS)	(mg/L)	(mg/L)
NH_4NO_3	825	825
KNO_3	950	950
$CaCl_2 \times 2H_2O$	220	220 CaCl2
$MgSO_4 \times 7H_2O$	185 MgSO4	185
KH_2PO	85 KH2PO4	85
Microelements (1/2 MS)	(mg/L)	(mg/L)
H_3BO_3	3.1	3.1
$MnSO_4 \times 1H_2O$	8.45	8.45
$ZnSO_4 \times 7H_2O$	5.3	5.3
KJ	0.415	0.415
$Na_2MoO_4 \times 2H_2O$	0.125	0.125
$CuSO_4 \times 5H_2O$	0.0125	0.0125
$CoCl_2 \times 6H_2O$	0.0125	0.0125
	(mg/L)	(mg/L)
Fe EDTA	34.51	34.51
	(mg/L)	(mg/L)
Peptone	250 Peptone	250 Peptone
myo-Inositol	100	100
Glycine	2	2
Nicotinic acid	0.5	0.5
Thiamine HCl	0.1	0.1
Pyridoxine HCl	0.5	0.5
	(g/L)	(g/L)
Sucrose	30	30
D(+)-glucose monohydrate	2	2
	(mg/L)	(mg/L)
2,4-D	2.0	–
2iP	0.8	–
	(g/L)	(g/L)
Gelrite® (Duchefa)	3.7	3.7
pH	5.5–5.6	5.5–5.6

2. Differentiation medium for regeneration and germination of somatic embryos and culture of plantlets (Table 26.1). Store at room temperature in darkness for 2 months max.
3. Both media can be prepared as liquid media by just omitting the Gelrite®.
4. Autoclavable sieves of 500 and 1,000 μm mesh size (*see* Note 14).

2.3. Transfer to Soil

1. Lukewarm tap water.
2. Pricking soil mixed 2:1 with perlite.

3. Methods

3.1. Surface Sterilization and Explant Preparation

1. Harvest flower buds about 2 days before anthesis from healthy, well-nourished plants (*see* Note 2).
2. Flower buds are surface sterilized on a magnetic stirrer for 30 s in ethanol (70%), then 20 min in NaOCl (2.4–2.8%) followed by three rinses in sterile water for 5 min each. They should remain in the last water bath until the use for ovule preparation (*see* Note 3).
3. From the outside to the centre, the whirls are cut off step-wise (Fig. 26.1a), until the ovary is reached. The ovary coat is removed by a flat section, so that the ovules become visible (Fig. 26.1a, right).
4. Preparation of ovules is best done with two preparation needles, one to fix the peduncle (which should have 8–10-mm length now), and the other one to smoothly touch the ovules individually. They will easily stick to the needle and can be put onto callus induction medium (*see* Note 4).
5. 25 ovules are placed in one 6-cm Petri dish containing 12 ml of callus induction medium, which is then sealed with Parafilm® (*see* Note 5).
6. The Petri dishes with the ovules are incubated at 22–25°C in complete darkness for 8 weeks. We normally prepare 50 ovules per flower bud and two buds per genotype. This is repeated three times on three different dates, so 300 ovules are prepared in total per genotype, in order to be able to evaluate the regeneration ability.

3.2. Callus Culture and Maintenance

1. Usually, after 8 weeks, callus has formed usually on 60–80% of the ovules (Fig. 26.1b). Different types of callus can be distinguished regarding colour and consistency, ranging from dark brown to yellow and from hard and compact to soft.

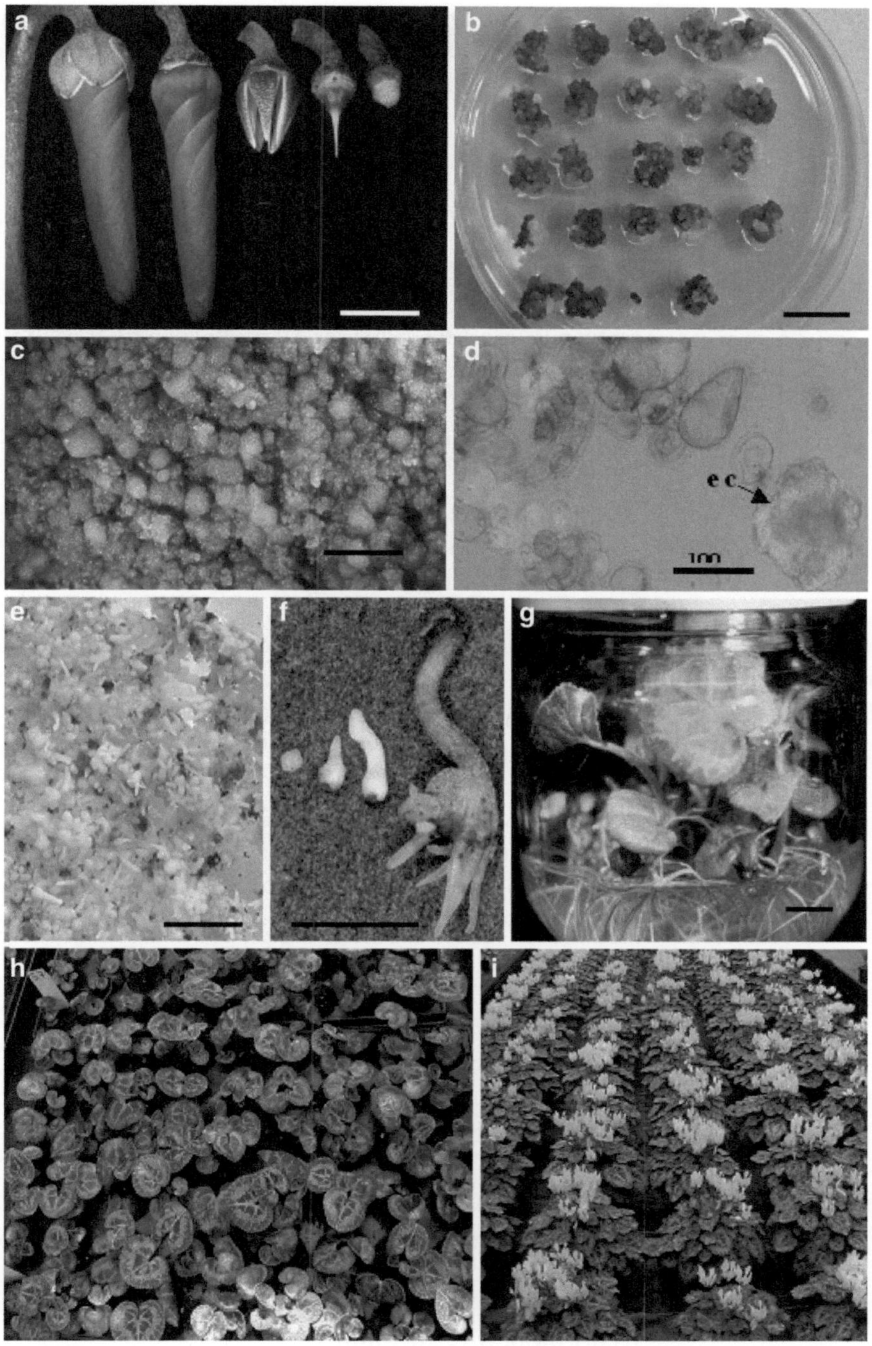

Fig. 26.1. Somatic embryogenesis in *Cyclamen persicum*. (**a**) Preparation of ovules, (**b**) Callus induction after 8 weeks of culture, (**c**) Embryogenic callus (*top view*), (**d**) Embryogenic callus stained with fluorescein diacetate under fluorescent light (microscopic view, e c = embryogenic cells), (**e**) Differentiation of somatic embryos after 4 weeks on hormone-free medium, (**f**) Typical developmental stages of somatic embryos, (**g**) Young plants ready for acclimatization, (**h**) Acclimatized young plants 2 months after transfer to soil, (**i**) Flowering plants regenerated via somatic embryogenesis (bar represents 1 cm, except for **d**).

Embryogenic callus normally is very soft and often light brown (Fig. 26.1c). Under the microscope, it can be recognized easily by the presence of small, dense embryogenic cells (Fig. 26.1d).

2. Calluses of more than 3 mm in diameter are transferred to fresh callus induction medium. Then 12 calluses per 200 ml vessel filled with about 30 ml medium are still cultivated in the dark at 22–25°C for another 4 weeks.
3. From now on, the length of culture passages should be 4 weeks. Sixteen weeks after ovule preparation, calluses are homogenized for the first time. If further propagation is sought, callus should be spread on callus induction medium as a thin layer (*see* Note 6). A small portion of cells should be plated onto differentiation medium to check for embryogenic capacity.
4. The embryogenic callus can also be cultivated as a suspension culture in liquid callus induction medium (11) enabling faster growth and scaling-up (*see* Subheading 26.3.5).

3.3. Differentiation and Germination of Somatic Embryos

1. At the end of a culture passage of 4 weeks, cell material can be transferred to differentiation medium. A very thin layer of cells corresponding to 0.3–0.5 g fresh mass per 5 cm circular area results in mainly single somatic embryos of good quality (Fig. 26.1e, *see* Note 7).
2. After 4–6 weeks of culture at 22–25°C in the dark, somatic embryos develop through characteristic stages (Fig. 26.1f). Within this time, they reach about 3–5 mm and should consist of a small white tuber and the cotyledon. At this stage (third stage from the right in Fig. 26.1f), somatic embryos can be separated and cultured individually (10–20 per vessel) still on hormone-free differentiation medium (*see* Note 8).
3. If possible, the temperature should be lowered to 18–20°C, and the germinating embryos kept in dark until the cotyledon has reached 2 cm long. Then they can be transferred to light (25–40 μmol/m^2/s from cool fluorescent lamps for 16 h per day).
4. Subculture to fresh medium should take place every 8 weeks until the plantlets have developed a small tuber of about 1 cm in diameter, the first true leaves and a good root system (Fig. 26.1g). This usually takes 2–4 months from the transfer to hormone-free medium.

3.4. Acclimatization

1. Well-developed plants (*see* Subheading 26.3.3, step 4) can be adapted to greenhouse conditions. First the plantlets are taken out of the culture medium, washed carefully in lukewarm water, and roots are shortened to 2 cm long if necessary (*see* Note 9).
2. They are pricked out into substrate in trays or multicell plates, taking care that one third of the tuber is situated above the ground.

3. After thorough watering, the plantlets are placed under a foil tunnel (*see* Note 10) at 95–98% relative humidity and 22–24°C. Prevent direct sunlight in the first 2–3 weeks (*see* Note 11).
4. After 7–10 days, the foil tunnel may be opened gradually starting from few hours per day up to 12 h within 7 days.
5. Six to eight weeks after the transfer to soil newly formed leaves indicate successful acclimatization (Fig. 26.1h), and with 3–5 true leaves, the plants can be potted according to their growth type and further cultured to flowering (Fig. 26.1i) as in common horticultural practice.
6. At the flowering stage, they should be checked for morphological and genetic stability. We found the amount of somaclonal variation low (about 3%) and dependent on the genotype (*see* Note 12).

3.5. Liquid Culture Systems

Two phases of the protocol described above can also be performed in liquid culture to save labour and to achieve faster growth. Within this paragraph, the establishment and maintenance of embryogenic suspension cultures and the differentiation of globular-staged somatic embryos will be described.

1. To initiate embryogenic suspension cultures 2.5 g embryogenic callus (*see* Subheading 26.3.2) is suspended in 10 ml liquid callus induction medium (Table 26.1) lacking Gelrite® and incubated at 22–25°C in the dark on a rotary shaker at 100 rpm.
2. After one, 3 and 7 days 10, 15, and 20 ml, respectively, of this medium are added (*see* Note 13). After 14 days, the cultures are diluted 1:2 and, at day 21, 20 ml liquid callus induction medium is added again.
3. After 4 weeks, regular subculturing starts after sieving the suspensions through 1,000 µm mesh size (*see* Note 14). In the filtrate, the packed cell volume (pcv) is measured and the new subculture is started with 4% pcv (*see* Note 15). Suspension cultures are subcultured every 14 days.
4. One millilitre of the embryogenic suspension plated onto solid hormone-free differentiation medium leads to hundreds of somatic embryos after 4 weeks of culture in the dark.
5. For the differentiation of somatic embryos in liquid culture 500–1,000 µm cell aggregates are collected in liquid hormone-free medium (Table 26.1) and adjusted to 0.5–1% pcv. Globular somatic embryos can also be realized in liquid culture, by cultivating this suspension at 22–25°C in the dark on a rotary shaker at 100 rpm (12). Within 3 weeks, somatic embryos regenerate, but thereafter they need to be plated onto solid differentiation medium, as they will not develop proper polarity in shaken culture.

4. Notes

1. With prolonged callus or suspension phase the risk of somaclonal variation increases (13), so it is recommended to establish new embryogenic cell lines every 1–2 years or to cryopreserve them.
2. Flower buds should not start to unroll their petals (that would be too late), which should be fully coloured. Optimal stages are in between and buds should be 2.5–3 cm in length for standard cyclamen, and about 2 cm for mini cyclamen.
3. In order to be able to sterilize a range of different genotypes together, the peduncles can be cut to different lengths or scored with different patterns.
4. The ovules are sometimes hidden by cotton-like placental emergences. If so, a starting point has to be found and some ovules loosened and, if they are wounded, discarded. Thereafter, ovules can easily be taken singly and placed on the medium surface.
5. To simplify matters, we found it beneficial to place a pattern of 25 black spots drawn on a piece of paper matching the size of the Petri dish beneath the dish. This helps check if the ovule is successfully deposited on the medium and saves counting.
6. For homogenization of callus, broad forceps are recommended. With these, the callus material is scraped off from the medium and mashed carefully. About 0.5 g fresh mass should then be distributed onto fresh medium. After a culture passage of 4 weeks, 3–5 g callus can be yielded.
7. If more than 0.5 g embryogenic callus per Petri dish or culture vessel is spread on the medium, the numbers of fused and malformed somatic embryos increase. If premature root formation is observed, it might help increase the sucrose concentration to 40–45 g/L.
8. Somatic embryos can be separated with fine forceps, transferring them individually and carefully to fresh differentiation medium (still in 200 ml jars filled with 30 ml medium). Best development is observed, if the somatic embryos are placed in an upright position.
9. Washed plants can be stored overnight between wet tissues or in humidified boxes.
10. We obtained very good hardening results by heating mats on the greenhouse tables, covered with wet sand, on which the trays with the plants are placed. This is covered with transparent foil and kept close for the first 7–10 days.

11. Spraying or watering with antimicrobial substances like 0.1% 8-hydroxychinolin or a fungicide once at the beginning of acclimatization may be helpful to prevent losses during the first days.
12. Most commonly observed types of phenotypic alterations were polyploidy, which can easily be recognized by thicker and larger leaves and flowers, changes in flower colour and morphology and changes in leaf pattern.
13. Wide neck Erlenmeyer flasks are ideal vessels for suspension cultures. We use them sealed with two layers of aluminium foil and fill them maximum with 25% of the total volume.
14. Sieves can easily be made by sections of polypropylene tubes (5 cm in diameter, 10 cm long), which are slightly melted at one end on a heating plate and then pressed on nylon mesh of the desired size until the tube is cooled and the mesh is tightly attached to it. Ideally, these sieves are autoclaved in glass vessels with twist-off lids, in which the filtrate can be collected.
15. The packed cell volume can be measured by taking three 1 ml samples in 1.5 ml graduated Eppendorf tubes each and spinning at 720 × *g* for 5 min. The Eppendorf tubes (Fig. 26.2) can be graduated by scoring with a scalpel in 1% steps (score the tube along the surface of 10 µl (= 1% of 1,000 µl) samples).

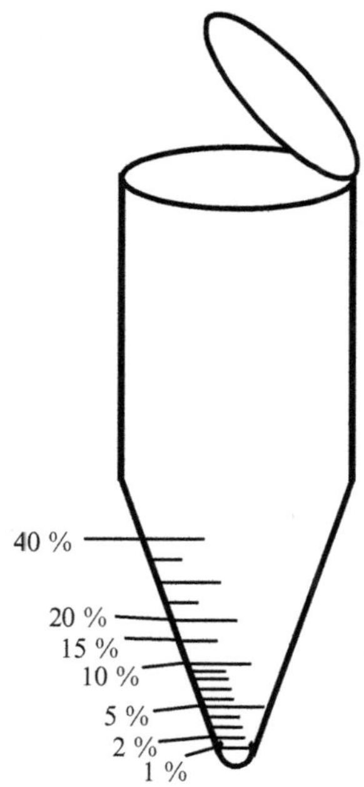

Fig. 26.2. Graduated Eppendorf tubes for determination of the packed cell volume.

Acknowledgments

The author is very grateful to Dr. Hans-Georg Schwenkel who was the first to develop somatic embryogenesis in cyclamen starting from ovules and who introduced me into the system. Furthermore, many thanks go to my students working on cyclamen, mainly Dr. Annette Hohe, Dr. Anke Pueschel, Dr. Viola Mußmann, Dr. Lara Meyer, Dr. Agnieszka Ilczuk, Antje Doil, Anika Prange, Janine Specht, and Christina Rode.

References

1. Schwenkel H-G, Winkelmann T (1998) Plant regeneration *via* somatic embryogenesis from ovules of *Cyclamen persicum* Mill. Plant Tiss Cult Biotechnol 4:28–34
2. Winkelmann T, Hohe A, Pueschel A-K, Schwenkel H-G (2000) Somatic embryogenesis in *Cyclamen persicum* Mill. Curr Topics Plant Biol 2:51–62
3. Winkelmann T, Serek M (2005) Genotypic differences in callus formation and regeneration of somatic embryos in *Cyclamen persicum* Mill. Euphytica 144:109–116
4. Winkelmann T, Mußmann V, Serek M (2004) Cryopreservation of embryogenic suspension cultures of *Cyclamen persicum* Mill. Plant Cell Rep 23:1–8
5. Winkelmann T, Specht J, Serek M (2006) Efficient plant regeneration from protoplasts isolated from embryogenic suspension cultures of *Cyclamen persicum* Mill. Plant Cell Tiss Org Cult 86:337–347
6. Winkelmann T, Prange ANS, Specht J, Serek M (2008) Morphological characterization of plants regenerated from protoplasts of *Cyclamen persicum* Mill. J Prop Ornam Plants 8:9–12
7. Winkelmann T, Meyer L, Serek M (2004) Desiccation of somatic embryos of *Cyclamen persicum* Mill. J Hortic Sci Biotechnol 79:479–483
8. Winkelmann T, Meyer L, Serek M (2004) Germination of encapsulated somatic embryos of *Cyclamen persicum*. HortScience 39:1093–1097
9. Seyring M, Hohe A (2005) Induction of desiccation tolerance in somatic embryos of *C. persicum* Mill. J Hortic Sci Biotechnol 80:65–69
10. Winkelmann T, Heintz D, van Dorsselaer A, Serek M, Braun H-P (2006) Proteomic analyses of somatic and zygotic embryos of *Cyclamen persicum* Mill. reveal new insights into seed and germination physiology. Planta 224:508–519
11. Winkelmann T, Hohe A, Schwenkel H-G (1998) Establishing embryogenic suspension cultures in *Cyclamen persicum* 'Purple Flamed'. Adv Hort Sci 12:25–30
12. Hohe A, Winkelmann T, Schwenkel H-G (2001) Development of somatic embryos of *Cyclamen persicum* Mill. in liquid culture. Gartenbauwissenschaft 66:219–224
13. Borchert T, Fuchs J, Winkelmann T, Hohe A (2007) Variable DNA content of *Cyclamen persicum* regenerated via somatic embryogenesis: rethinking the concept of long-term callus and suspension cultures. Plant Cell Tiss Org Cult 90:255–263
14. Murashige T, Skoog F (1962) A revised medium for rapid growth and bio assays with tobacco tissue cultures. Physiol Plant 15:473–497

Section B

Protocols for In Vitro Propagation of Ornamental Plants

Chapter 27

In Vitro Production of Sweet Peas (*Lathyrus odoratus* L.) via Axillary Shoots

Sergio J. Ochatt, C. Conreux, and L. Jacas

Abstract

The genus *Lathyrus* is best known because it includes a number of wild relatives of the protein pea which, despite being generally neglected and under-utilised, hold considerable potential as a useful genetic resource for the acquisition of interesting stress resistant traits important for a sustainable agriculture. However, also included in this genus are important commercially produced species with a significant ornamental value, among which the sweet pea (*Lathyrus odoratus* L.). Surprisingly though, there are no formal reports on the in vitro propagation of this species and, generally, these are scanty for in vitro approaches with all species of *Lathyrus*. Here, we describe simple, yet reliable strategies for the culture and multiplication of several landraces and species of *Lathyrus* including sweet peas.

Key words: *Lathyrus odoratus* L., Axillary bud proliferation, In vitro flowering

1. Introduction

Sweet pea (*Lathyrus odoratus* L.) is a flowering plant in the genus *Lathyrus* and a member of the Family *Fabaceae*, Subfamily *Faboideae*, Tribe *Vicieae* (which includes all grain legumes), native to the eastern Mediterranean region from Sicily east to Crete. Other *Lathyrus* species are sometimes also commonly called sweet peas, such as *L. annuus*, *L. belinensis*, *L. chilensis*, *L. chloranthus*, *L. latifolius*, *L. nervosus*, *L. sativus*, *L. tuberosus*, *L. vernus* (1) but, with most commercial varieties in the market, *L. odoratus* is by far the species of the genus most frequently called "sweet pea" commonly.

Lathyrus odoratus is an annual climbing plant, growing to a height of 1–2 m where suitable support is available. The leaves are pinnate with two leaflets and a terminal tendril, this twinning round supporting plant to help it climb. The flowers are purple,

S.M. Jain and S.J. Ochatt (eds.), *Protocols for In Vitro Propagation of Ornamental Plants*, Methods in Molecular Biology, vol. 589, DOI 10.1007/978-1-60327-114-1_27, © Humana Press, a part of Springer Science+Business Media, LLC 2010

2–3.5 cm broad, in the wild plant, larger and very variable in colour in the many cultivars. Unlike most peas, the seeds of the sweet pea are poisonous as they contain a neurotoxin, and should not be eaten (2). The illness caused by the ingestion of sweet peas is known as odoratism, or sweet pea lathyrism. Sweet peas have been cultivated since the seventeenth century and a vast number of cultivars are commercially available. They are often grown by gardeners for their bright colours and the sweet fragrance that gives them their name. Indeed, sweet peas give a sensational show of exquisitely coloured flowers on floppy, scrambling stems that need supporting. The best have a rich scent that carries right across the garden.

2. Biotechnological Tools with *Lathyrus* Species

The same as Gregor Mendel's work with cross breeding of pea plants (*Pisum sativum* L.), sweet pea has also been the subject of studies of genetic linkage in the early years of last century by Punnett (3), because of its ability to self-pollinate and easily observable Mendelian traits such as colour, height and petal form while, even earlier, complementary factor inheritance was also elucidated in sweet pea, from the cross of two pure-breeding white strains which gave rise to a blue hybrid, the blue colour requiring two genes, derived independently from the two white parents (4). Despite such a promising start, when it comes to biotechnological tools, studies with *Lathyrus* in general are few and far between, hence their inclusion among the neglected and under-utilised crops (2, 5).

Interspecific hybridization in the genus *Lathyrus* has been attempted by many researchers, but most attempts have failed (1, 5) and, in the few successful ones viable, *Fl* hybrids presented low fertility (6). In these cases, the rate of success can be increased through the utilization of embryo rescue techniques after fertilization (1, 2, 5). At intergeneric level, the incompatibility problems are even more serious (7, 8). Thus, from the information available in the literature on crossing, fertility and chromosome behaviour of hybrids, it is apparent that breeding strategies involving alien genetic transfer for the improvement of grass pea are only possible through the readily crossable species *L. cicera* and *L. amphicarpos* (5), while crosses involving other species will have to be assisted by biotechnology tools (2, 7–10).

As far as other biotechnology approaches is concerned, McCutchan et al. (8) established suspension cultures and isolated non-dividing protoplasts, while Durieu and Ochatt (9) induced sustained divisions in leaf protoplasts and fused them with pea protoplasts, to produce somatic hybrid microcalluses (10) and then calluses (2). They have obtained large number of viable protoplasts of pea and grass pea from leaves, and fused them by using electrofusion or chemical fusion. Several hundreds of somatic hybrid calluses have been obtained (2, 10), but regeneration has proven elusive to date. These could be the first steps into the introduction of ascochyta blight resistance from grass pea into pea (2, 7–10), of broomrape resistance from *L. ochrus* into grass pea (5), or of powdery mildew from sweet pea into other *Lathyrus* species (11). On the other hand, it is significant to this chapter that the only other report in the literature on the use of protoplast technology with a *Lathyrus* species, concerns precisely the sweet pea, where Razdan et al. (12) reported on the recovery of non-regenerating calluses from seedling-derived leaf protoplasts of this species.

Mutagenesis (13) and biotechnology (2, 7–9) via in vitro selection and gene transfer (14), would permit to produce novel genotypes and also to introduce interesting traits into the cross-incompatible species of the genus *Lathyrus* but also in other related genera (2).

Mutation breeding can be a valuable supplement to conventional plant breeding methods, aimed at creating additional genetic variability that may be utilised by the plant breeder in the development of cultivars for specific purposes or with specific adaptabilities (13, 15). Most mutation studies in this genus have concerned *L. sativus*, and a wide spectrum of morphological mutations has been found, affecting plant habit, maturity, branching, stem shape, leaf size, stipule shape, flower colour and structure, pod size, and seed size and colour (13, 15, 16).

In vitro selection through tissue culture for interesting traits can become routine when protocols become available to efficiently regenerate and multiply large numbers of plants after the selection of mutant cells or after the genetic transformation. High amounts of somaclonal variation for several traits (plant growth habit, ODAP content) has been found in regenerated plants (17–22), and this might be exploited in the production of desirable types for a faster improvement than via conventional crossing and backcrossing methods (1, 2).

Genetic transformation using *Agrobacterium tumefaciens* or *A. rhizogenes* as vectors or by using biolistic techniques has been tried repeatedly, but with very limited success. Barik et al. (14) are the only authors having developed a reproducible procedure for genetic transformation of *Lathyrus* (*L. sativus*, grasspea) using epicotyl segment cocultivation with *Agrobacterium*. Southern hybridization of genomic DNA of the kanamycin-resistant GUS-expressive shoots to a *gus* intron probe substantiated the integration of the transgene, further confirmed by progeny analysis, where among T1 seedlings of most transgenic plant lines, kanamycin-resistant and -sensitive plants segregated in a ratio close to 3:1.

Somaclones derived from various genotypes have been studied, mainly selected on the basis of β-ODAP content of the seeds and, while lines with nearly 1% β-ODAP level of the original genotype have been reported, no β-ODAP free lines have been found (23). Other agronomic, physiological, cytological and molecular characteristics of somaclones were also investigated (24–30).

On the other hand, various *Lathyrus* genotypes served to unravel the cytogenetic mechanisms underlying the occurrence of hyperhydricity during in vitro culture, a phenomenon commonly observed with many both legume and non-legume tissue cultures, for which only empirically developed solutions and no real explanation had been provided before (31). Finally, an in vitro protocol for fast the production of advanced progenies drastically shortening generation cycles has also been developed in *L. sativus* (32), where over three generations per year can be obtained instead of the normal two, allowing a faster progress in single-seed-descent (SSD) breeding strategies.

The regeneration protocols for *Lathyrus* (hormonal combination, hyperhydricity probability) have proven to be species- and, often, even genotype-specific (2, 8, 31), and there is a need to develop more generally applicable protocols. Interestingly, no general protocol for the propagation of *Lathyrus* species can be found in the literature.

Propagation by sexual means is the predominant technique in the genus *Lathyrus*, and sweet peas are no exception to this rule. However, in our laboratory, we have developed a very simple protocol for the maintenance and propagation in vitro of a number of *Lathyrus* species (2), including *L. odoratus*, which can even be coupled with the acceleration of generation strategies (32) as evoked above, and which ensures a constant supply of clean and fast-growing tissues that can be used for protoplast isolation purposes and also for multiplication of the genotypes studied.

3. Experimental Protocol for Propagation In Vitro

3.1. Disinfection and Excision of Initial Explants

1. Dry seeds are disinfected by immersion in 70°C ethanol for 30–45 s, followed by 15 min in calcium hypochlorite (70 g/l), and rinsed 3–4 times in sterile deionised water.
2. Disinfected seeds are kept in sterile water overnight for imbibition.
3. On the following morning, always under the laminar flow cabinet, embryo axes are excised from the seed by carefully removing the integuments and then detaching them from the cotyledons. Leaving the cotyledons may interfere with the subsequent multiplication rate obtained and/or with the initial growth of shoots, plus they are prone to callusing on the

medium whereby there is an inherent risk of genetic instability of cultures. Also in this context, particular care should be taken not to wound the embryo axis during excision to reduce the risks of callus developing from the wound.

3.2. Culture Initiation and Multiplication from Embryo Axes

1. Place excised embryos in a 10 cm diameter plastic Petri dish with 20-ml hormone-free modified B5 medium (33), with an additional 10 mM NH_4Cl, 3% sugar and 0.6 % agar (pH 5.6), as reported elsewhere for pea (34) and various *Lathyrus* species (9, 10, 31).
2. Plates with 10 axes each are then placed at 24/22°C, under a 16/8-h light/dark photoperiod 90 μE/m^2/s from warm white fluorescent tubes for 10–15 days.
3. The cotyledonary nodes from germinated axes are then dissected and transferred individually into tubes with 10-ml Murashige and Skoog (35) medium supplemented with 3% sugar, 1.0 mg/l BAP (N^6-Benzylaminopurine).
4. On this medium, an average of five shoots per explant will proliferate every 3–4 weeks, under the culture conditions as above.
5. For rooting, shoots are separated individually when they are at least 3–4 cm tall and they are explanted onto half-strength MS medium (35) with 0.5–1.0 mg/l (depending on genotype) NAA (1-naphthaleneacetic acid) for 3 weeks.
6. If the roots induced during the previous passage are not big enough (i.e. a minimum of 2-cm long and, preferably, ramified), a further 3 weeks on a half-strength hormone-free MS medium (35) will be needed for additional root elongation.

3.3. Multiplication from Shoot Nodes

1. Individual shoots are excised from the clusters produced from embryo axes when they are at least 1–2 cm tall.
2. Such shoots are transferred to MS medium (35) with 0.5–1.0 mg/l BAP as above, where axillary shoots will grow, particularly at the basal nodes of the shoots.
3. This step is repeated as required dependent on the number of shoots aimed at, by cutting each time the shoot bases and transferring individual nodes onto fresh medium of the same composition. Care should be taken to transfer single-node explants, as *Lathyrus* in general tends to exhibit a rather strong apical dominance that would otherwise interfere with the growth of axillary shoots and hence with multiplication in vitro.

3.4. Induction of Flowering and/or Rooting In Vitro

1. It is possible to induce flowering and seed set on un-rooted shoots taken from the multiplication medium in 1, by following the protocol as detailed in (32). Briefly, this is carried out as follows:

(a) Individual shoots (either origin, or even also for shoots regenerated from calluses or other sources as well) are identified within the clusters and separated when they are at least 1-cm tall and comprise two internodes.

(b) Such shoots are transferred onto half-strength hormone-free MS medium where they first elongate, then eventually flower and ultimately set seed. An average of up to seven flowers per shoot can be obtained, of which one or two will produce pods with seeds suitable to obtain subsequent generations.

(c) This procedure, which sidesteps the rooting phase, permits to save time, accelerates generations and is useful for SSD studies. Multiplied shoots produced can also be rooted prior to the induction of flowering and seed set in vitro, but this is not a prerequisite for fructification.

2. Alternatively, rooting of node-derived shoots is performed as described above.

3.5. Ex Vitro Acclimatization of In Vitro Produced Plants

Weaning of the in vitro produced plants is rather straightforward, provided plants with a sufficiently developed root system are transferred ex vitro. This, routinely, consists of:

1. Taking the cultures out from the culture chamber and leaving them (overnight to 24 h) on the laboratory bench, exposed to standard room temperature and illumination.
2. Thereafter, the rooted plants are taken out of the tubes, the roots rinsed under running tap water to wash away agar, and they are then planted on a mixture of perlite and soil (1:1) or standard commercial compost in small (9 cm) pots and kept in a mini glasshouse adapted to permit partial opening.
3. Such mini-glasshouses are then either transferred directly to the greenhouse at 25°C and natural daylight or kept under controlled conditions for the first week and then moved to the greenhouse.
4. Within about 10 days, the weaning period is complete, and the plants can be transferred to bigger pots and either kept in the greenhouse or, season permitting, cultivated outdoors. Figure 27.1 depicts several of the stages of in vitro multiplication and flowering with *Lathyrus*.

Plants of *Lathyrus* species produced in vitro in our laboratory to date with this strategy have proven true-to-type and normally reached flowering and set seeds. It is however important to point out that, after prolonged sub-culturing in both the multiplication (BAP-containing) or flowering (hormone-free) media used the shoots tend to age with time, whereby the multiplication ratios decrease and flowering is also diminished. In addition, such shoots

Fig. 27.1. Some of the steps for in vitro multiplication of *Lathyrus*: (**a**) Embryo axes cultured on hormone-free modified B5 medium for sprouting; (**b**) Detail of embryo axis at the time of culture initiation; (**c**) Embryo axis after 10–14 days of growth, *bars* indicate the cotyledonary node used as initial explant for multiplication; (**d**) A multiplying node on MS medium plus 1.0 mg/l BAP; (**e**) Shoots flowering in vitro on hormone-free MS medium; (**f**) Acclimatized plants in the greenhouse with tutorage suitable to support early flowering.

having undergone more than 12 months in culture are more difficult to root and the resulting plants tend to be rather enfeebled.

On the other hand, as with any tissue culture-based approach for the production of plants, it is advisable to check the plants obtained in terms of true-to-typeness. For commercial purposes, phenotype checks (36) might sometimes be suffice but, when in vitro plants were produced for other uses, if possible, they should also be characterised by cytological (37, 38), cytogenetic (2), isoenzymatic (39–41), and molecular markers (5, 42, 43).

Finally, as already mentioned at the beginning of this chapter, even if the protocol described is applicable to a large range of genotypes within sweet pea and also within *Lathyrus* species in general, optimum responses remain genotype-dependent.

References

1. Campbell CG (1997) Grass pea. *Lathyrus sativus* L. Promoting the conservation and use of under-utilized and neglected crops, vol 18. Institute of Plant Genetics and Crop Plant Research, Gatersleben/International Plant Genetic Resources Institute, Rome, Italy
2. Ochatt SJ, Abirached-Darmency M, Marget P, Aubert G (2007) The *Lathyrus* paradox: "poor men's diet" or a remarkable genetic resource for protein legume breeding? In: Ochatt SJ, Mohan Jain S (eds) Underutilised and neglected crops, herbs and spices. Science Press, Enfield, CT, pp 41–60
3. Punnett RC (1923) Linkage in the sweet pea (*Lathyrus odoratus*). J Genet 13:101–123
4. Bateson W, Saunders ER, Punnett RC (1906) Experimental studies in the physiology of heredity. Reports to the Evolution Committee. Proc R Soc B 77:236–238
5. Vaz Patto MC, Skiba B, Pang ECK, Ochatt SJ, Lambein F, Rubiales D (2006) *Lathyrus* improvement for resistance against biotic and abiotic stresses: From classical breeding to marker assisted selection. Euphytica 147: 133–147
6. Yunus AG, Jackson MT (1991) The gene pools of the grasspea (*Lathyrus sativus* L.). Plant Breed 106:319–328
7. Ochatt SJ, Benabdelmouna A, Marget P, Aubert G, Moussy F, Pontécaille C, Jacas L (2004) Overcoming hybridisation barriers between pea and some of its wild relatives. Euphytica 137:353–359
8. McCutchan JS, Larkin PJ, Stoutjesdijk PA, Morgan ER, Taylor PWJ (1999) Establishment of shoot and suspension cultures for protoplast isolation in *Lathyrus sativus* L. J Breed Genet 31:43–50
9. Durieu P, Ochatt SJ (2000) Efficient intergeneric fusion of pea (*Pisum sativum* L.) and grass pea (*Lathyrus sativus* L.) protoplasts. J Exp Bot 51:1237–1242
10. Ochatt SJ, Durieu P, Jacas L, Pontécaille C (2001) Protoplast, cell and tissue cultures forthe biotechnological breeding of grass peas (*Lathyrus sativus* L.). Lathyrus Neurolathyrism Newslett 2:35–38
11. Poulter R, Harvey L, Burritt DJ (2003) Qualitative resistance to powdery mildew in hybrid sweet peas. Euphytica 133:349–358
12. Razdan MK, Cocking EC, Power JB (1980) Callus regeneration from mesophyll protoplasts of sweet pea (*Lathyrus odoratus* L.). Z Pflanzenphysiol 96:181–183
13. Biswas AK (2007) Induced mutation in grasspea (*Lathyrus sativus* L.). In: Ochatt SJ, Mohan Jain S (eds) Underutilised and neglected crops, herbs and spices. Science Press, Enfield, CT, pp 29–39
14. Barik DP, Mohapatra U, Chand PK (2005) Transgenic grasspea (*Lathyrus sativus* L.): Factors influencing *Agrobacterium*-mediated transformation and regeneration. Plant Cell Rep 24:523–531
15. Nerkar YS (1976) Mutation studies in *Lathyrus sativus*. Indian J Genet Plant Breed 36: 223–229
16. Rybínski W (2003) Mutagenesis as a tool for improvement of traits in grasspea (*Lathyrus sativus* L.). Lathyrus Lathyrism Newslett 3:27–31
17. Mehta SL (1997) Plant biotechnology for removal of ODAP from *Lathyrus*. In: Haimanot RT, Lambein F (eds) Lathyrus and Lathyrism, a decade of progress. University of Ghent, Belgium, pp 103–1104
18. Roy PK, Singh B, Mehta SL, Barat GK, Gupta N, Kirti PB, Chopra VL (1991) Plant regeneration from leaf discs of *Lathyrus sativus*. Indian J Exp Biol 29:327–330
19. Misra M, Addis G, Narayan RKJ (1994) Methods for callus induction and differentiation of *Lathyrus sativus* and embryo rescue in interspecific crosses by tissue culture. J Univ Wales Agric Soc 74:129–144
20. Patil VD, Kulkarni UG, Harkal SR, Nerkar YS (1997) Studies on in vitro plant regeneration in grass pea (*Lathyrus sativus* L.). In: Haimanot RT, Lambein F (eds) Lathyrus and Lathyrism, a Decade of Progress. University of Ghent, Belgium, pp 125–127
21. Abd El-Moneim AM, Cocks PS (1993) Adaptation and yield stability of selected lines of *Lathyrus* spp. under rainfed conditions in West Asia. Euphytica 66:89–97
22. Zambre M, Chowdhury B, Kuo YH, van Montagu M, Angenon G, Lambein F (2002) Prolific regeneration of fertile plants from green nodular callus induced from meristematic tissues in *Lathyrus sativus* L. (grass pea). Plant Sci 163:1107–1112
23. Lambein F, Kuo YH (2004) Somaclonale variatie in *Lathyrus sativus*. In: Abstract, autumn meeting Netherlands society for plant biotechnology and tissue culture NVPW, Leiden, 5 November. http://www.nvpw.nl/pdf/NVPW04-2.pdf
24. Roy PK, Ali K, Gupta A, Barat GK, Mehta SL (1993) β-Noxalyl-L-α, β-diaminopropionic acid in somaclones derived from internode

explants of *Lathyrus sativus*. J Plant Biochem Biotechnol 2:9–13

25. Campbell CG, Mehra RB, Agrawal SK, Chen YZ, El Ali AMA, Khawaja HIT, Yadav CR, Toy J, Araya WA (1994) Current status and future strategy in breeding grass pea (*Lathyrus sativus*). Euphytica 73:167–175
26. Tiwari N, Bansai KC, Ali K, Sachdev A, Santha IM, Mehta SL (1995) Physiological and biochemical analysis of somaclones derived from leaf explants of *Lathyrus sativus*. J Plant Biochem Biotechnol 4:85–89
27. Singh S, Johari RP, Mehta SL (1997) Biochemical and cytological studies during seed development in somaclones of *Lathyrus sativus*. J Plant Biochem Biotechnol 6:101–104
28. Chakrabarti A, Santha IM, Mehta SL (1999) Molecular characterization of low ODAP somaclones of *Lathyrus sativus*. J Plant Biochem Biotechnol 8:25–29
29. Ali K, Bansal KC, Sachdev A, Tiwari N, Santha IM, Mehta SL (2000) Physiological activity and biochemical analysis of *Lathyrus sativus* somaclones derived from root and internode explants. J Plant Biochem Biotechnol 9:27–30
30. Kuo YH, Bau HM, Rozan P, Chowdhury B, Lambein F (2000) Reduction efficiency of the Neurotoxin β-ODAP in low toxin varieties of *Lathyrus sativus* seeds by solid-state fermentation with *Aspergillus oryzae* and *Rhizopus microsporus* var. *chinensis*. J Sci Food Agric 80:2209–2215
31. Ochatt SJ, Muneaux E, Machado C, Jacas L, Pontécaille C (2002) The hyperhydricity of in vitro regenerants is linked with an abnormal DNA content in grass pea (*Lathyrus sativus* L.). J Plant Physiol 159:1021–1028
32. Ochatt SJ, Sangwan RS, Marget P, Assoumou Ndong Y, Rancillac M, Perney P (2002) New approaches towards the shortening of generation cycles for faster breeding of protein legumes. Plant Breed 121:436–440
33. Gamborg OL, Miller RA, Ojima K (1968) Nutrient requirements for suspension cultures of soybean root cells. Exp Cell Res 50:151–158
34. Ochatt SJ, Pontécaille C, Rancillac M (2000) The growth regulators used for bud regeneration and shoot rooting affect the competence for flowering and seed set in regenerated plants of protein pea. In Vitro Cell Dev Biol Plants 36:188–193
35. Murashige T, Skoog F (1962) A revised medium for rapid growth and bioassays with tobacco tissue cultures. Physiol Plant 15:473–497
36. IPGRI (2000) Descriptors for *Lathyrus* spp. International Plant Genetic Resources Institute, Rome, Italy
37. Yamamoto OK, Fujiwara T, Blumenreich ID (1984) Karyotypes and morphological characteristics of some species in the genus *Lathyrus* L. Jpn J Breed 34:273–284
38. Ochatt SJ (2008) Flow cytometry in plant breeding. Cytometry A 73A:581–598
39. Chowdhury MA, Slinkard AE (1999) Linkage of random amplified polymorphic DNA, isozyme and morphological markers in grasspea (*Lathyrus sativus*). J Agric Sci 133:389–395
40. Alba E, Polignano GB, De Carlo D, Mincione A (2001) Electrophoretic phenotypes of different enzymes in some entries of *Lathyrus sativus* L. Lathyrus Lathyrism Newslett 2:15–20
41. Ben Brahim N, Salhi A, Chtourou N, Combes D, Marrakchi M (2002) Isozymic polymorphism and phylogeny of 10 *Lathyrus* species. Genet Res Crop Evol 49:427–436
42. Chtourou-Ghorbel N, Lauga B, Ben Brahim N, Combes D, Marrakchi M (2002) Genetic variation analysis in the genus *Lathyrus* using RAPD markers. Genet Res Crop Evol 49:363–370
43. Bard A, El Shazly H, El Rabey H, Watson LE (2002) Systematic relationships in *Lathyrus* sect. *Lathyrus* (Fabaceae) based on amplified fragment length polymorphism (AFLP) data. Can J Bot 80:962–969

Chapter 28

In Vitro Conservation and Cryopreservation of Ornamental Plants

Elif Aylin Ozudogru, Alberto Previati, and Maurizio Lambardi

Abstract

Today, the conservation of ornamental germplasm can take advantage of innovative techniques which allow preservation in vitro (slow growth storage) or in liquid nitrogen (cryopreservation) of plant material. Slow growth storage refers to the techniques enabling the in vitro conservation of shoot cultures in aseptic conditions by reducing markedly the frequency of periodic subculturing, without affecting the viability and regrowth of shoot cultures. Cryopreservation refers to the storage of explants from tissue culture at ultra-low temperature (−196°C). At such temperature, all the biological reactions within the cells are hampered, hence the technique makes available the storage of plant material for theoretically unlimited periods of time. An exhaustive review of papers dealing with the slow growth storage and the cryopreservation of ornamental species is reported here. Step-by-step protocols for the slow growth storage of rose germplasm, the production of synthetic seeds for the in vitro conservation of ornamentals, and the cryopreservation of *Chrysanthemum morifolium* are included.

Key words: Cryopreservation, In vitro conservation, Liquid nitrogen, Slow growth storage, Synthetic seeds

1. Introduction

The earliest recorded evidence of plant cultivation for ornamental purposes dates back to about 1500 B.C., when lotus ponds surrounded by acacias and palms were depicted in the tomb paintings of Egyptians. "Paradise gardens" and "Gardens of Babylon", among others, were early examples of magnificent gardens. The Italian villa gardens of the early Renaissance, the French parterres of the sixteenth century, and the English landscape gardens of the eighteenth century were important moments of diffusion and success of ornamentals in Europe. From the nineteenth century,

S.M. Jain and S.J. Ochatt (eds.), *Protocols for In Vitro Propagation of Ornamental Plants*, Methods in Molecular Biology, vol. 589, DOI 10.1007/978-1-60327-114-1_28, flower gardens became significant in city planning (1). As a result of an intense breeding activity, which has produced over time thousands of new selections for landscape architecture and indoor decoration, a large amount of germplasm is available today. In spite of this, little coordinated effort has been made up to date to conserve and protect the genetic resources of ornamental plants.

One exception worthy of mention is the genebank of the "Ornamental Plant Germplasm Center" (OPGC) of Ohio, USA, where the germplasm of herbaceous ornamentals is conserved, in conjunction with the USDA National Plant Germplasm Center at Ohio State University (http://opgc.osu.edu). Seed banks are a common way of conserving plant genetic resources. In these, orthodox and suborthodox seeds are stored at temperatures of either –15/–20°C (cold tolerant species), or 0/–5°C (temperate and tropical species). However, as the seeds do not represent the genetic profile of the mother plant, this approach cannot be used when endangered clonal germplasm is to be preserved, e.g., that of ancient cultivars. For vegetatively-propagated species, the conservation of clonal germplasm is made in field (clonal collections). However, genotypes preserved only this way run the risks of strong abiotic stresses and pathogen attacks which have often caused the loss of precious germplasm, e.g., in fruit species. Today, important complementary approaches to seedbanks and clonal orchards are offered by biotechnology with the possibility to preserve in vitro (slow growth storage) or at –196°C (cryopreservation) plant material.

Slow growth storage refers to the techniques enabling the maintenance in vitro of shoot cultures in aseptic conditions by reducing markedly the frequency of periodic subculturing, without affecting the viability and regrowth potential of shoot cultures. Manual labour and costs of tissue culture, as well as the risk of contamination during the transfers of material, are reduced too. Depending on the species, subculturing can be decreased to once in every several months (sometimes even to 1–2 years) and, because of this, the technique is considered a "medium-term conservation" method. By far, the most widely used approach to slow growth storage of shoot cultures is the coupling of a low temperature (generally, 2–5°C for temperate species, 15–25°C for tropical species; *see* (2, 3)) with the culture in the dark or low light intensity. Low temperature and light intensity have physiological consequences, such as the reduction of respiration, water loss, wilting, and ethylene production, which allow for safe conservation of shoot cultures. In general, cold storage of shoot cultures is held in the same medium composition as in the proliferation stage, though there are examples of a better conservation of cultures when made on hormone-free media (4). In addition to cold storage, in vitro conservation of shoot cultures can be achieved by modifying the medium compositions, i.e., by (i) reducing the sugar and/or mineral concentration, (ii) using growth retardants (e.g., chlorocholine chloride, abscisic acid) or osmotically active compounds (e.g., mannitol), and (iii) covering the explants with a layer of liquid medium or mineral oil to reduce the oxygen available to the plants (5). However, it should be noted that these methods, especially the first two, can lead to the appearance of somaclonal variations and thus are not so advisable when true-to-type material is required (6). As for ornamentals, no examples are available for studies approaching slow growth storage of shoot cultures this way.

Cryopreservation is the storage of explants from tissue culture at the ultra-low temperature of liquid nitrogen (−196°C). At such a temperature, all the biological reactions within the cells are practically stopped, hence the technique makes available the storage of the material for theoretically unlimited periods of time (long-term conservation) without genetic alterations. However, several precautions are required to achieve successful cryopreservation. Water content of cells during the immersion in the liquid nitrogen (LN) is the most critical factor affecting the success of the technique. It should be low enough to prevent lethal intra-cellular ice crystallization by inducing the "vitrification" of cytosol, and yet high enough to enable explant regrowth following post-thaw recovery (7). The term "vitrification" refers to a physical state of water solutions, in which the solidification of water molecules at ultra-low temperature is obtained through their transition to an amorphous (glassy) state, instead of crystallization (8). Traditionally, cell vitrification was achieved by imposing cryodehydration of explants, i.e., by a gradual decrease of their temperature (at a rate of 0.5–1°C/min) up to −40°C before the immersion in LN (slow cooling). Recently-developed cryopreservation techniques allow the direct immersion of samples in LN (one-step freezing) and have already been successfully applied to a wide range of species and explants, such as shoot tips (9), embryogenic cultures (10), seeds (11, 12), embryos and embryonic axes (13) dormant buds (14). Here, cell vitrification can be induced basically in two ways, i.e., (1) by the treatment with highly concentrated vitrification solutions (such as PVS2, (15)) of naked (vitrification method) or encapsulated explants (encapsulation-vitrification), and (2) by the physical dehydration of encapsulated explants either under sterile air flow or on silica gel (encapsulation-dehydration; (16)). Several additional steps contribute to the effectiveness of a cryopreservation protocol, such as the prehardening of stock cultures and excised shoot tips at 4°C, the preculturing of explants in media containing high sucrose concentrations (0.25–1 M), the thawing temperature, and the composition of regrowth medium (17). Moreover, in the encapsulation-based techniques, the production of high-quality "synthetic seeds" is required. Although the technique was initially proposed to encapsulate single somatic embryos inside an artificial seed coat (18), today various other explants (such as shoot tips, nodal segments, bulblets, and even callus samples) are used to produce synthetic seeds (19). The encapsulation method developed by Redenbaugh et al. (20) is still most widely used to produce synthetic seeds. In short, the method involves the incubation of explants in a Na-alginate solution (2–5%) and their subsequent release (immersed in a drop of alginate) into a complexing agent (usually, 100 mM $CaCl_2$ solution) where bead hardening occurs in 20–30 min.

In this chapter, the application of slow growth storage and cryopreservation techniques for the medium- and long-term conservation of ornamental plants is discussed. Example protocols of cold storage, synthetic seed production, and cryopreservation are also reported.

2. Slow Growth Storage of Ornamental Plants

To date, only a limited number of reports deal with the in vitro conservation of ornamental plants, all based on the cold storage approach (Table 28.1). Like for fruit species, ornamental plants are generally stored just a few degrees above the freezing (mainly, at 4–5°C). Storage in total dark conditions is preferred by most of the Authors for a better slowing down of cell metabolism; however, storage under low light intensity (from 3 to 8 μmol/m^2/s) showed to be effective for shoot cultures of *Camellia* (21) and *Humulus* spp. (22). In the latter study, the low light intensity was also coupled to a photoperiod (12 h) shorter than the one commonly applied during the proliferation stage of cultures in standard culture conditions (16 h). In comparison with the other reports, best conservation of *Cedrela fissilis* synthetic seeds was achieved with "unusual" storage conditions, i.e., at 25°C under 25 μmol/m^2/s of light intensity (23).

In general, maximum time of conservation of shoot cultures in cold conditions ranges from some months to 1 year, and ornamental species are not exception. Reed et al. (22) reported of 3 years of cold storage for *Humulus* spp., though the Authors do not report the percentage of shoot regrowth after this exceptionally long time of storage. As for synthetic seeds of ornamentals, they have a shorter time of conservability, ranging from 1.5 to 9 months. But, again, two exceptions are found, in *Hibiscus moscheutos* (almost 20 months of storage; (24)) and in *Splachnum ampullaceum* (30 months; (25)).

When optimizing a procedure of in vitro conservation, other factors, in addition to temperature and light conditions, can affect the maximum time of storage, such as the age, the size, and the physiological state of shoot cultures at the time of being

Table 28.1
Slow growth storage of shoot cultures and synthetic seeds (containing axillary buds or shoot tips, unless otherwise stated) of ornamental species

Species	Temp (°C)	Light conditions[a]	Months	Regrowth (%)	Ref.
Shoot cultures					
Camellia japonica (cv Alba Plena)	2–4	16 h (8 µmol/m²/s)	12	100	(21)
Camellia reticulata (cv Mouchang)	2–4	16 h (8 µmol/m²/s)	12	100	(21)
Gerbera (cv Marleen)	4	Darkness	3	96	(27)
Humulus spp.	4	12 h (3 µmol/m²/s)	36	NR	(22)
Rosa cvs[b]	4	Darkness	6	>80	(28)
Synthetic seeds					
Camellia japonica[c]	4	Darkness	2	30	(29)
Camellia japonica	2–4	NR	2.5	10	(21)
Cedrela fissilis	25	16 h (20–25 µmol/m²/s)	9	8	(23)
Hibiscus moscheutos (cv Lord Baltimore)	5	Darkness	19.5	80	(24)
Lilium longiflorum[d]	4	Darkness	6	92	(35)
Nerium oleander	4	Darkness	3	30	(31)
Morus spp.	5	Darkness	3	18	(30)
Paulownia elongate[e]	4	Darkness	2	32	(36)
Photinia fraseri	4	Darkness	3	91	(31)
Splachnum ampullaceum	5	Darkness	30	50	(25)
Syringa vulgaris	5	Darkness	1.5	83	(37)

"Regrowth" refers to the percentage of (1) shoots resuming proliferation or (2) synthetic seeds "germinating" (i.e., showing shoots emerging from the beads) after the maximum time of conservation reported by the authors (*NR* not reported)

[a]Photoperiod and light intensity (when reported)

[b]Cv Rosa Sant'Antonio di Padova (Tea hybrid) and cv Domenica (City Flor® Roses)

[c]Synthetic seeds contained somatic embryos at the cotyledonary stage

[d]Synthetic seeds contained bulblets

[e]Synthetic seeds contained somatic embryos (1.5–2 mm)

placed at low temperature, perhaps the latter being the most important (26). Indeed, Ballester et al. (21), in their study aiming to develop cold storage procedures for eight clones and cultivars of *Camellia japonica* and *Camellia reticulata*, showed that shoot clones with relatively thinner and less-cutinized leaves required to be subcultured at least every 6 months, while others could be left in cold storage for a year or more with no sign of deterioration.

Another important factor is the kind of container used for culture storage. When Hempel and Hempel (27) stored at 4°C gerbera shoots in 250-ml conical flasks, in which the plants had been rooted, 96% of the shoots could be successfully preserved up to 3 months. In contrast, when the shoots were taken out of the rooting flasks and placed in 40-ml plastic boxes prior to cold storage, the root tips died only after one month and the shoots could not be stored any longer. In rose, three different containers were compared during the storage at 4°C in the dark of shoot cultures from two Italian cultivars, the cvs. Rosa Sant'Antonio di Padova and Domenica ((28); *see* protocol in Subheading 28.4.1). The containers were gas-tight 500-cc glass jars, semi-permeable 80-cc plastic cylinders, and gas-permeable StarPac™ plastic bags. In 6 months of storage, the concentration of CO_2 detected in both the glass jars and the plastic cylinders was never exceeded 0.4%, which is a clear indicator of an effective reduction of shoot metabolism. For both the cultivars, after six months, best conservation was achieved in glass jars, with over 80% of healthy shoot recovery, no sign of hyperhydricity and a prompt regaining of normal proliferation after the return of shoot cultures to standard culture conditions. After the same time of storage in both plastic cylinders and StarPac™ plastic bags, the quality of shoot cultures was consistently lower. An interesting finding of this study was also the observation of the progressive reduction of cell metabolism during the conservation at 4°C. For instance, the cv Domenica exhibited a steady CO_2 concentration (0.35%) until the end of the 4th month of cold storage, after which it started to decrease, going below 0.25% after 6 months.

Encapsulating the explants within alginate beads for the production of synthetic seeds is another approach to the in vitro conservation of ornamental plants. In the various studies, the alginate concentration of beads ranged from 1% to 5%, 3% being the most used. Some Authors (29, 30) underlined that the synthetic seeds formed with the lower concentrations were fragile, difficult to handle, and thus not suitable for conservation, while at the highest concentration were too rigid and hindered the emergence of the shoots and roots. Concentration of the complexing agent ($CaCl_2$) also affects the quality of the synthetic seeds and varies between 50 and 100 mM, with the lower concentrations prolonging the ion-exchange duration (30). The conservation of synthetic seeds is generally made inside Petri dishes; however, this way the alginate beads can risk dehydrating too much. For this reason, Ozden-Tokatli et al. (31) compared three different conditions for the cold storage in Petri dishes of *Nerium oleander* and *Photinia fraseri* synthetic seeds, prepared with apical or axillary buds (*see* protocol in Subheading 28.4.2). The first set of synthetic seeds was kept in empty Petri dishes, tightly sealed with parafilm; the second set was placed over sterile filter papers, wetted with 2 ml dH_2O; while the third set was kept on gelled MS (32) hormone-free medium. For both the species, the highest germination rate after 3 months at 4°C in darkness was achieved from synthetic seeds stored on gelled MS medium. The same report also showed the importance of the medium composition to promote the "germination" of synthetic seeds after cold storage. Indeed, although shoot cultures of *P. fraseri* were successfully multiplied on QL (33) medium, supplemented with 4.4 μM BA, after the conservation at 4°C, best synthetic seed's "germination" was observed on hormone-free MS medium.

As for the production of synthetic seeds, Mallón et al. (25) compared the method described by Redenbaugh et al. (20) with the one developed by Patel et al. (34), consisting of placing moss buds in drops of a $CaCl_2$ solution containing CMC (carboxymethylcellulose), and then transferring them drop-wise into a 0.8% Na-alginate solution. The presence of CMC, which has a high capacity to absorb water, led to the opening of big pores in the gel matrix and the excessive uptake of water during cold storage in 10 ml dH_2O, which in turn resulted in the loss of sphericity of the bead and damage to the explant.

It is well known that, working with large collections, the development of effective protocols of slow growth storage for each genus, or even each species, can be extremely time consuming. However, Reed et al. (22) suggested the direct application of a well-tested standard procedure, with few modifications, also to genera different from the one for which it was developed. Indeed, these Authors proved that a procedure of in vitro conservation, originally developed for strawberry and mint shoot cultures, could be directly applied to a large collection of *Humulus* germplasm (a total of 70 species and cultivars), with satisfactory results. Subsequently, some improvements were made to the standard protocol to increase the time of storage of some specific *Humulus* accessions.

West et al. (24) suggested that, depending on the length of the storage period, the recovery from the stress imposed by low temperatures may take 1–3 months after the cultures are moved back to standard culture conditions, and this fact should be carefully considered when evaluating the effectiveness of an in vitro conservation procedure.

28.3. Cryopreservation of Ornamental Plants

Today cryopreservation is an important tool for long-term storage of valuable plant germplasm and a valid alternative to the traditional approaches of ex situ conservation. Working with ornamentals, the "slow cooling" approach is still used to induce cell tolerance to storage at −196°C; however, more innovative and simple "one-step freezing" techniques (such as "vitrification", "encapsulation-vitrification", and "encapsulation-dehydration") are widely preferred today. Effective cryoprotocols have been developed over time, though with varying results in terms of explant survival after recovery from cryostorage (Table 28.2).

Whatever the chosen technique, a step-by-step optimization of protocols (from explant preparation for LN to regrowth after thawing and plating) is fundamental to cryopreservation. For the cryopreservation of *Chrysanthemum morifolium* shoot tips, Halmagyi et al. (38) compared different cryo-techniques, i.e., "slow cooling", "encapsulation-dehydration", "droplet method", and "droplet-vitrification" (*see* protocols in Subheading 28.4.3). The latter techniques are based on the ultra-rapid freezing of explants included in droplets of cryoprotectant, as originally proposed for potato apices (39). In *Chrysanthemum*, the method was also combined with the use of PVS2 to produce 4-μl drops, containing the shoot tips, on aluminium strips ("droplet-vitrification"), being by far the most effective in terms of explant regrowth after cryopreservation (60%).

Preculturing explants with sucrose prior to LN is a fundamental step in many cryopreservation protocols developed for ornamentals. Protocorm-like bodies (PLBs) of *Oncidium*, precultured with high concentrations of sucrose and glycerol, maintained their cell shape and subcellular components and remained intact after desiccation and freezing, while the cellular structures of PLBs which were not precultured were severely damaged (40). Halmagyi and Pinker (41) tested four different sugars or sugar alcohols (i.e., sucrose, glucose, mannitol, and sorbitol) in preculture and observed that the tolerance to freezing of rose shoot tips was highest when sucrose was used.

In addition to sugar preculture, pre-treating explants or shoot cultures (from which shoot tips are to be excised) at 10°C or below (cold hardening) is another common approach for inducing freezing tolerance. Examples of cold hardening of different materials before use in cryopreservation trials are (i) nodal segments of *Chrysanthemum* stored at 10°C and 10 μmol/m^2/s of light intensity for 3 weeks (42), (ii) shoot cultures of *Photinia fraseri* stored at 4°C in darkness for 2–3 weeks (31), (iii) in vitro-grown gentian plants stored up to 50 days at 5°C and 26 μmol/m^2/s (43), and (iv) embryogenic callus of *Aesculus hippocastanum* stored at 4°C for 5 days in darkness (44).

The treatment inducing cell vitrification during ultra-rapid freezing in LN is the central step in every cryopreservation protocol. Here, the time of loading explants in PVS2, as well as of dehydrating beads under air flow or on silica gel, should be carefully determined for each species/material combination.

Table 28.2
Cryopreservation (–196°C) of ornamentals by "slow cooling" or "one-step freezing"

Species	Exlant type	Pre-freezing treatments	Thawing	Survival (%)	Ref.
Slow cooling					
Betula pendula	In vivo buds	18 h, 0°C; 0.17°C/min to -38°C	37°C	7	(57)
Caryophyllaceae[a]	Shoot tips (0.5–0.7 mm)	10% DMSO + 3% glu (1 h, 0°C); 0.5°C/min to -40°C	25–30°C	20–100	(49)
Chrysanthemum spp.	Shoot tips	5% DMSO (2d); 10% DMSO + 3% glu (1 h, 0°C); 0.2°C/min to -40°C	20 min, 25–30°C	9–100	(48)
Chrysanthemum grandiflora	Shoot tips	5% DMSO (2d); 10% DMSO + 3% glu (1 h, 0°C); 0.2°C/min to -40°C	20 min, 25–30°C	NR	(53)
Cyclamen persicum	Suspension cultures	0.6 M suc (2d, 24°C, D); 0.6 M suc + 10% DMSO (1 h); -20°C (2 h)	3 min, 38°C	<1	(58)
Humulus spp.	Shoot tips (0.8–1 mm)	0.1°C/min to -40°C	1 min, 45°C	41	(22)
Melia azedarach	Encapsulated som. embryos	0.5 M (1d), 0.75 M (1d), 1.0 M (1d) suc; SG (3 h); 1°C/min to -30°C	2 min, 30°C	36	(59)
Picea sitchensis	Embryogenic masses	0.4 M sorbitol (48 h); 5% DMSO (30 min, 0°C); -0.5°C/min to -50°C	2–3 min, 40°C	100	(60)
Salix spp.	Twigs	0.21°C/h to -35°C	1d, 2°C	0–95	(61)
Vitrification					
Acer mono	Suspended cells	MS + 0.3 M suc (16 h, 25°C); PVS2 (60 min, 0°C)	5 min, 40°C	81	(55)
Aesculus hippocastanum	Embryogenic callus	2 M gly + 0.4 M suc (30 min, 25°C); PVS2 (90 min, 0°C)	50 s, 45°C	94	(44)
Chrysanthemum grandiflora	Shoot tips	0.5 M suc (1d); PVS2 (5 min) ;PVS2 (4 µl drops)[b]	RT	60	(38)

(continued)

Table 28.2 (continued)

Species	Exlant type	Pre-freezing treatments	Thawing	Survival (%)	Ref.
Colocasia esculenta	Shoot tips	0.3 M suc (16 h, 25°C, D); 2 M gly + 0.4 M suc (20 min, 25°C); PVS2 (12 min, 25°C)	1.5 min, 40°C	30	(62)
Doritis pulcherrima	Mature seeds	PVS2 (50 min, 25°C)	2 min, 40°C	62	(63)
Elateriospermum tapos	Zygotic embryos	PVS2 (stepwise; 50–60–80–100–50–100%, 5 min each, 0 or 25°C)	NR	100	(64)
Limonium	Shoot tips (1 mm)	0.3 M suc (1d, 25°C); 2 M gly + 0.4 M suc (20 min, 25°C); PVS2 (15 min, 25°C)	90 sec, 40°C	76	(65)
Melia azedarach	Shoot tips	0.7 M suc (1d, 25°C); 2 M gly + 0.4 M suc (20 min, 25°C); PVS2 (60 min, 0°C)	1 min, 40°C	90	(66)
Ponerorchis graminifolia	Immature seeds	0.3 M suc (3d, 25°C, D); 2 M gly + 0.4 M suc (15 min, 25°C); PVS2 (60 min, 0°C)	2 min, 38°C	48	(67)
Rosa spp.	Shoot tips (1–2 mm)	0.5 M suc (2d, 24°C); PVS2 (20 min, RT)	RT	66–69	(41)
Rosa x hybrida	Shoot tips (3–4 mm)	0.5 M suc (1 or 2d, 24°C); PVS2 (20 min, RT)	RT	50–61	(52)
Sterculia cordata	Zygotic embryos	0.35–0.75 M suc (3d); 2 M gly + 0.4 M suc (1–3 h, 25°C); PVS2 (1–3 h, 0°C)	10–15 min, 38°C	≥80	(68)
Encapsulation-vitrification					
Acer mono	Suspended cells	0.3 M suc (16 h, 25°C); PVS2 (40 min, 0°C)	5 min, 40°C	81	(55)
Dianthus caryophyllus	Shoot tips (2–3 mm)	0.5–0.75 M suc (24 h, 25°C); PVS2 (3 h, 0°C)	25°C	73	(45)
Elateriospermum tapos	Zygotic embryos	PVS2 (10 min, 0°C)	NR	100	(64)
Gentiana spp.	Shoot tips (2 mm)	PVS2 (120 min, 25°C)	1 min, 37°C	74	(43)

Dehydration					
Camelia sinensis	Embryonic axes	LF (4 h)	2 min, 37°C	80	(69)
Paeonia lactiflora	Shoot tips	LF (5 h)	5 min, 40°C	74	(56)
Encapsulation-dehydration					
Anacamptis morio	Seeds	LF (16 h)	RT	26	(70)
Dactylorhiza fuchsii	Seeds	0.75 M suc (18 h); LF (16 h)	RT	85	(70)
Dendranthema grandiflorum	Shoot tips (2 mm)	0.8 M suc (16 h); SG (5 h)	3 min, 35°C	64	(42)
Dianthus hybridus	Nodal segments	0.4–1.0 M suc (2d); LF (11–13 h)	25–30°C	100	(71)
Oncidium	Protocorm-like body	0.8 M suc + 0.1 M gly (65 h, 25°C); SG (5 h)	90 s, 40°C	30	(40)
Oncidium bifolium	Protocorms	0.15 M (1d), 0.25 M (2d), 0.5 M (1d), 0.75 M (1d) suc; SG (5 h)	2 min, 30°C	11	(72)
Photinia fraseri	Shoot tips	0.5 M suc (2d); LF (8 h)	3 min, 40°C	13	(31)
Ribes spp.	Shoot tips (2–3 mm)	0.75 M suc (20 h); LF (4 h)	20 min, 20°C	30–70	(47)
Rhododendron simsii	Shoot tips (2–5 mm)	LF (6 h)	2 min, 38°C	42	(73)
Rosa multiflora	Shoot tips	0.5 M suc (24 h, 18°C, D); SG (2 h)	40°C	25	(46)

"Survival" refers to the maximum percentage of shoot tip regrowth, cell/callus regeneration, or seed/embryo/synthetic seed germination. When various species were investigated, the range among them is reported (*D* darkness, *glu* glucose, *gly* glycerol, *LF* laminar flow, *NR* not reported, *RT* room temperature, *SG* silica gel, *suc* sucrose)

[a]The study investigated species belonging to the following genera: *Dianthus, Gypsophila, Lychnis, Silene* and *Vaccaria*

[b]"Droplet-vitrification" technique (*see* Protocol in Subheading 28.4.3)

In reports dealing with ornamental plants, the PVS2 treatment ranges from 5 min (shoot tips of *Chrysanthemum grandiflora*; (38)) to 3 h (encapsulated shoot tips of *Dianthus caryophyllus*; (45)). As for dehydration of encapsulated shoot tips, Lynch et al. (46) reported that 2 h on silica gel were sufficient to induce 25% of "germination" of *Rosa multiflora* synthetic seeds after storage in LN; on the other hand, synthetic seeds of *Photinia fraseri* needed to be dehydrated 8 h under the sterile air of a laminar flow hood to achieve tolerance to ultra-rapid freezing in LN (31).

Following the storage in LN, a rapid warming in a waterbath can avoid recrystallization and ensure a proper recovery of vitrified material. However, how fast this "rapid warming" should be is still controversial; indeed, warming temperatures ranging from a minimum of 20°C (47) to 45°C (22, 44) have been proposed for ornamental species. Cryoprotocols reporting on naked or encapsulated explants rewarmed at room temperature are not unusual either.

Once a cryo-protocol has been developed, it is not unusual to come across different behaviors to freezing in LN of species from the same genus or, even, of the various cells composing the explant. Fukai et al. (48) reported that shoot tips of different *Chrysanthemum* species showed great differences in their ability to survive LN; on the other hand, the same Authors developed a cryoprotocol which was effective for a wide range of *Caryophyllaceae* species (49). Morphological observations of frozen shoot tips of *C. grandiflora* showed that recoveries were mainly from leaf primordia, and only rarely from the apical dome (50). In contrast, Siebert and Wetherbee (51) reported that, in *D. caryophyllus* shoot tips, cells of the leaf primordia died, while those in the meristem dome remained viable after freezing. More recently, Halmagyi and Pinker (52) observed that the small meristematic cells of rose shoot tips, containing only few and small vacuoles, had a low water content and exhibited high tolerance to ultra-rapid freezing in LN, while the cells of the stem, having large vacuoles, collapsed and did not survive.

Chimerism is an additional problem in the cryopreservation of ornamental species, as cultivars possessing chimeral structures in the meristem can have them broken after recovery from LN. Fukai et al. (53) tried to determine whether true-to-typeness of chimeric *Chrysanthemum* plants, cv Apricot Marble, was maintained after shoot-tip cryopreservation by "slow cooling". The study revealed that, although all the plants showed complete uniformity of growth, flowering date, flower and leaf shape, they showed also a great variability in the color of flowers. Indeed, 99% of the plants derived from untreated and unfrozen shoot tips and 94% of plants derived from treated (with DMSO and glucose) but unfrozen shoot tips maintained their apricot color; differently, as much as 70% of the plants derived from slow cooled and cryostored shoot tips evidenced an altered pink color. These results suggested that explant manipulation and cryoprotection did not modify the chimeral structure of shoot tips; however, they suffered from ultra-freezing and/or thawing, which led to the production of altered adventitious shoots. A possible adventitious origin of shoots recovered from cryopreservation has also been occasionally reported (38, 48).

The risk of genetic instability has always been a cause of concern when using tissue culture for plant regeneration. However, as for cryopreservation, the literature is overall "positive" regarding the outcome of stability assessments from cryopreservation, a large number of reports showing no evidence of morphological, cytological, biochemical, or molecular alterations in plants from storage at −196°C (54). As for ornamentals, the analysis with RAPD markers did not detect any genetic alteration in plants of *Acer mono* (55) and *Paeonia lactiflora* (56) coming from the cryostorage of suspended cells and shoot tips, respectively. Conversely, an occasional genetic alteration was observed with RAPD analysis in plants from chrysanthemum shoot tips, cryopreserved either by "vitrification" or "encapsulation-dehydration" (42). Using eight primers, which yielded 103 scorable bands ranging between 200 and 2,000 bp, only one of 46 samples (coming from the "encapsulation-dehydration" technique) showed a different band pattern with one specific primer. Even if the risk appears very limited, special attention should be given to the genetic assessment of plants from cryopreservation, also in consideration of the increasing use of the technique for the establishment of gene banks.

4. Sample Protocols

4.1. Slow Growth Storage of Virus-Free Rose Germplasm (Based on (28))

The following procedure was developed for two Italian selections of rose, i.e., "Rosa Sant'Antonio di Padova" (an ancient hybrid of Tea Rose) and "Domenica" (a selection of Rosa City Flor®). It includes protocols for the establishment of shoot culture lines, the recovery of virus-free lines by in vitro thermotherapy and shoot-tip culture and the conservation of lines at 4°C in the dark.

4.1.1. Introduction and Establishment In Vitro of Shoot Culture Lines

1. Uni-nodal segments (0.4–0.7 cm) are excised from potted plants and maintained under greenhouse conditions.
2. The excised segments are disinfected in a 0.75% Na-hypochlorite solution for 12 min and rinsed three times with sterile dH_2O.
3. The segments are introduced in glass tubes, on gelled MS medium, with 30 g/l sucrose, 1.5 µM BA (benzyladenine) and 0.05 µM NAA (naphthaleneacetic acid), and cultured at 20°C,

under a 16 h photoperiod and a light intensity of 45 μM/m²/s (standard culture conditions) (Fig. 28.1a).

4. After 4 weeks, developed shoots are transferred in 500-ml glass jars on fresh MS medium, with 2 μM BA (proliferation medium) and subcultured monthly (Fig. 28.1b).

4.1.2. In Vitro Thermotherapy and Shoot-Tip Culture

1. For establishment of virus-free lines, single shoots are transferred in 80-cc plastic cylinders on the proliferation medium; the tubes are then placed in a climatic chamber, where the initial temperature of 21°C is increased gradually to 37°C in 8 days (2°C per day), prolonging the thermotherapy at 37°C for further 2 days (Fig. 28.1c).
2. After in vitro thermotherapy, shoot tips (meristem plus the first leaf primordia; 0.6–0.8 mm) are aseptically excised, using a stereomicroscope under the laminar-flow hood, from healthy shoots showing no signs of damage due to the high-temperature treatment.
3. The shoot tips are placed again in plastic cylinders in the proliferation medium and kept in standard culture conditions; new shoot culture lines are then established as reported in **step 4** of Subheading 28.4.1.1 (Fig. 28.1d).
4. New shoot lines are subjected to ELISA test, and lines positive to ApMV (Apple Mosaic Virus), ArMV (Arabidopsis Mosaic Virus) and/or PNRSV (Prunus Necrotic Ring Spot Virus) viruses are discarded.
5. Rooting of virus-free lines is achieved by transferring single shoots on gelled MS medium, supplemented with 15 g/l sucrose, 2.5 μM IBA (indole-3-butyric acid) under standard culture conditions (Fig. 28.1e).
6. Rooted shoots are acclimatized by keeping them in plastic trays on a peat:perlite (4:1) substrate (Fig. 28.1f) for 4 weeks under plastic tunnels, progressively opened to reduce gradually the relative humidity.
7. Healthy plants (5 cm, on average) are then transferred to 1-l plastic pots on the above substrate and subjected to a second ELISA test, in order to confirm they are free from the above-mentioned viruses.

4.1.3. Slow Growth Storage of Virus-Free Shoot Culture Lines

1. Virus-free shoot cultures are transferred to 500-ml glass jars (*see* Note 1) on gelled hormone-free MS medium (15 shoot clusters per jar), wrapped with plastic film and kept in standard culture conditions.
2. One week later, the jars are placed at 4°C in darkness (Fig. 28.1g). The stored material is monitored periodically (every 2 weeks, at least) to check the maintenance of healthy

Fig. 28.1. Slow growth storage of virus-free rose lines. (**a**) Uninodal segment (0.4–0.7 cm long), used for introduction in vitro. *Arrow* indicates a shoot developed from an axillary bud in 3 weeks, (**b**) a shoot culture line proliferating on gelled MS medium with 30 g/l sucrose and 2 μM BA, (**c**) single shoots under thermotherapy treatment (37°C), (**d**) 0.6–0.8-mm long shoot tip (meristem plus the first leaf primordia) used for the establishment of virus-free culture lines, (**e**) rooted shoot clusters, after culturing on gelled MS medium with 15 g/l sucrose and 2.5 μM IBA, (**f**) acclimatization of rooted shoots in plastic trays, (**g**) conservation of shoot culture lines at 4°C in the dark inside different containers (500-cc glass jars, 80-cc plastic cylinders, semi-permeable "StarPac™" plastic bags), (**h**) shoot clusters recovered from 6-month storage at 4°C in the dark, (**i**) etiolated uninodal segments from 6-month cold storage, used to recover a culture line.

conditions; in particular, the appearance of etiolated, decayed, or hyperhydric shoots is monitored.

3. (*optional*) Accumulation of CO_2 and ethylene during the conservation is monitored by gas-chromatographic analysis (*see* Note 2).
4. Conservation is stopped when visual observation of jars evidences diffused decay, hyperhydricity and/or etiolation of shoot culture lines. Also high concentration of ethylene (over 1 μl/l) in the jar atmosphere is an indicator of shoot culture decay (*see* Note 3).

5. At the end of the conservation period, shoot cultures are transferred back to standard culture conditions (Fig. 28.1h). Subculturing on the proliferation medium (*see* step 4 of Subheading 28.4.1.1) is performed after 4 days, in order to evidence the presence of contamination in the jars coming from cold storage.
6. If necessary, etiolated shoots can be recovered by placing uninodal segments horizontally in the proliferation medium, under standard culture conditions (Fig. 28.1i). After 3 weeks, developed shoots are subcultured as stated in point 4.

4.2. Production of Synthetic Seeds for the In Vitro Conservation of Ornamental Plants (Based on (31))

This procedure for the production of synthetic seeds (containing apical or axillary buds) reported here was developed with three ornamental shrubs, i.e., lilac (*Syringa vulgaris* L.), oleander (*Nerium oleander* L.), and photinia (*Photinia fraseri* Dress.). The synthetic seeds were then used in slow growth storage and cryopreservation trials.

4.2.1. Shoot Culture Proliferation

1. Shoots of lilac are monthly subcultured on gelled DKW medium (74) supplemented with 8.8 μM BA
2. Shoots of oleander are monthly subcultured on gelled DKW medium, supplemented with 4.4 μM BA.
3. Shoots of photinia are monthly subcultured on gelled QL medium, supplemented with 4.4 μM BA.
4. Cultures are kept in 500-ml glass jars at 23°C, 16-h photoperiod (60 μM/m^2/s) until they are used for the preparation of synthetic seeds.

4.2.2. Preparation, Germination, and Conservation of Synthetic Seeds

1. For all species, aseptically excised apical or axillary buds are immersed in liquid MS medium (devoid of calcium), with 30 g/l sucrose and 3% Na-alginate.
2. The buds are then individually sucked by using a micropipette (with a tip trimmed to obtain a 2–4-ml hole) and released drop-wise (each drop containing a single bud) into 100 mM $CaCl_2 \cdot 2H_2O$ solution maintained under gentle agitation to avoid sticking of beads to each other.
3. Following 30 min of bead hardening, the synthetic seeds are collected with a sterile sieve, washed with sterile dH_2O.
4. They are then placed on gelled hormone-free MS medium in Petri dishes and transferred to 23°C at a 16-h photoperiod for germination, or stored at 4°C in darkness for conservation.

4.2.3. Cryopreservation of Photinia Synthetic Seeds by Encapsulation-Dehydration

1. Shoot cultures of photinia are hardened at 4°C in darkness for 2–3 weeks
2. Buds are aseptically excised and encapsulated as described above (**steps 1–3** of Subheading 28.4.2.2)

3. Synthetic seeds are incubated on gelled hormone-free MS medium, containing 0.5 M sucrose for 2 days at 23°C.
4. After sucrose preculture, beads are dehydrated under the sterile air of a laminar flow cabinet (at about 0.8 m/s) for 8 h (during dehydration, room temperature was 20°C and the relative humidity was 60%).
5. The beads are then inserted in 2-ml cryovials (eight beads per cyrovial) and directly immersed in liquid nitrogen.
6. Cryovials are thawed in water bath at 40°C for 3 min, after which the synthetic seeds are germinated as described in step 8 of Subheading 28.4.3.2.

4.3. Cryopreservation of *Chrysanthemum morifolium* Shoot Tips Using Different Techniques (Based on (38))

These authors compared various techniques of "slow cooling" or direct immersion in LN (one-step freezing) for cryopreservation of *C. morifolium* shoot tips.

4.3.1. Slow Cooling

1. Shoot tips (3–4-mm long, with two leaf primordia) are aseptically excised, under stereomicroscope, from in vitro-grown 4-week-old mother plants, cultured at 24°C on hormone-free MS medium, under a 16-h photoperiod and a light intensity of 25 $\mu M/m^2/s$.
2. The shoot tips are transferred on filter paper inside a Petri dish, soaked for 24 h with MS medium, containing 30 g/l sucrose, 1 mg/l BA and 0.1 mg/l NAA (shoot inducing medium), followed by a second 24-h incubation in the same medium, but containing 0.5 M sucrose.
3. For cryoprotection, shoot tips are incubated for 2 h at room temperature in MS medium containing 10% DMSO, and then placed with 1-ml medium into 2 ml cryovials.
4. Cryovials are placed in a programmable freezer and cooled to –40°C (cooling rate: –0.25°C/min) before being plunged into LN.
5. For thawing, cryovials are incubated in a 40°C water bath. The shoot tips are then transferred on the shoot-inducing medium.

4.3.2. Encapsulation-Dehydration

1. Shoot tips are prepared as described in steps 1 and 2 of Subheading 28.4.3.1.
2. Shoot tips are then immersed in Ca^{2+}-free medium, containing 3% Na-alginate.
3. The solution is added drop-wise (each drop containing a shoot tip) to a medium containing 0.1 M $CaCl_2$.
4. After 30 min of gentle stirring, the medium is poured off and the synthetic seeds are washed three times with MS medium containing 30 g/l sucrose, 1 mg/l BA, and 0.1 mg/l NAA.
5. The synthetic seeds are then incubated for 4 h in the same medium but containing 0.75 M sucrose.

6. The medium is then poured off, beads are dried on sterile filter paper to remove adherent medium, placed in an open Petri dish (Ø 90 mm) under the sterile air of a laminar flow hood and dehydrated for 6 h.
7. After dehydration, beads are placed in a cryovial and immersed in LN.
8. For thawing, the synthetic seeds are poured into 15-ml MS medium, containing 30 g/l sucrose, 1 mg/l BA, and 0.1 mg/l NAA.
9. The synthetic seeds are re-cultured on the same medium for the "germination".

4.3.3. Droplet-Freezing Method

1. Following steps 1 and 2 of Subheading 28.4.3.1, shoot tips are placed on filter paper soaked with shoot inducing medium, containing 7.5% DMSO for cryoprotection.
2. After a 2-h exposure, shoot tips are plunged directly into cryovials filled with LN using a hypodermic needle. Cryovials are cooled to LN temperature to prevent boiling of the LN inside the vials.
3. Rewarming is performed in MS medium with 50 g/l sucrose at 35°C for 30 min.
4. The shoot tips are then placed in small Petri dishes (Ø 30 mm) and covered with a liquid drop of low-melting 1% agarose.
5. Following solidification of the agarose, liquid shoot inducing medium is added, and the Petri dishes are kept in translucent plastic boxes for 4 weeks.

4.3.4. Droplet-Vitrification

1. Shoot tips are prepared as described in steps 1 and 2 of Subheading 28.4.3.1.
2. They are then precultured for 24 h in the same medium, but with 0.5 M sucrose.
3. Shoot tips are incubated in 100% PVS2 solution for 5 min.
4. Then, they are placed in 4-μl droplets of PVS2 solution on small strips of thin aluminium foils (2 × 0.8 mm). Each foil is then immersed in LN in 2-ml cryovial, cooled from underneath to LN temperature to prevent boiling of the LN inside the vial.
5. For thawing, the aluminium foils with the shoot tips are immersed in 15-ml MS medium at room temperature in a Petri dish (Ø 60 mm). The shoot tips are then recultured on shoot inducing medium.

5. Notes

1. Alternatively, the shoots can be stored inside the above-mentioned 80-cc plastic cylinders or semi-permeable plastic bags ("Star-Pac"™), one shoot cluster per cylinder and cell.

2. For the determination of CO_2 and ethylene accumulation, some glass jars are provided of a small hole in the cap, tightly closed with a rubber stopper. Gas samples (1 ml) are withdrawn weekly from the headspace of the jars, and injected into a gas chromatograph, equipped with thermal conductivity detector (TCD) and Poropaq N (80–100 Mesh) column (CO_2), and with flame ionization detector (FID) and Poropak Q (80–100 Mesh) column (ethylene). Column-injector-detector temperatures are set according to the instructions of the column producer. Carrier gases are He (CO_2) and N_2 (ethylene), both at 20 psi.
3. With the cvs Sant'Antonio di Padova and Domenica the storage of shoot cultures at 4°C in the dark can be safely prolonged up to 6 months.

References

1. Turner T (2005) Garden history : Philosophy and design 2000 BC–2000 AD. Abingdon, Spon 304 pp
2. Reed BM (1992) Cold storage of strawberries in vitro: A comparison of three storage systems. Fruit Varieties J 46:98–102
3. Lambardi M, De Carlo A (2003) Application of tissue culture to the germplasm conservation of temperate broad-leaf trees. In: Jain SM, Ishii K (eds) Micropropagation of woody trees and fruits. Kluwer, The Netherlands, pp 815–840
4. Lambardi M, Benelli C, De Carlo A, Fabbri A, Grassi S, Lynch PT (2002) Medium- and long-term in vitro conservation of olive germplasm (*Olea europea* L.). Acta Hortic 586:109–112
5. Withers LA, Engelmann F (1997) In vitro conservation of plant genetic resources. In: Altman A (ed) Biotechnology in agriculture. Marcel Dekker, New York, pp 57–88
6. Ashmore SE (1997) Status report on the development and application of in vitro techniques for the conservation and use of plant genetic resources. International Plant Genetic Resources Institute, Rome 67 pp
7. Benson EE (2008) Cryopreservation theory. In: Reed BM (ed) Plant cryopreservation: A practical guide. Springer, USA, pp 15–32
8. Fahy GM, MacFarlane DR, Angell CA, Meryman HT (1984) Vitrification as an approach to cryopreservation. Cryobiology 21:407–426
9. Sakai A, Hirai D, Niino T (2008) Development of PVS-based vitrification and encapsulation-vitrification protocols. In: Reed BM (ed) Plant cryopreservation: A practical guide. Springer, USA, pp 33–57
10. Lambardi M, Ozudogru EA, Benelli C (2008) Cryopreservation of embryonic cultures. In. Reed BM (ed) Plant cryopreservation: A practical guide. Springer, USA, pp 177–210
11. Walters C, Wesley-Smith J, Crane J, Hill LM, Chmielarz P, Pammenter NW, Berjak P (2008) Cryopreservation of recalcitrant (i.e. desiccation-sensitive) seeds. In: Reed BM (ed) Plant cryopreservation: A practical guide. Springer, USA, pp 465–484
12. Pritchard HW, Nadarajan J (2008) Cryopreservation of orthodox (desiccation tolerant) seeds. In: Reed BM (ed) Plant cryopreservation: A practical guide. Springer, USA, pp 485–501
13. Normah MN, Makeen AM (2008) Cryopreservation of excised embryos and embryonic axes. In: Reed BM (ed) Plant cryopreservation: A practical guide. Springer, USA, pp 211–240
14. Towill LE, Ellis DD (2008) Cryopreservation of dormant buds. In: Reed BM (ed) Plant cryopreservation: A practical guide. Springer, USA, pp 421–464
15. Sakai A, Kaboyashi S, Oiyama I (1990) Cryopreservation of nucellar cells of navel orange (*Citrus sinensis* Osb. var. brasiliensis Tanaka) by vitrification. Plant Cell Rep 9:30–33
16. Fabre J, Dereuddre J (1990) Encapsulation-dehydration: A new approach to cryopreservation of *Solanum* shoot tips. CryoLetters 11:413–426
17. Panis B, Lambardi M (2006) Status of cryopreservation technologies in plants (crops and forest trees). In: Ruane J, Sonnino A (eds)

The role of biotechnology in exploring and protecting agricultural genetic resources. Food and Agriculture Organization of the United Nations, Rome, Italy, pp 61–78

18. Murashige T (1977) Plant cell and organ cultures as horticultural practices. Acta Hortic 78:17–30
19. Lambardi M, Benelli C, Ozudogru EA, Ozden-Tokatli Y (2006) Synthetic seed technology in ornamental plants. In: Teixeira da Silva JA (ed) Floriculture, ornamental and plant biotechnology: Advances and topical issues, vol II. Global Science Books, UK, pp 347–354
20. Redenbaugh K, Fujii J, Slade D, Viss P, Kossler M (1991) Artificial seeds-Encapsulated somatic embryos. In: Bajaj YPS (ed) Biotechnology in agriculture and forestry, high-tech and micropropagation. Springer, Heidelberg, pp 395–416
21. Ballester A, Janeiro LV, Vieitez AM (1997) Cold storage of shoot cultures and alginate encapsulation of shoot tips of *Camellia japonica* L. and *Camellia reticulata* Lindley. Sci Hortic 71:67–78
22. Reed BM, Okut N, Achino JD, Narver L, DeNoma J (2003) Cold storage and cryopreservation of hops (*Humulus* L.) shoot cultures through application of standard protocols. CryoLetters 24:389–396
23. Nunes EDC, Benson EE, Oltramari AC, Araujo PS, Righetto M, Viana M (2003) In vitro conservation of *Cederla fissilis* Vellozo (Meliaceae), a native tree of the Brazilian Atlantic Forest. Biodivers Conserv 12:837–848
24. West TP, Ravindra MB, Preece JE (2006) Encapsulation, cold storage, and growth of *Hibiscus moscheutos* nodal segments. Plant Cell Tissue Organ Cult 87:223–231
25. Mallón R, Barros P, Luzardo A, González ML (2007) Encapsulation of moss buds: an efficient method for the in vitro conservation and regeneration of the endangered moss *Splachnum ampullaceum*. Plant Cell Tissue Organ Cult 88:41–49
26. Orlikowska T (1992) Effect of in vitro storage at 4°C on surviving and proliferation of two apple rootstocks. Plant Cell Tissue Organ Cult 31:1–7
27. Hempel T, Hempel M (1984) Storage of gerbera propagated in vitro. In: Proceedings of International Symposium on "Plant Tissue and Cell Culture Application to Crop Improvement", Czechoslovakia, pp 559–560
28. Previati A, Benelli C, Da Re F, Ozudogru A, Lambardi M (2008) Micropropagation and in vitro conservation of virus-free rose germplasm. Propagat Ornam Plants 8:93–98
29. Janeiro LV, Ballester A, Vieitez AM (1997) In vitro response of encapsulated somatic embryos of camellia. Plant Cell Tissue Organ Cult 51:119–125
30. Pattnaik S, Chand PK (2000) Morphogenic response of the alginate-encapsulated axillary buds from in vitro shoot cultures of six mulberries. Plant Cell Tissue Organ Cult 60:177–185
31. Ozden-Tokatli Y, De Carlo A, Gumusel F, Pignattelli S, Lambardi M (2008) Development of encapsulation techniques for the production and conservation of synthetic seeds in ornamental plants. Propagat Ornam Plants 8(1):17–22
32. Murashige T, Skoog F (1962) A revised medium for rapid growth and bioassays with tobacco tissue cultures. Physiol Plant 15:473–479
33. Quoirin M, LePoivre P (1977) Etude de milieux adaptés aux cultures in vitro de *Prunus*. Acta Hortic 78:437–442
34. Patel AV, Pusch I, Mix-Wagner G, Vorlop KD (2000) A novel encapsulation technique for the production of artificial seeds. Plant Cell Rep 19:868–874
35. Standardi A, Micheli M, Piccioni E (1995) Incapsulamento in alginato di espianti micropropagati. Italus Hortus 2(1–2):46–48
36. Ipekci Z, Gozukirmizi N (2003) Direct somatic embryogenesis and synthetic seed production from *Paulownia elongate*. Plant Cell Rep 22:16–24
37. Refouvelet E, Le Nours S, Tallon C, Daguin F (1998) A new method for in vitro propagation of lilac (*Syringa vulgaris* L.): Regrowth and storage conditions for axillary bud encapsulation in alginate beads, development of a preacclimatisation stage. Sci Hortic 74:233–241
38. Halmagyi A, Fischer-Klüver G, Mix-Wagner G, Schumacher HM (2004) Cryopreservation of *Chrysanthemum morifolium* (*Dendranthema grandiflora* Ramat.) using different approaches. Plant Cell Rep 22:371–375
39. Schäfer-Menuhr A, Mix-Wagner G, Schumacher HM (1997) Cryopreservation of potato cultivars-design of a method for routine application in genebanks. Acta Hortic 447: 447–483
40. Miao N-H, Kaneko Y, Sugawara Y (2005) Ultrastructural implications of pretreatment for successful cryopreservation of *Oncidium* protocorm-like body. CryoLetters 26:333–340
41. Halmagyi A, Pinker I (2006) Cryopreservation of *Rosa* shoot tips: importance of preculture conditions. Acta Hortic 725:351–356
42. Martin C, González-Benito ME (2005) Survival and genetic stability of *Dendranthema grandiflora* Tzvelev shoot apices after cryopreservation

by vitrification and encapsulation-dehydration. Cryobiology 51:281–289
43. Tanaka D, Niino T, Isuzugawa K, Hikage T, Uemura M (2004) Cryopreservation of shoot apices of in-vitro grown gentian plants: comparison of vitrification and encapsulation-vitrification protocols. CryoLetters 25:167–176
44. Lambardi M, De Carlo A, Capuana M (2005) Cryopreservation of embryogenic callus of *Aesculus hippocastanum* L. by vitrification/one-step freezing. CryoLetters 26:185–192
45. Halmagyi A, Deliu C (2007) Cryopreservation of carnation (*Dianthus caryophyllus* L.) shoot tips by encapsulation-vitrification. Sci Hortic 113:300–306
46. Lynch P, Harris WC, Chartier-Hollis JM (1996) The cryopreservation of shoot tips of *Rosa multiflora*. Plant Growth Regul 20:43–45
47. Johnson JW, Harding K, Benson EE (2007) Antioxidant status and genotypic tolerance of *Ribes* in vitro cultures to cryopreservation. Plant Sci 172:524–534
48. Fukai S, Goi M, Tanaka M (1991) Cryopreservation of shoot tips of *Chrysanthemum morifolium* and related species native to Japan. Euphytica 54:201–204
49. Fukai S, Goi M, Tanaka M (1991) Cryopreservation of shoot tips of *Carophyllaceae* ornamentals. Euphytica 56:149–153
50. Fukai S, Oe M (1990) Morphological observations of chrysanthemum shoot tips cultured after cryopreservation and freezing. J Jpn Soc Hort Sci 59:383–387
51. Siebert M, Wetherbee PJ (1977) Increased survival and differentiation of frozen herbaceous plant organ cultures through cold treatment. Plant Physiol 59:1043–1046
52. Halmagyi A, Pinker I (2006) Plant regeneration from *Rosa* shoot tips cryopreserved by a combined droplet vitrification method. Plant Cell Tissue Organ Cult 84:145–153
53. Fukai S, Goi M, Tanaka M (1994) The chimeric structure of the apical dome of chrysanthemum (*Dendranthema grandiflorum* (Ramat.) Kitam.) is affected by cryopreservation. Sci Hortic 57:347–351
54. Harding K (2004) Genetic integrity of cryopreserved plant cells: A review. CryoLetters 25:3–22
55. Park Y-G, Kwon G-S, Tay D (2005) Cryopreservation for gene conservation of *Acer mono* Max. Propagat Ornam Plants 5(2): 78–83
56. Seo MJ, Shin JH, Sohn JK (2007) Cryopreservation of dormant herbaceous peony (*Paeonia lactiflora* Pall.) shoot-tips by desiccation. CryoLetters 28:207–213
57. Ryyänen L, Aronen T (2007) Phenotypic expression of leaf variegation in two *Betula pendula* Roth genotypes following micropropagation, cryopreservation and grafting. Propagat Ornam Plants 7(1):23–28
58. Winkelmann T, Mußmann V, Serek M (2004) Cryopreservation of embryogenic suspension cultures of *Cyclamen persicum* Mill. Plant Cell Rep 23:1–8
59. Scocchi A, Vila S, Mroginski L, Engelmann F (2007) Cryopreservation of somatic embryos of paradise tree (*Melia azedarach* L.). CryoLetters 28:281–290
60. Gale S, John A, Benson EE (2007) Cryopreservation of *Picea sitchensis* (Sitka spruce) embryogenic suspensor masses. CryoLetters 28:225–239
61. Towill LE, Widrlechner M (2004) Cryopreservation of *Salix* species using sections from winter vegetative scions. CryoLetters 25:71–80
62. Sant R, Taylor M, Tyagi A (2006) Cryopreservation of in vitro-grown shoot-tips of tropical taro (*Colocasia esculenta* var. Esculenta) by vitrification. CryoLetters 27:133–142
63. Thammasiri K (2000) Cryopreservation of seeds of a Thai orchid (*Doritis pulcherrima* Lindl.) by vitrification. CryoLetters 21:237–244
64. Zaimah NA, Benson EE, Marzalina M (2007) Viability of *Elateriospermum tapos* (Perah) embryo after storage in liquid nitrogen. J Trop For Sci 19(1):1–5
65. Matsumoto T, Takahashi C, Sakai A, Nako Y (1998) Cryopreservation of in vitro-grown apical meristems of hybrid statice by three different procedures. Sci Hortic 76:105–114
66. Yang B-H, Kim H-T, Park J-Y, Park Y-G (2006) Cryopreservation of in vitro-cultured axillary shoot tips of Japanese bead tree (*Melia azedarach*) using vitrification technique. Korean J Plant Res 19(3):385–391
67. Hirano T, Ishikawa K, Mii M (2005) Cryopreservation of immature seeds of *Ponerorchis graminifolia* var. suzukiana by vitrification. CryoLetters 26:139–146
68. Nadarajan J, Staines HJ, Benson EE, Mansor M, Krishnapillay B, Harding K (2007) Optimization of cryopreservation for *Sterculia cordata* zygotic embryos using vitrification techniques. J Trop For Sci 19(2):79–85
69. Kim H-H, Cha Y-S, Baek H-J, Cho E-G, Chae Y-A, Engelmann F (2002) Cryopreservation of tea (*Camellia sinensis* L.) seeds and embryonic axes. CryoLetters 23:209–216
70. Wood CB, Pritchard HW, Miller AP (2000) Simultaneous preservation of orchid seed and its fungal symbiont using encapsulation-dehydration is dependent on moisture

content and storage temperature. CryoLetters 21:125–136

71. Fukai S, Togashi M, Goi M (1994) Cryopreservation of in vitro-grown *Dianthus* by encapsulation-dehydration. Tech Bull Fac Agric Kagawa Univ 46:101–107
72. Flachsland E, Terada G, Scocchi A, Rey H, Mroginski L, Engelmann F (2006) Cryopreservation of seeds and in vitro-cultured protocorms of *Oncidium bifolium* Sims. (*Orchidaceae*) by encapsulation-dehydration. CryoLetters 27:235–242
73. Verleysen H, Bockstaele EV, Debergh P (2005) An encapsulation-dehydration protocol for cryopreservation of the azalea cultivar 'Nordlicht' (*Rhododendron simsii* Planch.). Sci Hortic 106:402–414
74. Driwer JA, Kuniyuku AH (1984) In vitro propagation of paradox walnut rootstock. HortScience 19:507–509

Chapter 29

Genetic Engineering of Novel Flower Colors in Floricultural Plants: Recent Advances via Transgenic Approaches

Masahiro Nishihara and Takashi Nakatsuka

Abstract

Since the first successful genetic engineering of flower color in petunia, several new techniques have been developed and applied to modify flower color not only in model plants but also in floricultural plants. A typical example is the commercial violet-flowered carnation "Moondust series" developed by Suntry Ltd. and Florigene Ltd. More recently, blue-flowered roses have been successfully produced and are expected to be commercially available in the near future. In recent years, successful modification of flower color by sophisticated regulation of flower-pigment metabolic pathways has become possible. In this chapter, we review recent advances in flower color modification by genetic engineering, especially focusing on the methodology. We have included our own recent results on successful production of flower-color-modified transgenic plants in a model plant, tobacco and an ornamental plant, gentian. Based on these results, genetic engineering of flower color for improvement of floricultural plants is discussed.

Key words: Carotenoid, Flavonoid, Flower color, Genetic engineering, Gentian, RNAi, Transgenic plants

1. Introduction

There are more than 300,000 species of flowering plants, which are classified in more than 400 families. Many beautiful flowers of various shapes, colors, and fragrances are found in nature. Except for some modern domesticated or anemophilous flowers, they have evolved naturally, principally for attracting pollinators such as insects and small birds.

Among floral organs, the petals and in some cases sepals, are usually showy and colorful, depending on the species, and make the flower visually attractive. Even within a species, variations in flower pigmentation frequently exist. The most critical factor determining

S.M. Jain and S.J. Ochatt (eds.), *Protocols for In Vitro Propagation of Ornamental Plants*, Methods in Molecular Biology, vol. 589
DOI 10.1007/978-1-60327-114-1_29, © Humana Press, a part of Springer Science+Business Media, LLC 2010

flower color is the components of accumulated pigments. Flowers can accumulate secondary metabolites such as flavonoids, carotenoids, and betalains. These pigments are known to have multiple functions such as protection against ultraviolet (UV) radiation, defense against pathogen attack, attraction of pollinators, and signal compounds to symbionts (1–3). The chemical features of these pigments have been well studied and characterized (4–9). Along with the development of molecular biology, various metabolic pathways involved in the biosynthesis of these flower pigments have been revealed. An overview of the flavonoid biosynthetic pathway is shown in Fig. 29.1. The contribution of each pigment to plant flower pigmentation has been well reviewed by Grotewold (10). The elucidation and genetic engineering of their biosynthetic pathways have been attempted by many researchers, and their efforts have been summarized previously (11–15).

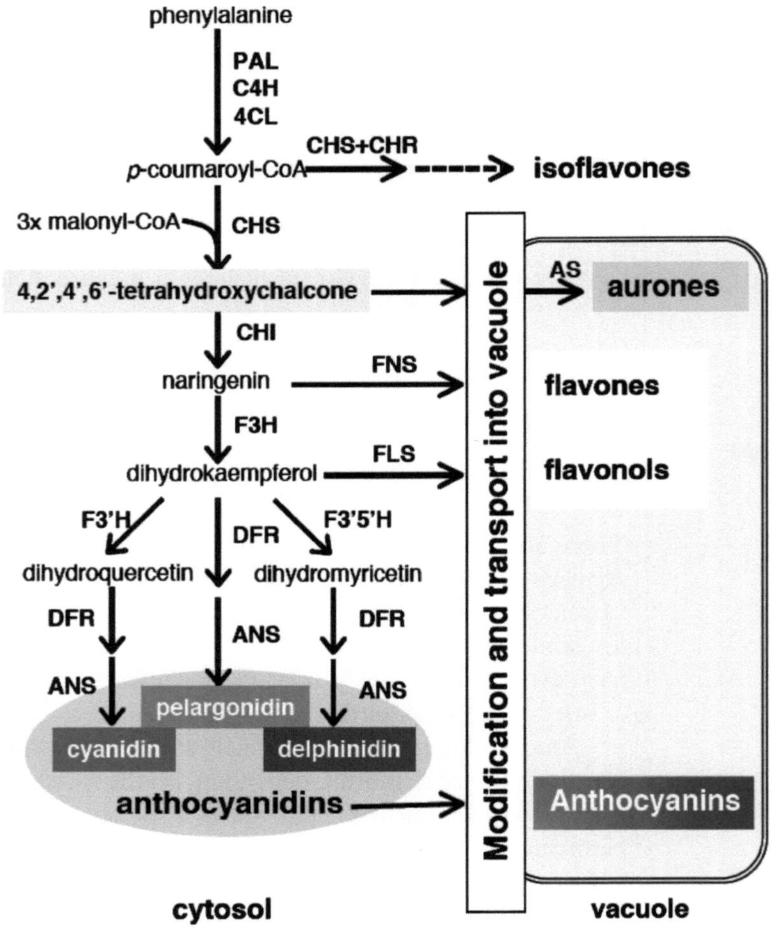

Fig. 29.1. General flavonoid biosynthetic pathway in higher plants. *PAL* phenylalanine ammonia-lyase, *C4H* cinnamate-4-hydroxylase, *4CL* 4-coumarate:CoA ligase, *CHS* chalcone synthase, *CHR* chalcone reductase, *CHI* chalcone isomerase, *F3H* flavanone 3-hydroxylase, *DFR* dihydroflavonol 4-reductase, *ANS* anthocyanidin synthase, *F3′H* flavonoid 3′-hydroxylase, *F3′5′H* flavonoid 3′, 5′-hydroxylase, *FNS* flavone synthase, *FLS* flavonol synthase, *AS* aureusidin synthase.

The number of plant species that can be genetically transformed is increasing every year, as development of reliable transformation methods is being extensively studied. Much information on this topic is readily available (16–18). For example, the procedure for transformation of chrysanthemum has been improved (19) and is now used for modifying many traits such as disease resistance, flowering time, and flower color. Our research groups have also obtained transgenic gentian plants by both particle bombardment and *Agrobacterium*-mediated transformation (20, 21 etc.). Thus, it has become possible to modify flower color by direct manipulation of the specific genes in many plant species. Among the secondary metabolic pathways so far studied, flavonoid biosynthesis is one of the best-characterized pathways in higher plants. The first genetic engineering of flower color reported was the modification of flavonoid biosynthesis in petunia (22). Subsequently, many researchers have attempted to modify flower color by regulating the flavonoid biosynthetic pathway. Such work has been well reviewed (23–27). Recently, engineering of carotenoid pigments to modify flower color was successfully achieved by several research groups, including our own (28, 29). Therefore, we will cover both flavonoid and carotenoid engineering in this review.

2. Advantage of Genetic Engineering Over Conventional Breeding

Conventional breeding by repeated artificial crossing is the oldest breeding strategy adopted by humans. This is a dependable method, but is laborious and requires a relatively long time. Recent advances in biotechnology enabled the use of various techniques such as tissue culture and mutation induction by UV, ionizing radiation, and also by chemical mutagens (e.g., EMS and sodium azide) to raise the breeding efficiency (30, 31). For example, employing a combination of tissue culture and mutation induction is very useful and indeed many successful results are obtained. Recently, mutagenesis using heavy-ion beam irradiation has been shown to be useful for flower color modification (32). In addition, "molecular breeding" by genetic transformation of specific genes of interest has become possible. Molecular biology techniques have become more widely adopted by researchers and plant breeders, therefore genetic engineering is now a viable method to generate new cultivars. To modify the flower color of a certain species by traditional breeding methods, many crosses and selections were often necessary to obtain the desired flower color. As traits other than flower color are simultaneously changed during hybridization, repeated backcrossing is necessary to recover original traits. Furthermore, this hybridization strategy cannot be performed if useful genetic resources are unavailable in related species or other populations. In contrast, a transgenic approach can generate a new color comparatively quickly without changing other traits. Flower-color-modified transgenic plants are very useful for breeding because the traits are dominantly inherited and fixed by a single cross (33). Of course, genetic stability should be carefully checked, because transgenes tend to undergo gene silencing or epigenetic changes.

3. Modification of Flower Color by Genetic Engineering

The breeding of floricultural plants requires paying attention to many traits, such as flower color, shape, size, and flowering time. In particular, flower color is one of the most prominent traits in breeding strategies, and flower markets and consumers demand a wide range of color variations in each crop. From ancient times, breeders have invested much effort into producing novel flower colors, and consequently most cultivated flowers have taken on various novel colors. For example, the wild prairie gentian or lisianthus (*Eustoma grandiflorum*) has purple flowers, but cultivars with white, pink, magenta, yellow, or orange flowers have been raised by conventional hybridization breeding. Fortunately, in the case of lisianthus, related wild species or mutants with various other flower colors existed and could be utilized for breeding. However, such genetic resources are not available for all floricultural plants. For example, despite concerted breeding during the last two centuries, roses, chrysanthemums, and lilies lack vivid blue flowers, and cyclamen and gentian lack yellow flowers. To overcome this problem, transgenic approaches are most promising, although many difficulties for their practical use remain to be solved. This is due not only to technical problems, but also to economic or social aspects. Other traits, such as flower shape, flowering time, disease resistance, and fragrance, are also important targets of transgenic approaches and are a major focus of ongoing research projects.

About ten years ago, the first genetically modified (GM) flower cultivar became commercially available, namely the purple-mauve mini carnation "Moondust" (17, 24, 34). GM cut flowers have been accepted by consumers without any problems and are used as bridal and gift items. The "Moondust" series was developed to accumulate delphinidin pigments in the petals, which are not synthesized in nontransgenic carnation by introducing petunia flavonoid 3′,5′-hydroxylase (*F3′5′H*) and petunia dihydroflavonol 4-reductase (*DFR*) genes into carnation. In this case, selection of host plants for transformation was important; a white-flowered host cultivar of carnation that lacked endogenous DFR enzyme activity was used for transformation with multiple genes (24).

Recently, a transgenic rose was developed by Suntry Ltd using a similar strategy (35) (see also below). These studies implied that transgenic approaches should be carefully applied to obtain the desired traits. There are three important aspects to be considered for the development of flower color-modified transgenic plants (1) Genetic engineering methods; (2) elucidation of pigment biosynthesis and use of suitable genes; and (3) evaluation of biosafety for genetically modified organisms (GMO). Among these, genetic engineering includes reliable methods to produce transgenic plants and to control the transgene. Plant transformation methods have been aggressively developed in many plant species (16, 17), and many protocols for a wide diversity of plant species are now available. Techniques to control the transgenes in plants are being improved every year, and this review covers only transgenic approaches in the field of flower color modification. With regard to pigment biosynthesis, there are many excellent reviews already mentioned and we will not elaborate this. Guidelines for risk assessment of GMOs differ among countries and plant species and are also beyond the scope of this review. However, the latter is the most important aspect for practical use of flower-color-modified plants produced using a transgenic approach, and the Japanese situation has been reviewed recently (36). Overall, the public is more accepting of GM floricultural plants, compared with GM food crops. However, assessment of the environmental effects should be similarly carefully handled in flowers as well as food crops to prevent genetic introgression with wild species. It should be noted that all of the transgenic carnations commercially available rarely produce viable pollen (36), and sterility is an essential trait for their commercial release.

In the following sections, we will briefly introduce the target pigments for flower color modification by transgenic approaches. Three main types of pigments are found in flowering plants and are responsible for flower coloration.

4. Targeted Plant Pigments for Modification of Flower Color

There are more than 200,000 kinds of compounds in higher plants, of which many include pigmented compounds (37). Pigmentation in flowers and fruits is derived from accumulation of certain visible compounds in these organs. For example, flavonoids, carotenoids, and betalains are the three major pigments that contribute to flower coloration (10). The chemical nature of each pigment is well-described elsewhere (e.g., (7–9)) and will be mentioned only briefly here.

Flavonoids and anthocyanins are responsible for a wide range of colors including pale yellow, orange, red, pink, and blue.

They are generally accumulated in vacuoles as water-soluble products and are widely distributed in plants. Carotenoids are isoprenoids, responsible for colors from yellow and orange to red, and are accumulated in chromoplasts as fat-soluble products. It should be noted that carotenoids are essential for plant survival because they play important roles in photosynthesis and also serve as substrates of plant hormones. Therefore, carotenoids are ubiquitously distributed in plants. Betalains are nitrogen-containing compounds, also showing yellow, orange to red coloration, and similar to flavonoids, are accumulated in vacuoles. Betalains are found only in the order Caryophyllales. Betalain biosynthesis is partly characterized, but remains to be elucidated fully, especially a tyrosine hydroxylation step that is probably catalyzed by a tyrosine hydroxylase such as tyrosinase and polyphenol oxidase (PPO). Anthocyanins and betalains do not coexist, and the reason for this is not yet fully understood. Shimada et al. (38, 39) reported that recombinant DFR and anthocyanidin synthase (ANS) proteins of *Spinacia oleracea* and *Phytolacca americana* expressed in *Escherichia coli* do exhibit enzymatic activity, namely conversion of dihydroquercetin and leucocyanidin into leucocyanidin and cyanidin, respectively. The evolutionary advantage for Caryophyllales in producing betalains instead of anthocyanins remains to be revealed in future studies. Genetic engineering of flavonoids, anthocyanins, and carotenoids has been achieved, but so far no successful modification of flower color with betalain pigments is reported. The possible manipulation of betalain biosynthesis is very interesting and a transgenic approach to modification of flower color with betalains might be successful in the near future. For that purpose, the key enzyme PPO-type tyrosinase involved in betalain synthesis must be isolated and characterized. In the orchid *Oncidium* "Gower Ramsey", both anthocyanin and carotenoid genes are shown to contribute to the color phenotype (40). Because the coexistence of flower pigments is useful to generate novel colors, engineering of flavonoids, carotenoids, and also betalains should be targeted for genetic manipulation.

Recent examples of flower-color-modified plants were summarized in Dr. Tanaka's recent review (26); readers can get useful information from this review. There is no successful example for monocotyledonous plants, because they are relatively difficult to transform compared to dicotyledonous plants. However, monocots contain many commercially important floricultural plants such as lilies and orchids. Recently, several effective transgenic protocols have been developed in these plants (e.g., (41–44)), therefore modification of flower colors in monocotyledonous plants is certain to increase in the future. For example, it has been announced that Suntry Ltd and Niigata prefecture are developing a blue-flowered lily in Japan.

In the following section, various methods to modify flower color are introduced. Among these, RNA interference (RNAi), which is a relatively recently developed method, is proven to be a most efficient and strong gene-silencing technique for flower color modification and is a particular focus of this review.

5. Development of Flower Color Modification Methods

5.1. Suppression of Endogenous Flavonoid Biosynthetic Genes

Most flower color modifications in transgenic plants have involved suppression of the particular flavonoid biosynthetic gene. Early reports used antisense and sense (cosuppression) technology to suppress the expression of an endogenous gene. The first successful engineering of flower color modification by gene suppression was the production of white-flowered petunia plants by the introduction of the antisense chalcone synthase *(CHS)* gene (45). About the same time, it was demonstrated that cosuppression technology is able to downregulate the expression of anthocyanin biosynthetic genes (46, 47). In the subsequent decade, RNAi technology has been frequently used as a gene-suppression method in plants, because its suppression is more efficient and stronger than former approaches, and RNAi has been applied to downregulate gene expression and modify flower colors (48–51).

Regarding color, most early studies produced white-flowered transgenic plants by either antisense or sense suppression of *CHS* genes in chrysanthemum (52), gerbera (53), lisianthus (54), torenia (55, 56) and gentian (33). In addition, *DFR* genes have been suppressed to modify flower color using molecular engineering (47, 55). Suppression of different steps in the flavonoid biosynthetic pathway resulted in different flower colors. For example, transgenic torenia plants introduced with antisense *CHS* and *DFR* genes exhibited white and pale-blue flower colors, respectively (57). Similarly, transgenic torenia and gentian plants in which the *ANS* gene was suppressed by RNAi also showed pale-blue flower colors (50, 58). The downregulation of genes encoding enzymes catalyzing later steps in the flavonoid biosynthetic pathway results in the reduction of anthocyanin production, but does not affect the composition of flavonols and/or flavones, which are colorless flavonoids contributing to copigmentation. We have also reported that *CHI*-suppressed tobacco plants (generated using RNAi) have yellowish-white flowers and yellow-colored pollen, accumulating chalcone derivatives (49). Generally, the suppression of genes encoding an enzyme catalyzing an early step in flavonoid biosysnthesis, such as *CHS*, can produce a purer-white flower than that resulting from the suppression of genes involved in later steps. As another example, the suppression of *F3′5′H*, encoding an enzyme that catalyzes the hydroxylation of

3′- and 5′-groups in the B-ring of dihydroflavonols, induced a change from blue to either pink or white flowers in torenia, petunia, *Nierembergia*, and gentian plants (56, 58–60).

Our data indicated that the antisense method seemed to be less efficient than RNAi for suppression of *CHS* in gentian plants. Both techniques could be applied to change the gentian flower color from blue to white, but the antisense efficiency (18%) (33) was lower than that with RNAi (70%) (58), although the promoters used were different (CaMV35S vs *rol*C). This may be partly due to the fact that the CaMV35S promoter underwent strong transcriptional gene silencing (TGS) in gentian (61). Undoubtedly, the promoter is one of the most important factors for the modification of flower color (see also below), and studies should focus more attention on the development of promoters used for transformation. Evidently, the CaMV35S promoter is not petal specific and influences gene expression in the whole plant. In addition, from the point of view of GMO biosafety, it is better to avoid using virus- or bacteria-derived promoters. Instead, use of plant (hopefully endogenous) promoters is preferable in future plant transgenic procedures.

When RNAi is applied to tobacco and gentian, at least 500 nt in length is required as a trigger for successful suppression of the target genes (M. Nishihara and T. Nakatsuka, unpublished data). This is in contrast to the case of *Arabidopsis*, in which introduction of hairpin RNA constructs with arm lengths ranging from 98 to 853 nt was effective to induce gene silencing (62). When transgenic torenia plants were produced by RNAi using the coding and 3′-untranslated regions of *CHS* mRNA as a trigger, the original blue flowers were converted to white and pale-blue flowers, respectively (48). Thus, the length and position of trigger sequences seem important to suppress targeted genes successfully and these factors should be tested in each plant species to which RNAi is applied. More recently, it was also demonstrated that transgenic petunia expressing dominant-negative *MjCHS*, harboring lysine instead of methionine at the 138th residue, exhibited reduced flower color intensity compared with the wild type (63). This is a unique approach not dependent on posttranscriptional gene silencing (PTGS) that seems worth trying in other plants.

5.2. Introduction of Additional Foreign Genes

It is known that pelargonidin (red to orange) and delphinidin (blue to purple) derivatives of anthocyanin pigments are synthesized in a restricted number of plant species. DFR that can reduce precursors efficiently and existence or nonexistence of F3′5′H activity are necessary to produce each anthocyanin (Fig. 29.1). Thus, molecular engineering of the genes encoding these enzymes might create solutions for the production of new flower colors. Successful engineering of pelargonidin synthesis was first demonstrated by expression of the maize *DFR* (*A1*) gene in petunia to produce brick-red flowers (22). DFRs from gerbera and rose can also reduce dihydrokaempferol (DHK) effectively, and produced red flowers when the genes were transformed into triple-deficient petunia accumulating DHK (64, 65).

Several *F3′5′H* genes isolated from blue-flowered plants were used to synthesize delphinidin derivatives (66, 67). However, in many previous studies, over-expression of *F3′5′H* alone did not lead to a change to blue flowers in the transgenic plants (see also below).

Anthocyanin modifications such as glycosylation, acylation, and methylation are partly responsible for the diversity of flower colors among plant species. Therefore, novel modifications of anthocyanin biosynthesis are considered useful for changing flower colors. Transgenic lisianthus, expressing *3GT* cDNA from snapdragon, synthesized novel 3-*O*-glucosylated anthocyanins accounting for approximately 30% of total petal anthocyanins (68). Heterologous expression of gentian *5GT* in tobacco plants led to additional accumulation of cyanidin 3-*O*-rutinoside-5-*O*-glucoside (69). Fukuchi-Mizutani et al. (70) produced transgenic petunia expressing two glucosyltransferase genes, *5GT* from torenia and *3′GT* from gentian. The petals of this transgenic petunia contained delphinidin 3,5,3′-*O*-tri-glucoside, which was absent in the wild type. Although these transgenic plants accumulated the additional anthocyanins, no noticeable change in flower colors was observed due to low levels of accumulation of the novel pigments. Therefore, further development of methods to enhance the synthesis and accumulation of modified anthocyanins may be needed.

Tobacco plants over-expressing gentian *F3′H* exhibited enhanced flower color intensity in comparison with the wild type (71). Because endogenous *F3′H* transcripts in tobacco plants are also expressed during flower development and decreased after anthesis, constitutive expression of *GtF3′H* using the CaMV35S promoter enhanced anthocyanin accumulation in tobacco flowers. Similarly, Muir et al. (72) reported that over-expression of the petunia chalcone isomerase (*CHI*) gene enhanced flavonol concentrations in transgenic tomato fruits. Therefore, it is implied that the capacity for anthocyanin biosynthesis in floricultural plants might be enhanced by reinforcement of the rate-limiting biosynthetic step.

Petunia and tobacco plants accumulate flavonols in their flowers, whereas torenia and gentian accumulate flavones. The over-expression of flavone synthase II (*FNSII*) led to additional flavone accumulation and the reduction of anthocyanins (59, 71). In contrast, the suppression of flavonol synthase (*FLS*) in petunia resulted in an increased quantity of anthocyanin derivatives (59), but the suppression of *FNSII* in torenia led to reduction in both flavone and anthocyanin amounts (73). Molecular engineering of flavonols/flavones might receive attention because copigmentation with anthocyanins is important for stability of flower color.

6′-deoxychalcones occur in a limited number of plant species, primarily in the Leguminosae, and their synthesis is catalyzed by chalcone reductase (CHR) together with CHS. By introducing *CHR* from *Medicago sativa* into acyanic-flowered petunia, the flower color was changed from white to pale yellow (74). Shimada et al. (75) also demonstrated that transgenic petunia expressing the gene encoding polyketide reductase (*PKR*) from *Lotus japonicus* exhibited reduced anthocyanin quantities and additional formation of isoliquiritigenin derivatives in their petals.

5.3. Metabolic Engineering of Flower Color by Regulating Multiple Genes

In some cases, deficient mutants accumulating suitable pigment precursors are necessary to achieve the above approach, but most floricultural plants do not have flower color mutants useful for metabolic engineering. Therefore, transgenic floricultural plants with novel flower colors must be produced by manipulating multiple genes. Not only additional expression of a foreign gene, but also simultaneous suppression of several endogenous genes is necessary to produce the desired pigments. Suppression of *F3′5′H* by RNAi and over-expression of DHK-reducible DFR resulted in pelargonidin accumulation in *Osteospermum hybrida* (76) and petunia (59). Transgenic torenia in which *F3′5′H* was downregulated by cosuppression exhibited redder flower colors when further transformed to induce over-expression of torenia *F3′H* (73).

Roses, carnations, chrysanthemums, lilies, and orchids are among the most important floricultural crops, but do not accumulate delphinidin-based anthocynanins. In carnation, over-expression of *F3′5′H* alone was insufficient to convert the metabolic flux fully toward delphinidin biosynthesis. A *DFR*-deficient white carnation transformed with petunia or viola *F3′5′H* in combination with petunia *DFR* resulted in violet flowers, accumulating a sufficient amount of delphinidin-based anthocyanins (34). However, a blue-flowered rose could not be produced with the same approach. In the case of rose, selection of the transformation host plant was important, focusing on the higher vacuole pH, lack of F3′H activity, and flavonol accumulation. Moreover, specific suppression of the endogenous *DFR* gene by RNAi and coexpression of iris *DFR* and viola *F3′5′H* were required to produce a blue-flowered rose (35).

Chalcone and aurone derivatives of flavonoids confer the yellow color in flowers of carnation, dahlia, and snapdragon. Snapdragon aureusidin synthase (AmAS1) was identified as a key enzyme that catalyzes aurone biosynthesis from chalcones (77), but transgenic flowers over-expressing the *AmAS1* gene failed to produce aurones. Because AmAS1 is localized in the vacuole, it was necessary to transport the substrate by glucosylation with snapdragon chalcone 4′-*O*-glucosyltransferase (Am4′CGT) to the vacuole.

Therefore, coexpression of *AmAS1* and *Am4'CGT* resulted in yellow flower colors in transgenic torenia plants in which flavanone 3-hydroxylase (*F3H*) or *DFR* were suppressed by RNAi (78).

As mentioned above, to downregulate two biosynthetic genes, the binary vectors usually harbor two independent expression cassettes containing the promoter and inverted repeats of each targeted gene. However, manipulation of these binary vectors is difficult due to the larger size of the binary plasmid itself and the limited restriction enzymes for the vector construct. In addition, the available promoters might be limited in many floricultural plants. Therefore, repeated usage of the same sequence frequently tends to induce gene silencing or recombination during the transformation process. To overcome this problem, Nakatsuka et al. (79) demonstrated that "chimeric-RNAi" is useful to produce red-flowered tobacco plants. This approach was first applied to perform a functional analysis of the *OsRac* gene family in rice (80). The chimeric RNAi construct, containing inverted repeats of the fused *FLS* and *F3'H* controlled by a single CaMV35S promoter, suppressed both endogenous genes simultaneously. The combination of both chimeric RNAi of the *F3'H* and *FLS* genes and the over-expression of gerbera *DFR* resulted in the production of transgenic tobacco plants with red flowers that accumulated pelargonidin derivatives (79). As shown in the these studies, it is important to modify the metabolic pathway by combining the introduction of foreign genes, the suppression of endogenous genes, and the selection of apposite transformed hosts.

5.4. Effect of Promoter Used for Flower Modification

In most of the transgenic plants described above, the CaMV35S promoter was used to control transgene expressions. The CaMV35S promoter is a useful promoter that can induce constitutive and strong gene expressions in higher plants, and this promoter is active in most plant organs. Thus, previous studies using TGS, PTGS, and RNAi mainly focused on changing the flower pigment composition itself and did not intend to modify flower pigmentation patterns. Transgenic petunia plants with *CHS* and *DFR* suppressed by antisense and cosuppression methods included rare plants with patterned corollas with irregular or no pigmentation, but it was impossible to control the flower patterns (46). Although RNAi has been successfully adapted using fruit-specific promoters in tomato (81), no study has compared the silencing ability of different promoters in the same target gene and tissues. Nakatsuka et al. (82) first attempted the suppression of targeted genes controlled by promoters with different tissue-specific activities, and could successfully design flower-color patterns. Inverted repeats of *CHS* driven by the flower-specific gentian *CHS* promoter (83) were transformed in tobacco plants, which produced flowers with a bicolor pattern with white periclinal petals. This observation is quite helpful to control the color pattern such as a bicolor pattern in transgenic flowers. Generally, the natural picotee flowers of petunia and lisianthus are considered to be inherited as a dominant genotype, and those of petunia were reported to be caused through PTGS (84). In contrast, the suppression of *CHS* by the *rolC* promoter from *Agrobacterium rhizogenes* resulted in pale-colored flowers in transgenic tobacco. Thus, the intensity and pattern of flower colors could be partly designed by using promoters with different tissue specificities or strengths instead of the CaMV35S promoter, even if it was used for driving the same gene as a trigger. This clearly indicates that the promoters used for RNAi are the vital factors for determining the suppression efficiency and inducing local suppression of the flower pigmentation. Thus, the promoters used for genetic engineering of flower color might be an important factor and should receive greater attention in future studies. In addition to flower colors, flavonoids are responsible for attraction of pollinators, plant–microorganism interactions, protection from harmful UV irradiation, and pollen-tube growth. Therefore, metabolic engineering of flavonoids would require precise regulation of transgene expression in the targeted tissues, in most cases petals. This problem would be partly solved by the development of new promoters with different tissue-specific activities.

The successful examples of modification of flower colors using the techniques mentioned above are shown in tobacco (Fig. 29.2) and gentian (Fig. 29.3). The procedural details for gentian were also reviewed by Nishihara et al. (85).

5.5. Flower Color Modification by Transcription Factors

Recent molecular analyses indicate that the pigment biosynthetic pathways are regulated at the transcriptional levels. Of the three major flower pigments, flavonoid biosynthesis has been well studied and three transcription factors, including R2R3-MYB, basic Helix-Loop-Helix (bHLH), and WD40 repeat (WDR), are known to be involved in the regulation of the flavonoid biosynthetic genes (86). These transcription factors form a complex and activate the different targeted biosynthetic gene sets in each plant. No key transcription factor involved in carotenoid and betalain biosynthesis has been identified.

In contrast to the biosynthetic structural genes, transcription factor genes that control the transcription of the structural genes tend to control multiple pathway steps and have emerged as powerful tools for the manipulation of complex metabolic pathways in plants, as reviewed by Broun (87). However, few studies have reported flower color modification using transcription factors related to anthocyanin biosynthesis. When the *Lc* gene encoding the anthocyanin biosynthetic regulator from maize, under the control of the CaMV35S promoter, was transformed in petunia

Fig. 29.2. Tobacco flowers of various colors produced by RNAi suppression of endogenous flavonoid biosynthetic genes and/or over-expression of foreign genes. (**a**) Host plant tobacco cv. SR1 (**b**) *CHS* suppression (**c**) *CHI* suppression (**d**) *F3H* suppression (**e**) *F3′H* suppression (**f**) *and* (**g**) CHS suppression (**h**) Over-expression of gentian *FNSII* (**i**) Over-expression of gentian *F3′H* (**j**) Over-expression of *Lotus japonicus* polyketide reductase (*PKR*) (**k**) Over-expression of *Arabidopsis PAP1* (**l**) Over-expression of gerbera *DFR* and suppression of *F3′H* and *FLS*. CaMV35S promoter was used except for (**f**) and (**g**). Gentian *CHS* and *Agrobacterium rhizogenes rol* C promoters were used in (**f**) and (**g**), respectively. Abbreviations are explained in Fig. 29.1

plants, the transgenic plants exhibited enhanced pigmentation in both vegetative and floral tissues (88). Similarly, tobacco and *Arabidopsis* plants expressing *Arabidopsis PAP1* and expressing maize *R* and *C1* also showed increased anthocyanin accumulation in the plant (89, 90). The over-expression of transcription factors can enhance flower color intensity, but results in ectopic anthocyanin

Fig. 29.3. Flower color modification by RNAi in gentian plants. (**a**) Host plant gentian cv. Albireo (**b**), (**c**) Suppression of *CHS* under control of the *rol*C promoter of *A. rhizogenes* (**d**) Suppression of *F3′5′H* under control of the *Arabidopsis* actin2 (*ACT2*) promoter (**e**), (**f**) Suppression of *ANS* under control of the *rol*C promoter of *A. rhizogenes* Abbreviations are explained in Fig. 29.1

accumulation. Therefore, it would be necessary to regulate the spatial expression of transcription factors with flower-specific promoters to overcome this problem.

Xie et al. (91) reported on metabolic engineering with the combination of anthocyanin regulatory and structural genes. Low levels of proanthocyanidins accumulate in tobacco leaves expressing an anthocyanidin reductase (*ANR*) gene, due to the lack of an anthocyanidin substrate (92). *ANR*-expressing tobacco with the anthocyanidin pathway enhanced by *PAP1* synthesized more proanthocyanidins than those expressing the *ANR* gene alone. Ben Zvi et al. (93) demonstrated that both increased metabolic flux and transcriptional activation of scent and color genes underlie the enhancement of petunia flower color and scent production by *PAP1*.

Recently, chimeric repressor gene-silencing technology (CRES-T) was developed as a method converting the transcriptional activator to repressor (94). When a transcription factor fused to the EAR-motif repression domain (SRDX), the chimeric transcription factor dominantly represses the transcription of its target genes. The over-expression of PAP1-SRDX suppressed the expression of flavonoid biosynthetic genes and inhibited the accumulation of anthocyanin in *Arabidopsis* (95). Moreover, this approach was also applied to produce sterile plants and to modify floral morphogenesis (96, 97). Since the CRES-T database for genetic engineering of horticultural pants, "FioreDB," is provided in a web interface (98), CRES-T will become an increasingly powerful tool to modulate secondary metabolism by the transcription factors including flower pigmentation.

6. Modification of Flower Color by Manipulation of Carotenoid Biosynthesis

Compared with green tissues, carotenoids are found quite variably in nongreen tissues, such as in roots, fruits, flowers, tubers, and seeds, depending on the plant species (99). For example, carotenoid composition was analyzed in petals of sandersonia (*Sandersonia aurantiaca*) (100), calendula (*Calendula officinalis* L.) (101), and chrysanthemum (102). In addition, for flavonoids, the carotenoid biosynthetic pathway has also been well-characterized and the genes involved have been cloned from marine bacteria and higher plants (11, 99, 103). Carotenoid biosynthesis during tomato fruit development is one of the best-characterized pathways at the molecular level in higher plants (104, 105). Because carotenoids are also important compounds for human health (106, 107), crop plants such as tomato and rice have been metabolically engineered to contain enhanced levels of beta-carotene, lutein, and lycopene. Golden rice is the best-known example that was engineered to accumulate beta-carotene (108). Recent research has resulted in even higher carotenoid accumulation by using maize phytoene synthase (PSY) instead of daffodil PSY (109). Potato tubers containing enhanced levels of beta-carotene and lutein were also genetically engineered by introducing *crt* genes from *Erwinia* sp. (110, 111).

Several research groups isolated carotenogenic genes that are involved in lutein biosynthesis in flowers and analyzed the gene expression in an ornamental plant, marigold (*Tagetes erecta*) (112, 113), and a pharmaceutical plant, great yellow gentian (*Gentiana lutea*) (114–116). Lutein is the most abundant carotenoid in the flowers of these plants. cDNAs responsible for lutein biosynthesis and plastid division genes have been isolated from marigold (112, 113). Zhu et al. (114–116) have also isolated cDNAs for lutein biosynthesis in *G. lutea* with yellow flowers. Many genes have been isolated and characterized in other plant species. For example, cDNAs of the carotenoid biosynthetic pathway, including carotenoid cleavage dioxygenase (CCD) family members, were isolated and their expression was analyzed in *Coffea canephora* (117). Three carotenogenic genes involved in formation of apocarotenoid during stigma development were isolated in *Crocus sativus* (118).

In the past decade or so, flavonoid biosynthesis has been the main target for modification of flower color, but recently regulation of carotenoid biosynthesis in flowers has been successfully achieved (28, 29, 119, 120). Metabolic engineering of astaxanthin production was first achieved using *crtO* (beta-carotene ketolase) from the alga *Haematococcus pluvialis*, under the control of the tomato phytoene desaturase (*PDS*) promoter in tobacco, in which the color of the nectary was changed from yellow to red (121). The first genetic engineering of flower color via carotenoid biosynthesis was achieved in chrysanthemum (28). It is reported that white petal color is dominant over yellow in chrysanthemum (122). Recently, a carotenoid cleavage dioxygenase (CCD) that cleaves carotenoid substrates was identified as a key enzyme expressed specifically in white petals in chrysanthemum and the suppression of *CmCCD4a* by RNAi resulted in yellow pigmentation in the petals (28). In this case, the endogenous gene was genetically engineered. Simkin et al. (123) reported that petunia PhCCD1 controls emission of beta-ionone, a volatile fragrance, at the transcript level. Thus, combinational metabolic engineering of flavor and color might be possible as well as flavonoid engineering as mentioned above. On the other hand, a *crt* gene from a marine bacterium (*Agrobacterium aurantiacum*) has been successfully used to change flower color in the model plant *Lotus japonicus*. The *A. aurantiacum crtW* gene encoding β-carotene ketolase (4,4′-β-oxygenase) was used to produce orange-flowered *L. japonicus* plants by Suzuki et al. (29). The leaves and flower petals of transgenic *L. japonicus* plants contained novel ketocarotenoids, including astaxanthin, adonixanthin, canthaxanthin, and echinenone. In another study, Zhu et al. (119) demonstrated additional formation of ketocarotenoid in leaves and flowers of *Nicotiana glauca* transformed with a cyanobacterial *crtO* gene. Development of chromoplasts might be a prerequisite for high levels of carotenoid accumulation in flower petals. For example, the *Or* gene, identified in cauliflower (*Brassica oleracea* var. *botrytis*) (124) seems to play an important role in the differentiation of plastids into chromoplasts (125). Several cDNA-encoding fibrillin proteins involved in carotenoid sequestration were also isolated (117) and these kinds of genes may be important to accumulate specific carotenoids to modify flower color in petals. Of course, tissue-specific regulation is quite important for controlling carotenoid biosynthesis in the petals. For example, tissue-specific regulation of carotenoid biosynthesis during tomato fruit development is reported (104, 105). Until now, no key transcription factor(s) of carotenoid biosynthesis has been identified, although the International Tomato Genome Sequencing Project is in progress (126). Future studies will reveal the transcription factors regulating the carotenoid biosynthesis and they might be useful for genetic engineering of carotenoids.

7. Perspectives

Flower color is one of the most important characters of floricultural plants. Consumers tend to request a variety of flower colors, even if some were once "dream" colors (e.g. a blue rose); they are no longer impossible today. Transgenic approaches have the potential to generate many previously unattainable flower colors that do not exist in nature. However, as we can see, throughout nature, there are many beautiful flowers with unusual and diverse colors and shapes. Modification of flower color by genetic engineering has graduated from an early stage of investigation and can now generate novel transgenic cultivars with direct commercial application. Other factors such as vacuole pH, copigmentation, and formation of metal complexes might be the targets for future molecular breeding. Flower color and patterns will come under increasingly sophisticated control along with improvement of the techniques using RNAi and tissue-specific promoters and other forthcoming techniques. We can look forward to the production of genetically modified flowers with an expanded range of colors and patterns in the near future, because this is currently the most active area of transgenic research in flowers.

Acknowledgments

This research was partially supported by the Ministry of Education, Science, Sports, and Culture, Grant-in-Aid for Scientific Research (C) (no. 18580010) and Young Scientists (B) (no. 18789005). We thank Drs S. Mohan Jain and S. J. Ochatt for providing us with the opportunity to write this review. We also thank Dr Y. Tanaka (Suntry Ltd, Japan) for providing information about literature.

References

1. Mol J, Grotewold E, Koes R (1998) How genes paint flowers and seeds. Trends Plant Sci 3:212–217
2. Winkel-Shirley B (2001) Flavonoid biosynthesis. A colorful model for genetics, biochemistry, cell biology, and biotechnology. Plant Physiol 126:485–493
3. Subramanian S, Stacey G, Yu O (2007) Distinct, crucial roles of flavonoids during legume nodulation. Trends Plant Sci 12: 282–285
4. Goodwin TW (ed) (1980) The biochemistry of the carotenoids, vol I, 2nd edn. Chapman and Hall, London

5. Sandmann G (2001) Carotenoid biosynthesis and biotechnological application. Arch Biochem Biophys 385:4–12
6. Grotewold E (ed) (2006) The science of flavonoids. Springer, London
7. Britton G, Liaaen-Jensen S, Pfander H (eds) (2004) Carotenoids handbook. Birkhäuser, Basel
8. Andersen OM, Markham KR (eds) (2006) Flavonoids: Chemistry, biochemistry and applications. CRC, Boca Raton, FL
9. Strack D, Vogt T, Schliemann W (2003) Recent advances in betalain research. Phytochem 62:247–269
10. Grotewold E (2006) The genetics and biochemistry of floral pigments. Annu Rev Plant Biol 57:761–780
11. Tanaka Y, Sasaki N, Ohmiya A (2008) Biosynthesis of plant pigments: anthocyanins, betalains and carotenoids. Plant J 54:733–749
12. Springob K, Nakajima J, Yamazaki M, Saito K (2003) Recent advances in the biosynthesis and accumulation of anthocyanins. Nat Prod Rep 20:288–303
13. Koes R, Verweij W, Quattrocchio F (2005) Flavonoids: a colorful model for the regulation and evolution of biochemical pathways. Trends Plant Sci 10:236–242
14. Dixon RA (2005) Engineering of plant natural product pathways. Curr Opin Plant Biol 8:329–336
15. Yu O, Matsuno M, Subramanian S (2006) Flavonoid compounds in flowers: genetics and biochemistry. In: Teixeira da Silva JA (ed) Floriculture, ornamental and plant biotechnology: Advances and topical. Global Science Books, London, pp 282–292
16. Jackson JF, Linskens HF (2003) Genetic transformation of plants. Springer, Berlin
17. Chandler S, Tanaka Y (2007) Genetic modification in floriculture. Crit Rev Plant Sci 26:169–197
18. Shibata M (2008) Importance of genetic transformation in ornamental plant breeding. Plant Biotech 25:3–8
19. Teixeira da Silva JA (2003) Chrysanthemum: advances in tissue culture, cryopreservation, postharvest technology, genetics and transgenic biotechnology. Biotech Adv 21:715–766
20. Hosokawa K, Matsuki R, Oikawa Y, Yamamura S (2000) Production of transgenic gentian plants by particle bombardment of suspension-culture cells. Plant Cell Rep 19:454–458
21. Mishiba K, Nishihara M, Abe Y, Nakatsuka T, Kawamura H, Kodama K, Takesawa T, Abe J, Yamamura S (2006) Production of dwarf potted gentian using wild-type *Agrobacterium rhizogenes*. Plant Biotech 23:33–38
22. Meyer P, Heidmann I, Forkmann G, Saedler H (1987) A new petunia flower colour generated by transformation of a mutant with a maize gene. Nature 330:677–678
23. Forkmann G, Martens S (2001) Metabolic engineering and applications of flavonoids. Curr Opin Biotech 12:155–160
24. Tanaka Y, Tsuda S, Kusumi T (1998) Metabolic engineering to modify flower color. Plant Cell Physiol 39:1119–1126
25. Tanaka Y, Katsumoto Y, Brugliera F, Mason J (2005) Genetic engineering in floriculture. Plant Cell Tiss Org Cult 80:1–24
26. Tanaka Y, Ohmiya A (2008) Seeing is believing: engineering anthocyanin and carotenoid biosynthetic pathways. Curr Opin Biotech 19:190–197
27. To KY, Wang CK (2006) Molecular breeding of flower color. In: Teixeira da Silva JA (ed) Floriculture, ornamental and plant biotechnology: Advances and topical. Global Science Books, London, pp 300–310
28. Ohmiya A, Kishimoto S, Aida R, Yoshioka S, Sumitomo K (2006) Carotenoid cleavage dioxygenase (CmCCD4a) contributes to white color formation in chrysanthemum petals. Plant Physiol 142:1193–1201
29. Suzuki S, Nishihara M, Nakatsuka T, Misawa N, Ogiwara I, Yamamura S (2007) Flower color alteration in Lotus japonicus by modification of the carotenoid biosynthetic pathway. Plant Cell Rep 26:951–959
30. Chahal GS, Gosal SS (eds) (2002) Principles and procedures of plant breeding: biotechnological and conventional approaches. Alpha Science International Ltd, Pangbourne
31. Brown J, Caligari P (eds) (2008) An introduction to plant breeding. Blackwell Publishing, Oxford
32. Sasaki K, Aida R, Niki T, Yamaguchi H, Narumi T, Nishijima T, Hayashi T, Ryuto H, Fukunishi T, Abe T, Ohtsubo N (2008) High-efficiency improvement of transgenic torenia flowers by ion beam irradiation. Plant Biotech 25:81–89
33. Nishihara M, Nakatsuka T, Hosokawa K, Yokoi T, Abe Y, Mishiba K, Yamamura S (2006) Dominant inheritance of white-flowered and herbicide-resistant traits in trangenic gentian plants. Plant Biotech 23:25–31
34. Fukui Y, Tanaka Y, Kusumi T, Iwashita T, Nomoto K (2003) A rationale for the shift in colour towards blue in transgenic carnation flowers expressing the flavonoid 3', 5'-hydroxylase gene. Phytochem 63:15–23

35. Katsumoto Y, Fukuchi-Mizutani M, Fukui Y, Brugliera F, Holton TA, Karan M, Nakamura N, Yonekura-Sakakibara K, Togami J, Pigeaire A, Tao GQ, Nehra NS, Lu CY, Dyson BK, Tsuda S, Ashikari T, Kusumi T, Mason JG, Tanaka Y (2007) Engineering of the rose flavonoid biosynthetic pathway successfully generated blue-hued flowers accumulating delphinidin. Plant Cell Physiol 48:1589–1600
36. Kikuchi A, Watanabe K, Tanaka Y, Kamada H (2008) Recent progress on environmental biosafety assessment of genetically modified trees and floricultural plants in Japan. Plant Biotech 25:9–15
37. Fiehn O (2002) Metabolomics – the link between genotypes and phenotypes. Plant Mol Biol 48:155–171
38. Shimada S, Takahashi K, Sato Y, Sakuta M (2004) Dihydroflavonol 4-reductase cDNA from non-anthocyanin-producing species in the Caryophyllales. Plant Cell Physiol 45: 1290–1298
39. Shimada S, Inoue YT, Sakuta M (2005) Anthocyanidin synthase in non-anthocyanin-producing Caryophyllales species. Plant J 44:950–959
40. Hieber AD, Mudalige-Jayawickrama RG, Kuehnle AR (2006) Color genes in the orchid Oncidium Gower Ramsey: identification, expression, and potential genetic instability in an interspecific cross. Planta 223:521–531
41. Hoshi Y, Kondo M, Mori S, Adachi Y, Nakano M, Kobayashi H (2004) Production of transgenic lily plants by *Agrobacterium*-mediated transformation. Plant Cell Rep 22:359–364
42. Ogaki M, Furuichi Y, Kuroda K, Chin DP, Ogawa Y, Mii M (2008) Importance of co-cultivation medium pH for successful *Agrobacterium*-mediated transformation of *Lilium x formolongi*. Plant Cell Rep 27:699–705
43. Liau CH, You SJ, Prasad V, Hsiao HH, Lu JC, Yang NS, Chan MT (2003) *Agrobacterium tumefaciens*-mediated transformation of an *Oncidium* orchid. Plant Cell Rep 21:993–998
44. Chan MT, Chan YL, Sanjaya (2006) Orchids (*Cymbidium* spp., *Oncidium*, and *Phalaenopsis*). Methods Mol Biol 344:331–338
45. van der Krol AR, Lentlng PJ, Veenstra JG, van der Meer IM, Koes RE, Gerats AGM, MOI JNM, Stuitje AR (1988) An antisense chalcone synthase gene in transgenic plants inhibits flower pigmentation. Nature 333:866–869
46. Napoli C, Lemieux C, Jorgensen R (1990) Introduction of a chimeric chalcone synthase gene into petunia results in reversible co-suppression of homologous genes *in trans*. Plant Cell 2: 279–289
47. van der Krol AR, Mur LA, Beld M, Mol JN, Stuitje AR (1990) Flavonoid genes in petunia: addition of a limited number of gene copies may lead to a suppression of gene expression. Plant Cell 2:291–299
48. Fukusaki E, Kawasaki K, Kajiyama S, An CI, Suzuki K, Tanaka Y, Kobayashi A (2004) Flower color modulations of *Torenia hybrida* by downregulation of chalcone synthase genes with RNA interference. J Biotech 111: 229–240
49. Nishihara M, Nakatsuka T, Yamamura S (2005) Flavonoid components and flower color change in transgenic tobacco plants by suppression of chalcone isomerase gene. FEBS Lett 579:6074–6078
50. Nakamura N, Fukuchi-Mizutani M, Miyazaki K, Suzuki K, Tanaka Y (2006) RNAi suppression of the anthocyanidin synthase gene in *Torenia hybrida* yields white flowers with higher frequency and better stability than antisense and sense suppression. Plant Biotech 23:13–18
51. Tanaka Y, Nakamura N, Togami J (2008) Altering flower color in transgenic plants by RNAi-mediated engineering of flavonoid biosynthetic pathway. Methods Mol Biol 442: 245–257
52. Courtney-Gutterson N, Napoli C, Lemieux C, Morgan A, Firoozabady E, Robinson KE (1994) Modification of flower color in florist's chrysanthemum: production of a white-flowering variety through molecular genetics. Bio/technology 12:268–271
53. Elomaa P, Honkanen J, Puska R, Seppanen P, Helariutta P, Mehto M, Kotilainen M, Nevalainen L, Teeri TH (1993) *Agrobacterium*-mediated transfer of antisense chalcone synthase cDNA to *Gerbera hybrida* inhibits flower pigmentation. Bio/Technology 11:508–511
54. Deroles S, Bradley M, Davies KM, Schwinn KE, Manson DG (1995) Generation of novel patterns in lisianthus flowers using an antisense chalcone synthase gene. Acta Hortic 420:26–28
55. Aida R, Kishimoto S, Tanaka Y, Shibata M (2000) Modification of flower colour in torenia (*Torenia fournieri* Lind.) by genetic transformation. Plant Sci 153:33–42
56. Suzuki K, Xue H, Tanaka Y, Fukui Y, Fukuchi-Mizutani M, Murakami Y, Katsumoto Y, Tsuda S, Kusumi T (2000) Flower color modifications of *Torenia hybrida* by cosuppression of anthocyanin biosynthesis genes. Mol Breed 6:239–246
57. Aida R, Yoshida K, Kondo T, Kishimoto S, Sibata, M (2000) Copigmentation gives bluer

flowers on transgenic torenia plants with the antisense dihydroflavonol-4-redutase gene. Plant Sci 160:49–56

58. Nakatsuka T, Mishiba K, Abe Y, Kubota A, Kakizaki Y, Yamamura S, Nishihara M (2008) Flower color modification of gentian plants by RNAi-mediated gene silencing. Plant Biotech 25:61–68
59. Tsuda S, Fukui Y, Nakamura N, Katsumoto Y, Yonekura-Sakakibara K, Fukuchi-Mizutani M, Ohira K, Ueyama Y, Ohkawa H, Holton TA, Kusumi T, Tanaka Y (2004) Flower color modification of *Petunia hybrida* commercial varieties by metabolic engineering. Plant Biotech 21:377–386
60. Ueyama Y, Katsumoto Y, Fukui Y, Fukuchi-Mizutani M, Ohkawa H, Kusumi T, Iwashita T, Takahata Y (2006) Molecular characterization of the flavonoid biosynthetic pathway and flower color modification of *Nierembergia* sp. Plant Biotech 23:19–24
61. Mishiba K, Nishihara M, Nakatsuka T, Abe Y, Hirano H, Yokoi T, Kikuchi A, Yamamura S (2005) Consistent transcriptional silencing of 35 S-driven transgenes in gentian. Plant J 44:541–556
62. Wesley SV, Helliwell CA, Smith NA, Wang MB, Rouse DT, Liu Q, Gooding PS, Singh SP, Abbott D, Stoutjesdijk PA, Robinson SP, Gleave AP, Green AG, Waterhouse PM (2001) Construct design for efficient, effective and high-throughput gene silencing in plants. Plant J 27:581–590
63. Hanumappa M, Choi G, Ryu S, Choi G (2007) Modulation of flower colour by rationally designed dominant-negative chalcone synthase. J Exp Bot 58:2471–2478
64. Helariutta Y, Elomaa P, Kotilainen M, Seppanen P, Teeri TH (1993) Cloning of cDNA coding for dihydroflavonol-4-reductase (DFR) and characterization of *dfr* expression in the corollas of *Gerbera hybrida* var. Regina (*Compositae*). Plant Mol Biol 22:183–193
65. Tanaka Y, Fukui Y, Fukuchi-Mizutani M, Holton TA, Higgins E, Kusumi T (1995) Molecular cloning and characterization of Rosa hybrida dihydroflavonol 4-reductase gene. Plant Cell Physiol 36:1023–1031
66. Shimada Y, Nakano-Shimada R, Ohbayashi M, Okinaka Y, Kiyokawa S, Kikuchi Y (1999) Expression of chimeric P450 genes encoding flavonoid-3', 5'-hydroxylase in transgenic tobacco and petunia plants. FEBS Lett 461:241–245
67. Okinaka Y, Shimada Y, Nakano-Shimada R, Ohbayashi M, Kiyokawa S, Kikuchi Y (2003) Selective accumulation of delphinidin derivatives in tobacco using a putative flavonoid 3', 5'-hydroxylase cDNA from *Campanula medium*. Biosci Biotech Biochem 67:161–165
68. Schwinn KE, Davies KM, Deroles SC, Markham KR, Miller RM, Bradley JM, Manson DG, Given NK (1997) Expression of an *Antirrhinum majus* UDP-glucose: flavonoid-3-*O*-glucosyltransferase transgene alters flavonoid glycosylation and acylation in lisianthus (*Eustoma grandiflorum* Grise.). Plant Sci 125:53–61
69. Nakatsuka T, Sato K, Takahashi H, Yamamura S, Nishihara M (2008) Cloning and characterization of the UDP-glucose:anthocyanin 5-*O*-glucosyltransferase gene from blue-flowered gentian. J Exp Bot 59:1241–1252
70. Fukuchi-Mizutani M, Okuhara H, Fukui Y, Nakao M, Katsumoto Y, Yonekura-Sakakibara K, Kusumi T, Hase T, Tanaka Y (2003) Biochemical and molecular characterization of a novel UDP-glucose:anthocyanin 3'-*O*-glucosyltransferase, a key enzyme for blue anthocyanin biosynthesis, from gentian. Plant Physiol 132:1652–1663
71. Nakatsuka T, Nishihara M, Mishiba K, Yamamura S (2006) Heterologous expression of two gentian cytochrome P450 genes can modulate the intensity of flower pigmentation in transgenic tobacco plants. Mol Breed 17: 91–99
72. Muir SR, Collins GJ, Robinson S, Hughes S, Bovy A, Ric De Vos CH, van Tunen AJ, Verhoeyen ME (2001) Overexpression of petunia chalcone isomerase in tomato results in fruit containing increased levels of flavonols. Nat Biotech 19:470–474
73. Ueyama Y, Suzuki K, Fukuchi-Mizutani M, Fukui Y, Miyazaki K, Ohkawa H, Kusumi T, Tanaka Y (2002) Molecular and biochemical characterization of torenia flavonoid 3'-hydroxylase and flavone synthase II and modification of flower color by modulating the expression of these genes. Plant Sci 163: 253–263
74. Davies KM, Bloor SJ, Spiller DB, Deroles SS (1998) Production of yellow colour in flowers: redirection of flavonoid biosynthesis in *Petunia*. Plant J 13:259–266
75. Shimada N, Nakatsuka T, Nishihara M, Yamamura S, Ayabe S, Aoki T (2006) Isolation and characterization of a cDNA encoding polyketide reductase in *Lotus japonicus*. Plant Biotech 23:509–513
76. Seitz C, Vitten M, Steinbach P, Hartl S, Hirsche J, Rathje W, Treutter D, Forkmann G (2007) Redirection of anthocyanin synthesis in *Osteospermum hybrida* by a two-enzyme

manipulation strategy. Phytochemistry 68: 824–833

77. Nakayama T, Yonekura-Sakakibara K, Sato T, Kikuchi S, Fukui Y, Fukuchi-Mizutani M, Ueda T, Nakao M, Tanaka Y, Kusumi T, Nishino T (2000) Aureusidin synthase: a polyphenol oxidase homolog responsible for flower coloration. Science 290:1163–1166
78. Ono E, Fukuchi-Mizutani M, Nakamura N, Fukui Y, Yonekura-Sakakibara K, Yamaguchi M, Nakayama T, Tanaka T, Kusumi T, Tanaka Y (2006) Yellow flowers generated by expression of the aurone biosynthetic pathway. Proc Nat Acad Sci USA 103:11075–11080
79. Nakatsuka T, Abe Y, Kakizaki Y, Yamamura S, Nishihara M (2007) Production of red-flowered plants by genetic engineering of multiple flavonoid biosynthetic genes. Plant Cell Rep 26: 1951–1959
80. Miki D, Itoh R, Shimamoto K (2005) RNA silencing of single and multiple members in a gene family of rice. Plant Physiol 138: 1903–1913
81. Davuluri GR, van Tuinen A, Fraser PD, Manfredonia A, Newman R, Burgess D, Brummell DA, King SR, Palys J, Uhlig J, Bramley PM, Pennings HM, Bowler C (2005) Fruit-specific RNAi-mediated suppression of *DET1* enhances carotenoid and flavonoid content in tomatoes. Nat Biotech 23:890–895
82. Nakatsuka T, Pitaksutheepong C, Yamamura S, Nishihara M (2007) Induction of differential flower pigmentation patterns by RNAi using promoters with distinct tissue-specific activity. Plant Biotech Rep 1:251–257
83. Kobayashi H, Oikawa Y, Koiwa H, Yamamura S (1998) Flower-specific expression directed by the promoter of a chalcone synthase gene from *Gentiana triflora* in *Petunia hybrida*. Plant Sci 131:173–180
84. Koseki M, Goto K, Masuta C, Kanazawa A (2005) The star-type color pattern in *Petunia hybrida* 'red Star' flowers is induced by sequence-specific degradation of chalcone synthase RNA. Plant Cell Physiol 46:1879–1883
85. Nishihara M, Nakatsuka T, Mizutani-Fukuchi M, Tanaka Y, Yamamura S (2008) Gentians: From gene cloning to molecular breeding. In: Teixeira da Silva JA (ed) Floricultural and ornamental biotechnology V. Global Science Books, London, pp 57–67
86. Quattrocchio F, Baudry A, Lepiniec L, Grotewold E (2006) The regulation of flavonoid biosynthesis. In: Grotewold E (ed) The science of flavonoids. Springer, New York, pp 97–122
87. Broun P (2004) Transcription factors as tools for metabolic engineering in plants. Curr Opin Plant Biol 7:202–209
88. Bradley JM, Davies KM, Deroles SC, Bloor SJ, Lewis DH (1998) The maize *Lc* regulatory gene up-regulates the flavonoid biosynthetic pathway of *Petunia*. Plant J 13:381–392
89. Borevitz JO, Xia Y, Blount J, Dixon RA, Lamb C (2000) Activation tagging identifies a conserved MYB regulator of phenylpropanoid biosynthesis. Plant Cell 12:2383–2394
90. Lloyd AM, Walbot V, Davis RW (1992) *Arabidopsis* and *Nicotiana* anthocyanin production activated by maize regulators R and C1. Science 258:1773–1775
91. Xie DY, Sharma SB, Wright E, Wang ZY, Dixon RA (2006) Metabolic engineering of proanthocyanidins through co-expression of anthocyanidin reductase and the PAP1 MYB transcription factor. Plant J 45:895–907
92. Xie DY, Sharma SB, Paiva NL, Ferreira D, Dixon RA (2003) Role of anthocyanidin reductase, encoded by *BANYULS* in plant flavonoid biosynthesis. Science 299:396–399
93. Ben Zvi MM, Negre-Zakharov F, Masci T, Ovadis M, Shklarman E, Ben-Meir H, Tzfira T, Dudareva N, Vainstein A (2008) Interlinking showy traits: co-engineering of scent and colour biosynthesis in flowers. Plant Biotechnol J 6:403–415
94. Hiratsu K, Matsui K, Koyama T, Ohme-Takagi M (2003) Dominant repression of target genes by chimeric repressors that include the EAR motif, a repression domain, in *Arabidopsis*. Plant J 34:733–739
95. Matsui K, Tanaka H, Ohme-Takagi M (2004) Suppression of the biosynthesis of proanthocyanidin in *Arabidopsis* by a chimeric PAP1 repressor. Plant Biotech J 2:487–493
96. Mitsuda N, Hiratsu K, Todaka D, Nakashima K, Yamaguchi-Shinozaki K, Ohme-Takagi M (2006) Efficient production of male and female sterile plants by expression of a chimeric repressor in *Arabidopsis* and rice. Plant Biotech J 4:325–332
97. Narumi T, Aida R, Niki T, Nishijima T, Mitsuda N, Hiratsu K, Ohme-Takagi M, Ohtsubo N (2008) Chimeric *AGAMOUS* repressor induces serrated petal phenotype in Torenia fournieri similar to that induced by cytokinin application. Plant Biotech 25: 45–54
98. Mitsuda N, Umemura Y, Ikeda M, Shikata M, Koyama T, Matsui K, Narumi T, Aida R, Sasaki K, Hiyama T, Higuchi T, Ono M, Isuzugawa K, Saitoh K, Endo R, Ikeda K, Nakatsuka T, Nishihara M, Yamamura S,

Yamamura T, Terakawa T, Ohtsubo N, Ohme-Takagi M (2008) FioreDB: a database of phenotypic information induced by the chimeric repressor silencing technology (CRES-T) in Arabidopsis and floricultural plants. Plant Biotech 25:37–44

99. Howitt CA, Pogson BJ (2006) Carotenoid accumulation and function in seeds and non-green tissues. Plant Cell Environ 29: 435–445
100. Nielsen KM, Lewis DH, Morgan ER (2003) Characterization of carotenoid pigments and their biosynthesis in two yellow flowered lines of *Sandersonia aurantiaca* (Hook). Euphytica 130:25–34
101. Kishimoto S, Maoka T, Sumitomo K, Ohmiya A (2005) Analysis of carotenoid composition in petals of calendula (*Calendula officinalis* L.). Biosci Biotech Biochem 69:2122–2128
102. Kishimoto S, Maoka T, Nakayama M, Ohmiya A (2004) Carotenoid composition in petals of chrysanthemum (*Dendranthema grandiflorum* (Ramat.) Kitamura). Phytochem 65: 2781–2787
103. Hirschberg J (2001) Carotenoid biosynthesis in flowering plants. Curr Opin Plant Biol 4: 210–218
104. Giuliano G, Bartley GE, Scolnik PA (1993) Regulation of carotenoid biosynthesis during tomato development. Plant Cell 5:379–387
105. Bramley PM (2002) Regulation of carotenoid formation during tomato fruit ripening and development. J Exp Bot 53:2107–2113
106. Yonekura-Sakakibara K, Saito K (2006) Review: Genetically modified plants for the promotion of human health. Biotech Lett 28:1983–1991
107. Sandmann G, Romer S, Fraser PD (2006) Understanding carotenoid metabolism as a necessity for genetic engineering of crop plants. Metab Eng 8:291–302
108. Ye X, Al-Babili S, Kloti A, Zhang J, Lucca P, Beyer P, Potrykus I (2000) Engineering the provitamin A (beta-carotene) biosynthetic pathway into (carotenoid-free) rice endosperm. Science 287:303–305
109. Paine JA, Shipton CA, Chaggar S, Howells RM, Kennedy MJ, Vernon G, Wright SY, Hinchliffe E, Adams JL, Silverstone AL, Drake R (2005) Improving the nutritional value of Golden Rice through increased pro-vitamin A content. Nat Biotech 23: 482–487
110. Ducreux LJ, Morris WL, Hedley PE, Shepherd T, Davies HV, Millam S, Taylor MA (2005) Metabolic engineering of high carotenoid potato tubers containing enhanced levels of beta-carotene and lutein. J Exp Bot 56:81–89
111. Diretto G, Al-Babili S, Tavazza R, Papacchioli V, Beyer P, Giuliano G (2007) Metabolic engineering of potato carotenoid content through tuber-specific overexpression of a bacterial mini-pathway. PLoS ONE 2:e350
112. Moehs CP, Tian L, Osteryoung KW, Dellapenna D (2001) Analysis of carotenoid biosynthetic gene expression during marigold petal development. Plant Mol Biol 45:281–293
113. Del Villar-Martinez AA, Garcia-Saucedo PA, Carabez-Trejo A, Cruz-Hernandez A, Paredes-Lopeza O (2005) Carotenogenic gene expression and ultrastructural changes during development in marigold. J Plant Physiol 162:1046–1056
114. Zhu C, Yamamura S, Koiwa H, Nishihara M, Sandmann G (2002) cDNA cloning and expression of carotenogenic genes during flower development in *Gentiana lutea*. Plant Mol Biol 48:277–285
115. Zhu C, Yamamura S, Nishihara M, Koiwa H, Sandmann G (2003) cDNAs for the synthesis of cyclic carotenoids in petals of *Gentiana lutea* and their regulation during flower development. Biochim Biophys Acta 1625:305–308
116. Zhu C, Kauder F, Romer S, Sandmann G (2006) Cloning of two individual cDNAs encoding 9-cis-epoxycarotenoid dioxygenase from *Gentiana lutea*, their tissue-specific expression and physiological effect in transgenic tobacco. J Plant Physiol 164:195–204
117. Simkin AJ, Moreau H, Kuntz M, Pagny G, Lin C, Tanksley S, McCarthy J (2008) An investigation of carotenoid biosynthesis in *Coffea canephora* and *Coffea arabica*. J Plant Physiol 165:1087–1106
118. Castillo R, Fernandez JA, Gomez-Gomez L (2005) Implications of carotenoid biosynthetic genes in apocarotenoid formation during the stigma development of *Crocus sativus* and its closer relatives. Plant Physiol 139:674–689
119. Zhu C, Gerjets T, Sandmann G (2007) *Nicotiana glauca* engineered for the production of ketocarotenoids in flowers and leaves by expressing the cyanobacterial *crtO* ketolase gene. Transgenic Res 16:813–821
120. Gerjets T, Sandmann M, Zhu C, Sandmann G (2007) Metabolic engineering of ketocarotenoid biosynthesis in leaves and flowers of tobacco species. Biotech J 2:1263–1269
121. Mann V, Harker M, Pecker I, Hirschberg J (2000) Metabolic engineering of astaxanthin

production in tobacco flowers. Nat Biotech 18:888–892

122. Hattori K (1991) Inheritance of carotenoid pigmentation in flower color of chrysanthemum. Jpn J Breed 41:1–9

123. Simkin AJ, Underwood BA, Auldridge M, Loucas HM, Shibuya K, Schmelz E, Clark DG, Klee HJ (2004) Circadian regulation of the PhCCD1 carotenoid cleavage dioxygenase controls emission of beta-ionone, a fragrance volatile of petunia flowers. Plant Physiol 136:3504–3514

124. Li L, Paolillo DJ, Parthasarathy MV, Dimuzio EM, Garvin DF (2001) A novel gene mutation that confers abnormal patterns of beta-carotene accumulation in cauliflower (*Brassica oleracea* var. botrytis). Plant J 26:59–67

125. Lu S, Van Eck J, Zhou X, Lopez AB, O'Halloran DM, Cosman KM, Conlin BJ, Paolillo DJ, Garvin DF, Vrebalov J, Kochian LV, Kupper H, Earle ED, Cao J, Li L (2006) The cauliflower *Or* gene encodes a DnaJ cysteine-rich domain-containing protein that mediates high levels of beta-carotene accumulation. Plant Cell 18:3594–3605

126. Yano K, Aoki K, Shibata D (2007) Genomic databases for tomato. Plant Biotech 24: 17–25

Chapter 30

Azalea Phylogeny Reconstructed by Means of Molecular Techniques

Ellen De Keyser, Valentina Scariot, Nobuo Kobayashi, Takashi Handa, and Jan De Riek

Abstract

Plants belonging to the *Rhododendron* subgenera *Pentanthera* (deciduous) and *Tsutsusi* and *Azaleastrum* (evergreen) are called azaleas. Concerning their mutual phylogenetic positions, the *Pentanthera* subgenus is closer to evergreen rhododendrons (subgenera *Rhododendron* and *Hymenanthes*) than to the *Tsutsusi* subgenus. Both azalea types are important ornamentals with a long breeding tradition. Different hybrid groups are often named after the supposed principal ancestor species. Molecular techniques for phylogenetic and kinship research have been evaluated to a great extent. First, some studies using comparative gene sequencing are presented; this approach was then widened to the use of molecular markers to reveal more detailed genetic relationships. Finally, the use of candidate genes as functional markers for the assessment of genetic diversity is presented. This opens new research lines to the genetic mapping of plant traits and azalea genomics.

Key words: Azalea, Phylogeny, Molecular markers, Genetic diversity

1. Introduction

The genus *Rhododendron* of the heather family (*Ericaceae*) is well known for the beauty and diversity of floral and vegetative form of its more than 1,000 species. Because of the many species within this genus, during history taxonomists have made several classifications mainly based on morphological data (i.e. flowers, leaves, hairs, etc.). The genus is divided into eight subgenera, the four most important subgenera being *Tsutsusi* (evergreen azaleas except *Brachycalyx* section), *Pentanthera* (deciduous azaleas), *Rhododendron* (lepidote rhododendrons) and *Hymenanthes* (elepidote rhododendrons) (1). *Hymenanthes* species are evergreen

S.M. Jain and S.J. Ochatt (eds.), *Protocols for In Vitro Propagation of Ornamental Plants*, Methods in Molecular Biology, vol. 589, DOI 10.1007/978-1-60327-114-1_30, © Humana Press, a part of Springer Science+Business Media, LLC 2010

and have large non-scaly leaves; Rhododendrons (subgenus) species have smaller scaly leaves, are usually evergreen, but occasionally semi-deciduous. These latter two subgenera comprise what gardeners loosely refer to as "rhododendrons," while *Pentanthera* and *Tsutsusi*, together with the *Azaleastrum* subgenus (evergreen), comprise the "azaleas."

Within the evergreen azaleas, at least four more or less well defined groups of cultivars are distinguished according to their traditionally recognized most important ancestor: pot or Belgian azaleas (*R. simsii* putative hybrids, formerly erroneously called *Azalea indica* referring to *R. indicum* (L.) Sweet), Hirado azaleas (*R. scabrum* G. Don putative hybrids), Kurume azaleas (putative hybrids of *R. kiusianum* Makino var. *kiusianum* and var. *sataense* (Nakai) D.F. Chamb. and *R. kaempferi* Planch.) and Satsuki azaleas (putative mutants or hybrids of *R. indicum* and *R. eriocarpum* (Hayata) Nakai). The group of Hirado, Kurume and Satsuki azaleas is often referred to as Japanese azaleas because of their geographical origin; a distinction is made with regard to hardiness. Belgian pot azaleas have been created from a relatively narrow genetic basis of collectors material, introduced in botanical gardens and private collections from the far east. The Ghent region in Flanders (Belgium) is now the main European production area of pot azaleas, accounting for approximately 40 million plants per year, generating a value of 45 million Euro. Nevertheless, very little was known about the origin and the ancestor species used for this important ornamental crop. *R. simsii*, the species that is accepted to be the main ancestor, originates from hilly areas in China (Chang Jiang valley), Thailand, Laos and Burma. Apart from *R. simsii*, at least three other species from the *Tsutsusi* subgenus, from South-Asia and Japan, might have contributed: *R. indicum*, *R.* X*mucronatum* (Blume) G. Don and *R. scabrum* (2).

The Ghent horticulturists in the past (from 1,800 on) also have been at the basis of the well-known Hardy Ghent hybrids, deciduous azaleas related to *R. luteum* (Europe), *R. molle* (Asia) and many North-American species belonging to the *Pentanthera* subgenus. These hybrids are characterized by many cultivars with appealing yellow and orange flowers. It has been a dream of many pot azalea growers to introduce these types of flowers in their crop. However, phylogenetic studies by Kron (3, 4) on the *Rhododendron* genus and the *Pentanthera* subgenus in special, do contain a major warning: the *Pentanthera* subgenus seems to be a very old lineage that has relatively recently diversified, especially in North-America. Concerning their mutual positions, the *Pentanthera* subgenus (deciduous azaleas) is closer to evergreen rhododendrons (subgenera *Rhododendron* and *Hymenanthes*) than to the *Tsutsusi* subgenus (evergreen azaleas).

2. Molecular Techniques for Phylogenetic and Kinship Research

Morphological data to support taxonomical classifications are relatively easy to obtain, but only a small number of traits can be evaluated. Published research focusing on evergreen azalea morphology concerns diversity of flower colors (5), the inheritance of the hose-in-hose character (sepals forming a corolla like structure (6), content and composition of flavonoids which determine flower color (7), corolla size, number of stamens and percentage of plants with petaloid stamens (8) and other taxonomic characters (9). Moreover, when handling seedlings or wild plants outside the flowering season, one is not always able to observe decisive taxonomical traits. For this reason, an alternative classification may be based on molecular data. Molecular data provide the possibility to study a tremendous number of genetic loci in three distinct genomes (nuclear, mitochondrial, and chloroplast DNA). In plant systematics, the chloroplast genome has proven to be very informative. Among many genes studied for phylogenetic reasons, two that have been very useful in constructing evolutionary trees in the *Ericaceae* are *rbc*L (rubisco large subunit) and *mat*K (maturase K).

The *mat*K gene is known to evolve approximately three times faster than *rbc*L, and is therefore a powerful tool for phylogenetic reconstruction within angiosperm families and genera (10). On the other hand, the use of molecular markers can be very effective to study diversity within species or the relationships between closely related species.

Rhododendron taxonomy has been tested in recent times by molecular phylogenies based on several DNA regions. Most of these studies have aimed at higher-level relationships, despite the importance of lower ranks, such as sections, to most workers on the genus. Since the reference papers on the phylogenetic relationships by Kron (3, 11), integrating previous molecular studies using *rbcL* and nr18s sequences and *mat*K sequences of 42 taxa from traditional *Rhododendroideae* and potentially related clades, several authors have used these techniques to clarify taxonomical positions of specific genera. So far, classification of *Rhododendron* species based on morphology has led to a consensus taxonomy recognizing the major subgenera *Azaleastrum*, *Hymenanthes*, *Pentanthera*, *Rhododendron*, *Tsutsusi*, and three minor ones. To determine whether these subgenera are monophyletic and to infer phylogenetic relationships between sections and species, Goetsch (12) carried out a cladistic analysis using molecular data, including all groups within the genus. They introduced the comparison of the nuclear gene RPB2-I, encoding a major RNA Polymerase II subunit. A phylogeny based on data analysis of 87 species by means of maximum parsimony, maximum likelihood, and Bayesian methods showed subgenera *Azaleastrum* and *Pentanthera* to be polyphyletic and grouped all *Rhododendron* species (except the two in section *Therorhodion*) into three large clades. Based upon these results, modifications in *Rhododendron* classification were proposed, which consolidate minor subgenera and recognize monophyletic subgenera and sections.

Two interesting studies because of their link to azaleas as an ornamental group concern the genetic relationships of section *Pentanthera* and *Vireya*. Scheiber (13) unraveled the genetic relationships within *Rhododendron* section *Pentanthera* based on sequences of the internal transcribed spacer (ITS) region. For specimens of the 15 currently recognized species in *Rhododendron* section *Pentanthera*, sequences of the entire ITS region including ITS1, ITS2, and the 5.8 S subunit were generated by direct sequencing of polymerase chain reaction (PCR) amplified fragments. *Rhododendron vaseyi* A. Gray section *Rhodora*, was used as an outgroup. Aligned sequences of the 16 taxa resulted in 688 characters. The region contained 38 variable sites and eight phylogenetically informative characters. A bootstrap analysis was performed and a dendrogram was constructed. Divergence values among the taxa were extremely low ranging from 0.00 to 3.51%, providing support to traditional views of section *Pentanthera* as a group of very closely related species. Brown (14, 15) studied the phylogenetic relationships of *Rhododendron* section *Vireya* inferred from the ITS nrDNA region and based on two noncoding regions of cpDNA. Section *Vireya* is one of the most morphologically diverse groups of the genus Rhododendron. *Vireyas* have a unique distribution, for the genus, is predominantly found throughout the Malaysian Archipelago. The alpha taxonomy of section *Vireya* is relatively well understood and taxa are easily distinguished from other rhododendrons by their general appearance. Results of phylogenetic analyses of the ITS region (ITS-1, 5.8 S and ITS-2) and of two cpDNA regions, psbA-trnH and trnT-trnL intergenic spacers for the genus *Rhododendron*, with sampling concentrated on section *Vireya*, were presented. Relationships between the species of section *Vireya* do not correspond to the traditional classification based on morphology, instead correlating strongly with geographic areas, with a disjunction between an Australian-New Guinea clade and clades of west and middle Malaysian taxa. The phylogeny also indicates that the ITS region may not undergo complete homogenization in all species of *Rhododendron*.

Our own research focused especially on section *Tsutsusi*. First, some studies using comparative gene sequencing are presented; this approach was then widened to the use of molecular markers to reveal more detailed genetic relationships.

2.1. Phylogenetic Analysis Based on MatK Sequencing for Different Species from the Tsutsusi Subgenus

Phylogenetic analysis based on the *mat*K sequences was used to confirm the taxonomic position of the different species of the *Tsutsusi* subgenus. For this purpose, some additional species belonging to the subgenera *Hymenanthes* and *Rhododendron* were also analyzed. A specific PCR to amplify the *mat*K gene was performed on 33 Chinese *Rhododendron* species (Fig. 30. 1) using four primer couples (16). The amplified PCR fragments were directly used for sequencing, using the PCR primers as sequencing primers (forward and reverse). After sequencing and alignment of the different contigs, a fragment of 2,529 bp was obtained covering *mat*K and its flanking *trn*K sequences. Analyzing these 2,529 characters by means of PAUP, 2,438 characters resulted to be constant, 30 variable and parsimony uninformative and 61 variable and parsimony informative. Trees were generated using 100 resampled bootstrap datasets, stepwise addition and tree bisection recognition as branch swapping algorithm. The nodes that separated the different subgenera and sections all received high bootstrap values. In Fig. 30.2, a parsimony cladogram is showing the genetic similarity between the assayed Chinese *Rhododendron* species and some *R. simsii* hybrids.

Five major clusters can be distinguished. The first group contains *R mariesii* Hemsl. & E. H. Wilson and *R. farrerae* Tate ex Sweet; the latter was used as an outgroup species to root the dendrogram. Both species belong to the section *Brachycalyx* of the subgenus *Tsutsusi.* The second cluster contains *R. yunnanense* Franch., *R. spiciferum* Franch., *R.* X*duclouxii* Levl., and two populations of *R. racemosum* Franch. All these species belong to the subgenus *Rhododendron.* A third cluster contains *R. fortunei* Lindl., *R. sinofalconeri* Balf., *R. argyrophyllum* Franch., *R. irroratum* Franch. and two populations of *R. decorum* Franch. All these species belong to the subgenus *Hymenanthes.* The fourth cluster contains *R. florulentum* P.C. Tam, *R. rufulum* P.C. Tam and two populations of *R. rivulare* Hand.-Mazz. All these species belong to the subgenus *Tsutsusi.* The last and largest cluster contains species all belonging to the section *Tsutsusi* and can be divided in two subclusters. A first subcluster contains four *R. simsii* populations, two *R. simsii* var. *mesembrinum* Rehder populations, *R. scabrum, R. kiusianum* and the Kurume hybrid "Cupido."

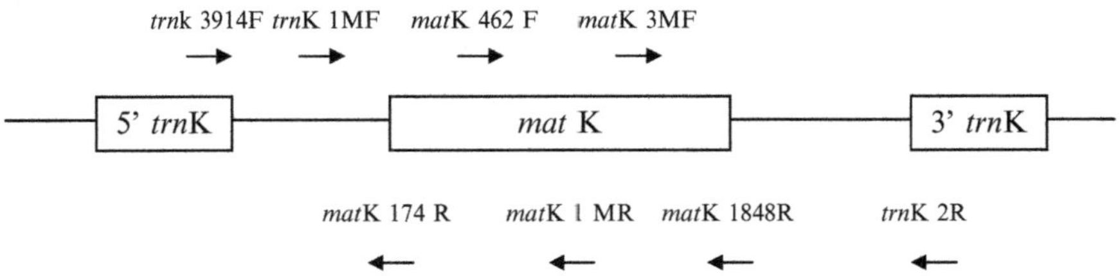

Fig. 30.1. MatK locus showing the position of the primers used for amplification of the partial sequences (16).

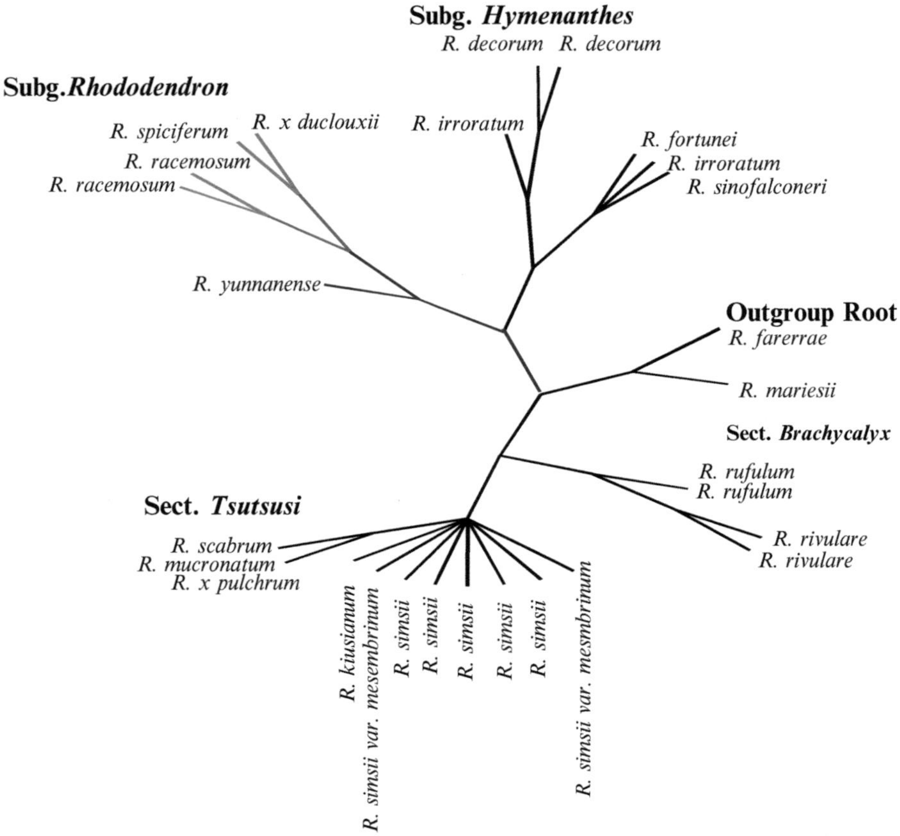

Fig. 30.2. Consensus parsimony tree based on matK data for a selection of Chinese Rhododendron species and cultivated R. simsii hybrids; length = 98; consistency index (excluding uninformative characters)= 0.9403; retention index = 0.9987.

A second subcluster contains five *R. simsii* hybrids, *R.* X*pulchrum* Sweet, *R.* X*mucronatum*, and *R. scabrum*.

Kurashige (16, 17) also analyzed *matK* with *trnK* intron sequences of chloroplast (cp) genome of the genus *Rhododendron* and its closely related genera, and elucidated a substantial part of the phylogenetic relationships (Fig. 30.3). Total DNA was extracted from fresh leaf tissues following the methods of Kobayashi (18) or Yukawa (19). All parsimony analyses were conducted with PAUP version 3.1 (20). The heuristic search option with 100 random replicates (21) was used to perform Fitch parsimony analyses (22). Branch lengths for trees were calculated by ACCTRAN optimization (23). For assessment of the relative robustness for clades found in each Fitch parsimony analysis, the bootstrap method (24) was used. In this study, all the species of genus *Menziesia* were nested within *Rhododendron* species.

Apart from above studies, some other authors also took section *Tsutsusi* as subject for their research. Hwang-ShihYing (25) investigated the molecular phylogeny of eight Taiwanese *Rhododendron* species based on sequence comparisons of the

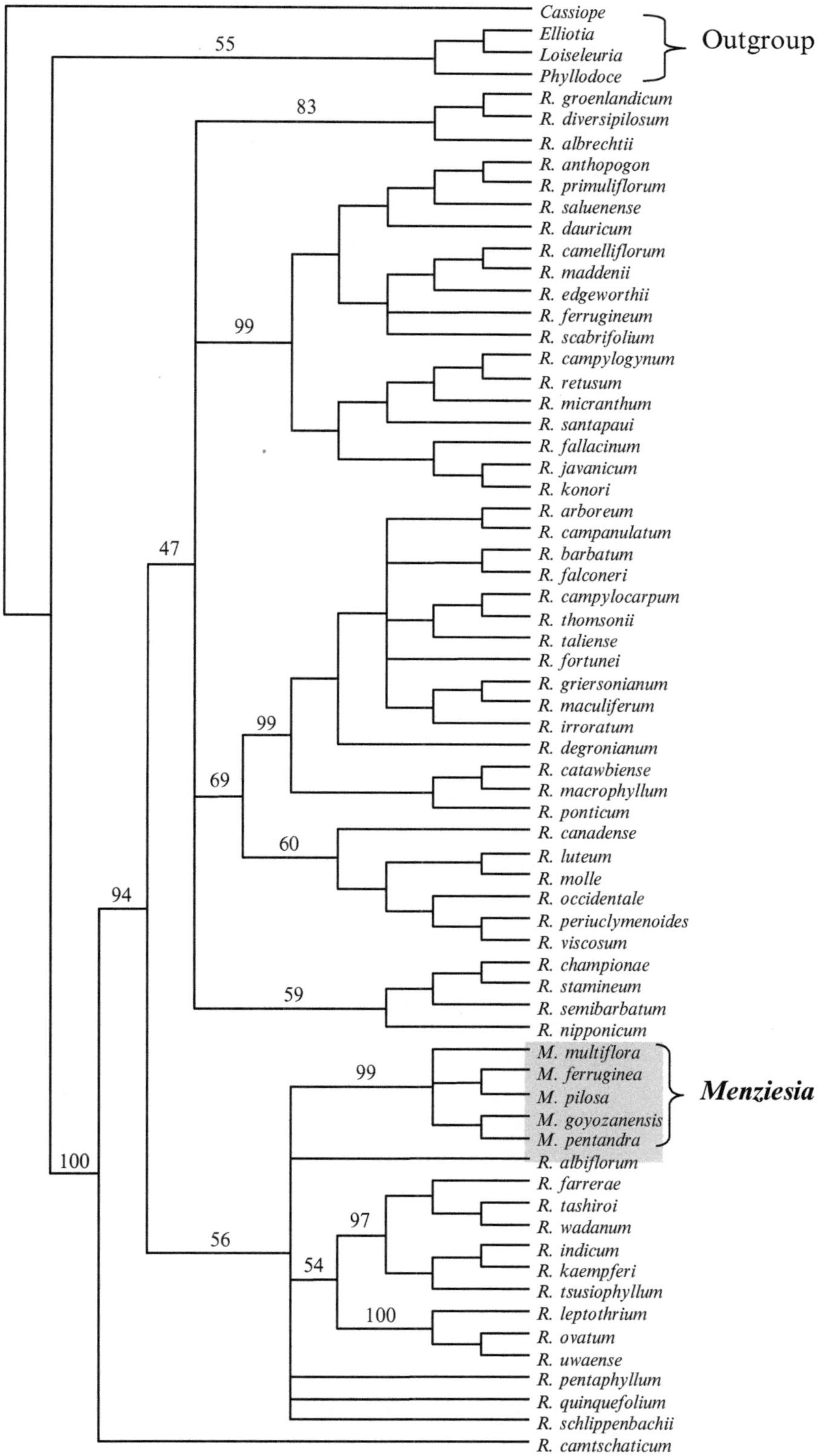

Fig. 30.3. Strict consensus of most parsimonious Fitch trees based on matK and trnK intron sequences for Rhododendron and Menziesia (17). Numbers above internodes indicate bootstrap values from 1,000 replicates.

chloroplast *trn*F-*trn*L intergenic spacer region. Aligned sequences of the eight *Rhododendron* taxa were composed by 460 base pairs. The region contained 16 variable sites, of which 10 were phylogenetically informative. Neighbor-joining and parsimony analyses were conducted. Identical topology with three major clades was obtained from parsimony and neighbor-joining trees. Clade 1 consisted of *R. pseudochrysanthum* Hayata, *R. rubropunctatum* Hayata, *R. morii* Hayata, *R. hyperythrum* Hayata and *R. formosanum* Hemsl. The latter was sister of the other four species in the clade 1. Clade 2 consisted of *R. oldhamii* Maxim. and *R. kanehirae* E.H. Wilson. *R. ellipticum* Maxim. formed clade 3 and was basal to all other *Rhododendron* species. Close phylogenetic relationships among *R. pseudochrysanthum*, *R. morii*, *R. rubropunctatum* and *R. hyperythrum* based on the cpDNA sequences agreed with those derived from morphological characters.

Gao-LianMing (26) compared the sequences of nuclear ribosomal DNA internal transcribed spacers (ITS) (including 5.8 S rDNA) of 12 species representing two sections of subgenus *Tsutsusi* and four species representing three sections of subgenus *Pentanthera*. To reconstruct the phylogeny of these two subgenera using parsimony methods, *R. redowskianum* (Maxim.) Hutch. of subgenus *Therorhodion* was included in the analysis as an outgroup. The size of ITS region within the three subgenera ranges from 642 to 645 bp. The aligned length is 653 positions, which provide 6.58% variable sites and 3.68% phylogenetic informative sites when gaps are treated as missing. Fifteen most parsimonious trees, with a length of 75 steps, consistency index of 0.9333 and retention index of 0.9515, were obtained. The topology of the strict consensus tree shows that: (1) subgenus *Tsutsusi* is a monophyletic group, the bootstrap value is 81%; (2) *R. tashiroi* is neither a member of section *Tsusiopsis* nor of section *Tsutsusi*, but it should be a member of section *Brachycalyx*; (3) *R. tsusiophyllum* Sugim. is a member of section *Tsutsusi*; and (4) the phylogenetic position of *R. schlippenbachii* Maxim. is not resolved in this study and needs further research.

2.2. The Usefulness of AFLP Markers for the Establishment of Phylogenetic Relationships in Evergreen Azaleas

To evaluate if the AFLP technique can be applied for the establishment of phylogenetic relationships and yields comparable results to common taxonomy and gene sequencing, 84 accessions of evergreen azaleas, including cultivars (19 Belgian azaleas, 6 Hirado azaleas, 14 Kurume azaleas, and 5 Satsuki azaleas) as well as the species and varieties that were most likely to have been involved in their origin were analyzed (27). Based on the pair wise Jaccard similarity matrix, cluster analysis with bootstrapping was applied on this dataset. The phenogram (Fig. 30.4) was compared to the taxonomic classification (9, 28) and previous phylogenetic studies based both on PCR-RFLP analysis of cpDNA (18) and on *mat*K and *trn*K intron sequences (16, 17). Subgenus *Pentanthera* was included as an outgroup.

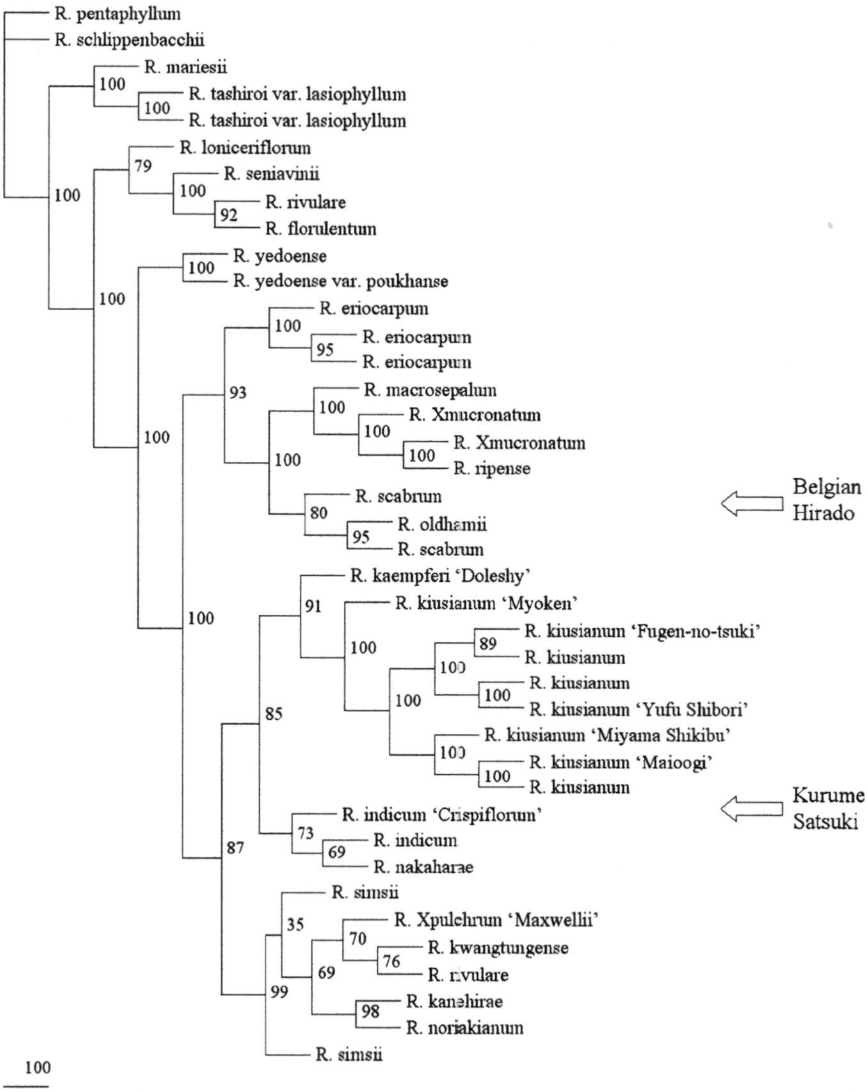

Fig. 30.4. Consensus tree between the species and varieties that were most likely to have been involved in the origin of the horticultural groups for the presence/absence of AFLP fragments. Numbers at the internodes indicate bootstrap values from 100 re-sampling cycles. The amount of replacement was fixed to 1% of the fragments per bootstrap. The arrows indicate the collocation of the 4 horticultural groups obtained by means of the same method (**27** with kind permission from Springer Science and Business Media).

Six distinct groups were supported by high bootstrap values (Fig. 30.4). The first group included the Japanese species classified by Yamazaki (28) in the subsection *Tsutsusi* and it is divided into two sub-clusters: one contains *R. kaempferi* "Doleshy" and *R. kiusianum* with its cultivars (series *Kaempferia*); the other one includes *R. indicum, R. indicum* "Crispiflorum" and *R. nakaharae* Hayata (series *Tsutsusi*). The second group consists of *R. simsii, R. Xpulchrum* "Maxwellii" and some Chinese species. The third group is formed by a sub-cluster enclosing *R. eriocarpum* belonging to the subsection *Tsutsusi*, series *Tsutsusi*, and a sub-cluster containing all the species belonging to the subsection *Scabra*, series *Scabra*, considered in this study, except for *R. yedoense* Maxim. and *R. yedoense* var. *poukhanense* (H. Lév.) Nakai which have a Korean origin as the fourth group. The fifth and sixth groups include *R. tashiroi* Maxim. var. *lasiophyllum* Hatus. ex T. Yamaz and Chinese species that are not or poorly involved in the origin of the Belgian, Hirado, Kurume and Satsuki azaleas.

The well-characterized horticultural groups were then brought into the cluster analysis (results not shown). As indicated in Fig. 30.4, the Hirado and Belgian azaleas, morphologically characterized by large flowers, grouped together with the species belonging to the subsection *Scabra*, series *Scabra*; the Kurume and the Satsuki azaleas, both with small flowers, clustered with the species belonging to the subsection *Tsutsusi* (series *Kaempferia* and *Tsutsusi*).

Results generated by AFLP fingerprinting are generally in good agreement with other phylogenetic studies, geographical distribution and horticultural classification. Deviations, if observed, are in line with open issues concerning the taxonomical position of certain species or hybrids. The close relationship of *R. kaempferi* and *R. kiusianum* revealed by the present analysis was previously established in their cpDNA (18). Likewise, the genetic closeness between *R. indicum* and *R. kaempferi* was formerly shown by Kurashige (16). The origin of *R.* X*pulchrum* is unclear. According to Heursel (29), this might be a Japanese natural garden hybrid. On the other hand, Chamberlain (1) consider *R.* X*pulchrum* originated from *R. indicum* x "Ledifolia." The present study seems to suggest that Chinese azaleas, including *R. simsii*, could be involved in its origin. Concerning *R.* X*mucronatum*, the hypothesis that it is an artificial hybrid of *R. ripense* Makino and its close ally *R. macrosepalum* Maxim. (syn. *R. stenopetalum* (Hogg) Mabb.) (9) seems to be confirmed. Based on morphological information, the taxonomic position of *R. tashiroi* has not been stabilized. It is considered an anomalous species that lies between section *Tsutsusi* Sweet and *Brachycalix* Sweet (9). In the present analysis it was nested within a clade with a species of subgenus *Tsutsusi* section *Brachycalyx* as in Kurashige (16, 17), supporting Yamazaki's (28) view.

Concerning the horticultural groups, it is accepted that Hirado and Belgian azaleas share at least one important common ancestor, *R. scabrum*, originating from some small southern Japanese islands, as the present study seems to confirm. *R. simsii* is generally considered to be the most important ancestor for Belgian azaleas due to the type of the flowers and the only ancestor exclusively distributed to China. In the present analysis, *R. simsii* seems to be more apart, in agreement with a previous research (30). Kurume azaleas are considered to be related to *R. kiusianum* and *R. kaempferi*, species native to Kyushu Island in Japan.

The Satsuki azaleas are believed to derive from *R. indicum* and *R. eriocarpum*. Both these statements look to find correspondence in this study.

3. Candidate Genes as Functional Markers for the Assessment of Genetic Diversity and Mapping of Plant Quality Traits

3.1. Development of EST Markers

Molecular markers can be very effective to study diversity within species or the relationships between closely related species and hybrids. DNA fingerprinting is usually based on techniques amplifying arbitrary sequences or specific loci. Among the arbitrary DNA fingerprinting systems, the multilocus AFLP technique (31) has already proven to be successful in assessing the genetic variation of a breeders' collection of evergreen azaleas (32) and for classifying an Italian gene pool of evergreen azaleas (27). Among the DNA fingerprinting techniques based on specific loci, microsatellite markers have the potential to provide reliable and highly informative genetic data. Molecular markers from the transcribed region of the genome have also great potential for applications in plant genotyping. Expressed Sequence Tag (EST) markers have already shown a good level of polymorphism in various plants but they were not previously developed in *Rhododendron*.

Starting from two cDNA libraries made from flowers of the *R. simsii* hybrid "Flamenco" and the species *R. luteum*, 323 cDNA fragments were randomly picked and sequenced. The putative functions of the cDNA fragments were determined by comparison of the sequences with EMBL accessions using FASTA33 (33). Reliable homologies were found for 31% of all fragments. Primers were developed on 127 cDNA fragments and used for PCR amplification on six different azalea cultivars and species. PAGE was used to separate the fragments and the presence of polymorphic bands was evaluated. In the end, this resulted in 45 polymorphic EST markers (34), 32 of which originated from "Flamenco" cDNA. Although markers were developed in less related species, markers from *R. luteum* could be applied in *R. simsii* hybrids and also "Flamenco" EST markers generated polymorphisms in *R. luteum*. Therefore the use of the markers will not be limited to the species in which they were developed, but they have a good chance to be useful in the whole *Rhododendron* genus as functional markers.

3.2. Comparison of AFLP, STMS, and EST Markers for the Assessment of Genetic Diversity

This study refers to the use of these EST markers in comparison to AFLP and STMS markers to define genetic differentiation among a set of cultivars. The material is representative of the four more or less well defined groups of evergreen azalea cultivars (Belgian, Hirado, Kurume and Satsuki azaleas) and of the species and varieties that were most likely to have been involved in their origin.

These techniques have shown different capacities in discriminating and assessing genetic relationships in evergreen azaleas (35). AFLPs revealed the highest polymorphism detection capacity while STMS and EST markers showed the highest discrimination capacity. This study evaluated the usefulness of AFLP, STMS and EST markers in (1) evaluation of the transferability of the different markers across the subgenus *Tsutsusi*; (2) differentiating between horticultural hybrid groups of evergreen azaleas; (3) assessing the extent of genetic variation within horticultural groups; and (4) revealing the genetic relationships between species and cultivars.

3.2.1. Marker Transferability Across Species and Horticultural Groups

At first, the transferability of ten EST markers (developed from the "Flamenco" cDNA library) and the ten STMS markers developed by Dendauw (36) across the species and cultivars genotyped was investigated. Concerning the STMS loci, four of ten proved to be not fully transferable across the sample set. Band profiles obtained using the other STMS loci were easily scored. A total of 99 alleles were detected at six STMS loci. The number of alleles per locus ranged from 12 to 29, with an average of 16.5 alleles per locus. The six STMS primer sets amplified across all the horticultural groups and almost all the different species included in this study, exhibited a great transferability.

Concerning the EST loci, one primer combination produced too many aspecific bands. Therefore, it was excluded from further analysis. At these nine loci, a total of 38 alleles could be scored. All primer sets allowed amplification across the Hirado and Belgian azaleas. Across the Kurume cultivars, no alleles were found in "Rex" and "Hino Crimson" by EST-04 neither in "Hatsugiri" by EST-10 nor in "Herbert" by both EST-04 and EST-10. No EST primer pair could amplify across all the Satsuki azaleas; only the cultivar "Shinkyo" could be typed by all of them. Also, in several accessions of *R. kiusianum*, "Myoken," and "Fugen-no-tsuki," "Miyama Shikibu", few primer sets amplified successfully; only by means of EST-02 and EST-06 all the *R. kiusianum* accessions showed amplified alleles. Apart from the *R. kiusianum* accessions, the nine EST primer sets exhibited a great transferability across the species. A few cases of no amplifications were also found (*R. macrosepalum* by EST-01, *R. yedoense* by EST-07, *R. simsii* and *R. schlippenbachii* by EST-09, *R. rivulare* and *R. loniceriflorum* P.C. Tam by EST-10). Primer EST-05 amplified only one allele in all but one sample, and in *R. loniceriflorum* no allele was obtained.

3.2.2. Genetic Variation Within Horticultural Groups

Knowledge of genetic diversity and relationships between genotypes is required for efficient rationalization and utilization of germplasm resources. Furthermore, it is important for planning future breeding programs. To attribute the distribution of the genetic variation in AFLP, STMS and EST patterns among and within horticultural groups (populations), AMOVA (37) was computed at one level from squared Euclidean distances with the number of permutations set at 1,000, using the Arlequin software (38). For all marker techniques used, most of the genetic diversity was attributable to differences among cultivars within horticultural groups (89.08%, 89.09% and 85.56% for AFLP, STMS and EST markers respectively, Table 30.1). The precise origin of horticultural evergreen azalea cultivars is still not well established. Recently, several studies have been undertaken on this subject (30, 32, 39, 40). Based on the genetic conformity as revealed by AFLP markers (27), a genetic continuum must be accepted, spanning a lot of species of the *Tsutsusi* subgenus. Common parental species could explain the low genetic diversity attributable to differences among horticultural groups. Besides, intensive breeding activities within groups, where seed is commonly mass selected on the basis of phenotypic characteristics, have yielded highly significant differentiation among cultivars within each

Table 30.1
AMOVA (analysis of molecular variance) significance tests (1,023 permutations) at one level based on AFLP, STMS, and EST data sets (binary Euclidian distance coefficient)

	Source of variation	Degree of freedom	Sum of squares	Variance components	Percentage of variation
AFLP	Among hybrid groups	3	33.900	0.623	10.92**
	Within hybrid groups	40	203.235	5.081	89.08
	Total	43	237.136	5.704	
	Fixation Index F_{st}: 0.10920[a]				
STMS	Among hybrid groups	3	11.942	0.219	10.91**
	Within hybrid groups	40	71.643	1.791	89.09
	Total	43	83.585	2.010	
	Fixation Index F_{st}: 0.10908[a]				
EST	Among hybrid groups	3	10.804	0.226	14.44**
	Within hybrid groups	40	53.651	1.341	85.56
	Total	43	64.455	1.567	
	Fixation Index F_{st}: 0.14438[a]				

The species group was not included in this analysis being, by its nature, artificial and comprising very different genotypes

[a] Significant with $p < 0.001$ as tested by 1,023 permutations

horticultural group. Comparing the marker techniques, EST markers outperformed AFLP and STMS markers concerning F_{st} values (approx. 0.15 vs. 0.11) indicating a low but significant differentiation among horticultural groups. This might be due to the more conservative nature of the polymorphism that EST markers can detect, that could allow better emphasis of differences among parental species.

3.3. Exploration of Segregation of Plant Quality Traits and the Construction of Genetic Maps

When breeding new Belgian pot azaleas, seedlings are at first selected based on their flower characteristics, it is only at in a later stage that other important plant quality traits such as leaf shape, leaf color, growth vigor, compactness, ... are evaluated. Attractive flowering plants are still too often rejected in a later stage of breeding because the plant as a whole does not comply with the present cultivation standard. The inheritance of flower color has been well studied (2), but information on the heritability of other plant quality traits is still lacking and needs to assure a more specific and consequently a more efficient selection when new crosses are made. For this purpose, different crosses are made for each trait between parents that reveal the extreme phenotypes of the range of the examined trait. All the siblings of a cross, segregating for one specific trait are then scored individually. To assure that these plants are evaluated under natural growth conditions, they receive no treatments with growth inhibitors or other sprays.

The phenotype of these (semi)-quantitative traits will be integrated on a genetic map of azalea by QTL-mapping. Flower color is inherited as a qualitative trait, so mapping is quite straightforward. For the construction of the genetic map of Belgian pot azalea, AFLP, and STMS markers are generated on 4 populations that are segregating for flower color and several plant quality traits under investigation. Since azalea has a rather high chromosome number ($2n=26$), integration of the separate population maps into a consensus genetic map of azalea will only be successful when there is at least one segregating co-dominant locus for each chromatide. STMS markers (36, 41), as well as EST markers (34), will be used for this purpose. Following the candidate gene approach, EST markers will also be developed for genes that are known to be involved in flower color biosynthesis or in literature that are reported to have a major impact on the generation of other complex plant quality traits. In this way, phenotypic and genetic data (candidate genes) will both be integrated on the genetic map of azalea. In case both are located at the same mapping position, these genes are proven to be directly involved in the creation of the phenotypic variation of the trait. Nevertheless, it is very likely that not the genes themselves, but transcription factors are the switches that regulate the phenotype of the trait. In that case, phenotype and genotype will be mapped at different positions, but phenotype is then expected to be mapped together with the true regulators, the transcription factors. To confirm this theory, gene expression profiles of the candidate genes will be generated in several populations using real-time RT-PCR (42) and eQTL mapping will integrate these data with the genetic map.

Acknowledgments

The authors wish to thank Romain Uytterhaegen for his hands-on experience in azalea breeding and cultivation, Veerle Buysens, Veerle Cools and Laurence Desmet for their efforts in doing DNA preparations and AFLP, STMS and EST-reactions. The whole staff of the biotech lab was very much appreciated for their support and skilful assistance.

References

1. Chamberlain DF, Hyam G, Argent G, Fairweather G, Walter K (1996) The genus Rhododendron. Its classification & synonymy. Royal Botanical Garden, Edinburgh
2. Heursel J (1999) Azalea's: oorsprong, veredeling en cultivars. Lannoo, Tielt
3. Kron KA (1997) Phylogenetic relationships of Rhododendroideae (Ericaceae). Am J Bot 84:973–980
4. Kron KA (2002) Phylogenetic relationships and major clades of Rhododendron (Rhodoreae, Ericoideae, Ericaceae) In: Argent G, McFarlane M (eds) Rhododendrons in horticulture and science. Royal Botanic Garden, Edinburgh, pp 79–85
5. Heursel J, De Roo R (1981) Polyploidy in evergreen azaleas. Hort Sci 16:765–766
6. Heursel J (1976) Die Vererbung des Merkmals "hose in hose" (Doppelkronigkeit) bei Azaleen. Gartenwelt 76:111–113
7. Heursel J, Horn W (1977) A hypothesis on the inheritance of flower colours and flavonoids in *Rhododendron simsii* Planch. Z Pflanzenzüchtung 79:238–249
8. Heursel J, Garretsen F (1989) Inheritance of corolla size, number of stamens and percentage of plants with petaloid stamens in evergreen azaleas (Rhododendron Subsect. obtusa). Plant Breed 103:304–309
9. Chamberlain DF, Rae SJ (1990) A revision of Rhododendron IV Subgenus Tsutsusi. Edinburgh J Bot 47:89–200
10. Hilu KW, Liang H (1997) The *matK* gene: sequence variation and application in plant systematics. Am J Bot 84:830–839
11. Kron KA (1993) A revision of Rhododendron section Pentanthera. Edinburgh J Bot 50: 249–363
12. Goetsch L, Eckert AJ, Hall BD (2005) The molecular systematics of Rhododendron (Ericaceae): a phylogeny based upon *PRB2* gene sequences. Syst Bot 30:616–626
13. Scheiber SM, Jarret RL, Robacker CD, Newman M (2000) Genetic relationships within Rhododendron L. section Pentanthera G. Don based on sequences of the internal transcribed spacer (ITS) region. Sci Hort 85:123–135
14. Brown GK, Craven LA, Udovicic F, Ladiges PY (2006) Phylogeny of Rhododendron section Vireya (Ericaceae) based on two non-coding regions of cpDNA. Plant Syst Evol 257: 57–93
15. Brown GK, Craven LA, Udovicic F, Ladiges PY (2006) Phylogenetic relationships of Rhododendron section Vireya (Ericaceae) inferred from the ITS nrDNA region. Aust Syst Bot 19:329–342
16. Kurashige Y, Mine M, Kobayashi N, Handa T, Takayanagi K, Yukawa T (1998) Investigation of sectional relationships in the genus Rhododendron (Ericaceae) based on matK sequences. J Jpn Bot 73:143–154
17. Kurashige Y, Etoh JI, Handa T, Takayanagi K, Yukawa T (2001) Sectional relationships in the genus Rhododendron (Ericaceae): evidence from matK and trnK sequences. Plant Syst Evol 228:1–14
18. Kobayashi N, Kawashima S, Handa T, Takayanagi K, Arisumi K (1998) Introgression

in Japanese evergreen azaleas (*Rhododendron kiusianum* and *R. kamepferi*). Acta Hortic 454:325–328

19. Yukawa T, Koga S, Handa T (2000) DNA uncovers paraphyly of Dendrobium (Orchidaceae). In: Andrews S et al (eds) Taxonomy of cultivated plants. Royal Botanic Gardens, Kew, London, pp 351–354
20. Swofford DL (1993) PAUP: phylogenetic analysis using parsimony, version 3.1. Computer program distributed by the Illinois Natural History Survey, Champaign, IL
21. Maddison DR (1991) The discovery and importance of multiple islands of most-parsimonious trees. Syst Zool 40:315–328
22. Fitch WM (1971) Toward defining the course of evolution: minimum change for a specific tree topology. Syst Zool 20:406–416
23. Swofford DL, Maddison WP (1987) Reconstructing ancestral character states under Wagner parsimony. Math Biosci 87:199–299
24. Felsenstein J (1985) Confidence limits on phylogenies: an approach using the bootstrap. Evolution 39:783–791
25. Hwang-ShihYing, Hsu-KuoKai (2001) Molecular phylogeny of eight Taiwanese Rhododendron species based on chloroplast trnF-trnL DNA sequences. Taiwan J For Sci 16:153–160
26. Gao-LianMing, Yang-JunBo, Zhang-ChangQin, Li-DeZhu (2002) Phylogenetic relationship of subgenus Tsutsusi (Rhododendron) based on ITS sequences. Acta Bot Yunnanica 24:313–320
27. Scariot V, Handa T, De Riek J (2007) A contribution to the classification of evergreen azalea cultivars located in the Lake Maggiore area (Italy) by means of AFLP markers. Euphytica 158:47–66
28. Yamazaki T (1996) A revision of the genus Rhododendron in Japan, Taiwan, Korea and Sakhalin. Tsumura Laboratory, Tokyo, pp 1–125
29. Heursel J (1987) Inheritance of flower colours and breeding of evergreen azaleas. J Am Rhododendron Soc 41:141–145
30. Dendauw J, De Riek J, De Loose M, Van Bockstaele E (2002) Identification of 33 Chinese *Rhododendron* species using matK sequences and AFLP data. Acta Hortic 572: 169–177
31. Vos P, Hogers R, Bleeker M, Reijans M, van de Lee T, Hornes M, Fijters A, Pot J, Peleman J, Kuiper M, Zabeau M (1995) AFLP: a new technique for DNA fingerprinting. Nucleic Acids Res 21:4407–4414
32. De Riek J, Dendauw J, Mertens M, De Loose M, Heursel J, Van Bockstaele E (1999) Validation of criteria for the selection of AFLP markers to assess the genetic variation of a breeders' collection of evergreen azaleas. Theor Appl Genet 99:1155–1165
33. Pearson WR, Lipman DJ (1988) Proc Natl Acad Sci USA 85:2444–2448
34. De Keyser E, De Riek J, Van Bockstaele E (2009) Discovery of species-wide EST-derived markers in *Rhododendron* by intron-flanking primer design. Mol Breed 23:171–178
35. Scariot V, De Keyser E, Handa T, De Riek J (2007) Comparative study of the discriminating capacity and effectiveness of AFLP, STMS and EST markers in assessing genetic relationships among evergreen azaleas. Plant Breed 126:207–212
36. Dendauw J, De Riek J, Arens P, Van Bockstaele E, Vosman B, De Loose M (2001) Development of sequenced tagged microsatellite site (STMS) markers in azalea. Acta Hortic 546: 193–197
37. Excoffier L, Smouse P, Quattro J (1992) Analysis of molecular variance inferred from metric distances among DNA haplotypes: application to human mitochondrial DNA restriction data. Genetics 131:479–491
38. Schneider S, Roessli D, Excoffier L (2000) Arlequin: a software for population genetics data analysis, version 2.000. Genetics and Biometry Laboratory, Department of Anthropology, University of Geneva, Switzerland
39. Handa T, Eto J, Kita K, Kobayashi N (2002) Genetic diversity of Japanese wild evergreen azaleas in Kyushu (south main island of Japan) characterized by AFLP. Acta Hortic 572: 159–161
40. Kobayashi N, Handa T, Yoshimura K, Tsumura Y, Arisumi K, Takayanagi K (2000) Evidence for introgressive hybridization based on chloroplast DNA polymorphisms and morphological variation in wild evergreen azalea populations in the Kirishima Mountains, Japan. Edinburgh J Bot 57:209–219
41. Dunemann F, Kahnau R, Stange I (1999) Analysis of complex leaf and flower characters in Rhododendron using a molecular linkage map. Theor Appl Genet 98:1146–1155
42. De Keyser E, De Riek J, Van Bockstaele E (2007) Gene expression profiling of key enzymes in azalea flower colour biosynthesis. Acta Hortic 743:107–113

Chapter 31

Status of Floriculture in Europe

Johan Van Huylenbroeck

Abstract

Europe is traditionally the largest producer of floricultural products in the world with an estimated production value of over 12 billion euro in 2006. The Netherlands, Italy, Germany, Spain, United Kingdom and France are the main centres of production. More recently, a significant growth in production area was observed in Poland also. The Dutch auctions remain the world's largest trading system for flowers and plants. Looking at the intra-European trade, Belgium and Denmark are also major exporting countries. The consumption of floricultural products increased strongly within Europe during the last years. Especially, Eastern Europe has a big potential as a new market. In these countries, the demand for ornamentals will increase due to the rise of income and the level of prosperity. In spite of the positive developments in consumption and production, increasing energy costs, growing environmental concern and globalisation of production, trade and markets will form the major challenges European floricultural industry has to deal with in the near future.

Key words: Europe, Floriculture, Cut flowers, Pot plants, Statistics

1. Introduction

Floriculture is widely spread throughout the world and is an important economic activity in many countries. The flower industry comprises cultivation and trade in cut flowers, cut foliage, potted plants and bedding plants. In some regions in the world as India and China, the industry is mainly organised in small companies which focus on the local market. In other countries as in Africa and South-America, cut flower production developed during the last decade into a major export oriented industry. Traditionally, Europe has been an important production and consumption centre for ornamental products. European floricultural production is characterised by a high diversity in products, production facilities and production areas. In Europe, the Netherlands are known for their leading role in the production of cut flowers, pot plants, as well as

S.M. Jain and S.J. Ochatt (eds.), *Protocols for in Vitro Propagation of Ornamental Plants,* Methods in Molecular Biology, vol 589,
DOI 10.1007/978-1-60327-114-1_31, © Humana Press, a part of Springer Science+Business Media, LLC 2010

bulbs, annuals and perennials. Germany focuses more on garden plants, while Italy produces lots of flowers and potted plants. Denmark is famous for flowering pot plants and Belgium for the production of azaleas. France and UK have a broad product pallet. Poland and Spain are new players in the market. In recent years, significant changes in European horticultural industry were observed. European integration, increasing consumption and new developing markets in Eastern Europe and some other parts of the world, liberalisation of trade and important shifts in production areas worldwide had a significant impact on the industry.

Compiling adequate and correct figures on European floricultural production, acreage and trade movements is a gigantic task, especially since some countries do not keep meticulous records of these figures. Furthermore, no uniform product definitions are used by the different countries, which make it hard to compare figures. Yearly the International Association of Horticultural Producers (IAPH) together with the Union Fleurs publishes a comprehensive handbook containing statistics of most countries. Data used in this review are mainly based on these publications (1).

2. Cut Flower and Pot Plant Production in Europe

Today, the total European production area for cut flowers and pot plants, both under protected cultures and in open air, is estimated to be almost 53,000 ha. Worldwide the production area is estimated on 609,000 ha, from which more then two-thirds is located in Asia (mainly open air production in China and India). Production area in Central and South America is about 48,000 ha, while in USA 25,000 ha are located. In Europe, about 40–45% of the production area is under glass or plastic coverage (Table 31.1). The Netherlands, Italy and Germany, traditionally the most important ornamental flowers/plants producing countries, have also today in absolute numbers the largest production areas within Europe, followed by the United Kingdom, Spain, France and Poland (Table 31. 1). In both the Netherlands and Italy more then 5,000 ha flowers and pot plants are produced under glass or plastic. This represents about 45% of the European protected production area. Germany and Spain have around 2,500 ha. Belgium and Denmark have both a relatively high production under glass compared to the country size (Table 31.1). In southern European countries, mainly plastic greenhouses are used for production, while in more northern regions, glass constructions are most common.

Over the last 10 years, total production area within Europe remained almost stable or even slightly increased. However, in

Table 31.1
Total area under production of cut flowers and pot plants in different European countries (ha under glass or plastic and open air)

	Protected (ha)	Open air (ha)	Total (ha)
Austria[a]	418	1,753	2,171
Belgium	625	1,027	1,652
Czech republic	126	89	215
Denmark	353	194	547
Finland	135	–[b]	135
France	1,961	2,503	4,464
Germany	2,524	5,116	7,640
Greece	–	–	944
Guernsey	55	–	55
Hungary	220	220	440
Ireland	–	–	62
Italy	–	–	7,976
Netherlands	5,365	2,784	8,149
Norway	118	–	118
Poland	1,417	3,176	4,593
Portugal	240	–	240
Spain	2,456	3,684	6,140
Sweden	159	–	159
Switzerland	229	288	517
United Kingdom	1,043	5,726	6,769

[a]Tree nursery included
[b]No separate data available
Source: AIPH (2007)

most West European countries, a decrease in especially cut flower production was observed. The largest part of cut flower production is nowadays located in countries as Colombia, Ecuador, Kenya, Ethiopia and Zimbabwe. On the otherhand, opposite, the Polish ornamental industry showed a boost in production area due to the lower production costs. Also, open air production in Spain grew significantly. Besides flowers and pot plant production, the area in Europe for ornamental tree production is about 110,000 ha, while almost 26,500 ha is used for bulb production.

A large total area of ornamental crops does not automatically imply a high production value. Differences in production and yield per hectare are considerable. In Table 31.2 an overview of the production value for the different countries is given. These figures

Table 31.2
Production values of flower and pot plants and number of nurseries in different European countries

	Value of production (million €)	Number of holdings
Austria	240[a]	855
Belgium	263	2,953
Czech republic	43	3,000
Denmark	349	683
Finland	96	825
France	956	7,663
Germany	1,289	9,561
Guernsey	45	64
Hungary	95	850
Ireland	19	–[b]
Italy	1,627	–
Netherlands	3,890	6,807
Norway	117	690
Poland	186	–
Portugal	457	704
Spain	1,757	5,454
Sweden	128	685
Switzerland	110	543
United Kingdom	454[c]	9,400[a]

Source: AIPH (2007)
[a]Tree nursery included
[b]No data available
[c]Flower bulbs included

clearly demonstrate the leading role of the Netherlands with a total production value of 3,900 million euro in 2006, representing 30% of the estimated total European production value. Production values of Italy, Germany and Spain are significantly lower, although the total production areas in these countries are comparable to the Netherlands. These big differences might be explained by the fact that in the Netherlands a greater share of production is carried out under glass. Another reason is the very intensive and highly automated production techniques used by Dutch growers, resulting in high productions per hectare.

The number of companies in the traditional production countries decreased continuously during the last decade. In combination with a stable production area, this results in an increasing average company size. In the economy of scale, increasing company size reduces production costs per unit which is necessary for profitable development. It might be expected that in future this tendency will continue or even accelerate.

3. European Trade and Market

The development of Europe's flower and plant market depends both on the medium and the long term general economic and demographic evolutions. The gross amount of Europe's production today is consumed within Europe. The United States and Japan are the most important over-seas clients for European products. On the other hand, Europe imports large numbers of especially cut flowers from Kenya, Colombia, Ecuador, Israel, Zimbabwe and some other minor flower producing countries. Total import of fresh cut flowers from non-European countries reached almost 800 million euro in 2006. From these, 54% came from Africa and 30% Latin-America. About 58% of these imported cut flowers are sold via the Dutch auctions. Compared to cut flowers, the import of ornamental pot plants from outside Europe is significantly lower (about 260 million euro in 2006).

When looking to import and export figures from individual European countries it is noticed that Germany, the United Kingdom, The Netherlands and France are the main importing countries (Table 31.3). For export, the dominant position of the Dutch producers is clear. They represent almost 70% of the total exported value for ornamentals within Europe, 83% for cut flowers and 53% for pot plants. For pot plants also, the prominent position of Belgian and Danish growers is remarkable (Table 31.4). These figures also demonstrate that in spite of the large production areas for ornamentals in some countries, they mainly produce for the local market, as is the case of i.e. Italy and France. Most European countries (except the Netherlands) have higher import then

Table 31. 3
Import values of cut flowers and pot plants

	Cut flowers (million €)	Pot plants (million €)	Sum (million €)
Germany	756	608	1,364
United Kingdom	791	273	1,064
France	399	370	769
Netherlands	501	301	802
Italy	162	152	314
Switzerland	130	169	299
Others	900	900	1,800
Total Europe	3,639	2,773	6,412

Source: AIPH (2007)

Table 31. 4
Export values of cut flowers and pot plants

	Cut flowers (million €)	Pot plants (million €)	Total (million €)
Netherlands	2,402	1,536	3,938
Belgium-Luxemburg	54	250	304
Denmark	4	289	293
Italy	60	215	275
Germany	37	190	227
Spain	37	63	100
Others	300	400	700
Total Europe	2,894	2,943	5,837

Source: AIPH (2007)

export figures, indicating that the domestic production capacity is not self-supplying.

In most European countries, there is a high consumption of floricultural products, stimulated by urbanization and prosperity. The average consumption of flowers and plants per capita increased

Table 31.5
Per capita consumption of cut flowers and plants in different European countries

	Cut flowers (€)	Plants (€)	Total (€)
Austria	45	38	83
Belgium	30	54	84
Croatia	7	7	14
Czech Republic	10	7	17
Denmark	44	44	88
Finland	34	44	78
France	31	20	51
Germany	36	50	86
Greece	16	8	24
Hungary	15	8	23
Ireland	35	13	48
Italy	23	12	35
Netherlands	55	33	88
Norway	59	56	115
Poland	8	3	11
Portugal	16	8	24
Russia	3	1	4
Slovakia	7	3	10
Slovenia	23	21	44
Spain	22	12	34
Sweden	38	46	84
Switzerland	80	42	122
United Kingdom	44	13	57
Europe	24	18	42

Source: AIPH (2007)

in most European countries during the last years and is estimated at €42 in 2006 (Table 31.5). Taking into account a population of more then 600 million people, this represents a total market value of 25 billion euro. At present, the spending potential of East European consumers is still far behind that in South, North or

West Europe. Per capita consumption of flowers and plants in these countries varies between €4 (Russia) and €17 (Czech Republic), compared to €30–€35 in Southern Europe and €80 in the Northern countries. Norway and Switzerland spend more then €100 per capita on flowers and plants (Table 31.5). Nevertheless, East Europe is an important growth market for flowers and plants. Today, Russian market represent already more then 725 million euro, while Poland with around 400 million euro and Hungary with around 240 million euro showed an increased demand. This positive development is expected to continue in the next years, since with a growing income, people will start to buy more ornamentals (2).

The most important trading places for ornamental plants in Europe are the Dutch auctions. The Dutch flower auctions are key institutions in the world flower industry and a base for the flower trade in the Netherlands (3). In 2007, the flower auctions collectively traded more than 4.2 billion euro in cut flowers and potted plants (Table 31.6). In total 12.7 billion flower stems and pot plants were sold. Total turnover of cut flowers was 2.5 billion euro and 1.3 billion euro for pot plants. In cut flowers, roses are the most important, followed by *Chrysanthemum* and tulips. In quantities, *Gerbera* also represents an important product. However,

Table 31.6
Turnover, quantities sold, and average price of cut flowers and pot plants on the Dutch auctions in 2007

	Turnover (×1,000 €)	Number of stems or plants	Average price (€)
Cut flowers			
Rosa	795,436	3,337,457	0.24
Chrysanthemum Ind Grp TR	296,040	1,314,354	0.23
Tulipa	204,608	1,468,932	0.14
Lilium	171,203	379,915	0.45
Gerbera	126,169	839,883	0.15
Cymbidium	71,174	34,667	2.05
Chrysanthemum Ind Grp	64,174	166,808	0.39
Freesia	56,375	340,495	0.17

(continued)

Table 31.6 (continued)

	Turnover (×1,000 €)	Number of stems or plants	Average price (€)
Anthurium	45,908	83,907	0.55
Alstroemeria	38,711	245,089	0.16
Eustoma russellianum	38,047	117,977	0.32
Hippeastrum	36,843	47,452	0.78
Zantedeschia	36,409	78,982	0.46
Gypsophila	31,471	179,871	0.18
Hypericum	29,743	190,612	0.16
Cut flowers total	2,547,694	11,372,903	0.23
Pot plants			
Phalaenopsis	221,423	45,523	4.86
Kalanchoe	48,382	67,771	0.72
Anthurium	47,268	12,983	3.64
Dracaena	40,201	23,360	1.72
Ficus	36,401	19,516	1.87
Rosa	32,467	32,882	0.99
Chrysanthemum	29,778	37,510	0.81
Hydrangea	26,148	9,551	2.74
Spathiphyllum	24,291	20,114	1.21
Hyacinthus	21,185	39,051	0.55
Hedera	20,204	31,114	0.66
Guzmania	20,162	12,108	1.67
Begonia	18,980	17,192	1.11
Cyclamen persicum	18,597	20,112	0.93
Euphorbia pulcherrima	18,494	19,211	0.99
Pot plants total	1,292,030	859,368	1.52

Source: VBN (2007)

when looking to turnover, lilies take the fourth place. In pot plants, flowering plants as *Phalaenopsis*, *Kalanchoe*, and *Anthurium* take the leading positions. For green pot plants *Dracaena*, *Ficus*, and *Hedera* are the most important ones (Table 31.6).

4. Some Future Trends and Challenges

Globalisation is set to have a major impact on world horticultural and floricultural production and distribution. Shifts in production areas as for cut flowers and markets for floricultural products were observed during the last decades. Today, plants are grown in places where the combinations of climatic conditions and cost price of labour is the most favourable. Outsourcing started with cut flowers, but nowadays starting material is also produced frequently outside Europe. Since long distance transport for pot plants is more expensive, production takes place closer to the market. The impact of globalisation on European horticulture is clearly seen in the reduction of the number of nurseries. European growers can only survive by specialisation and increasing the company size, by which expensive labour costs can be replaced by mechanisation. In future, rapid changes in information technology, packing and transport facilities will have an impact on international trade and distribution (2).

Another issue for European floriculture is the environmentally friendly production. Today, integrated production systems are a leading concept in Europe. Driving forces are the concern to reduce environmental impact on production and cost reduction by reducing the use of chemicals. MPS is an international certification organization, based in Holland, which owns and develops certificates for horticultural producers. It was set up by growers more than 10 years ago as an answer to the public's and media's inquiries surrounding the horticultural sector's environmental practices. Social issues can be addressed as a separate MPS certification. In Belgium a similar system called VMS exists. The goal is to assure the public and buyers that certified growers are using best practices, which include sustainable pest management, water use, fertilizing etc. In future it might be expected, that consumers are becoming more and more aware and sensitive of environmental, ecological and social issues. This will open new marketing opportunities for growers as they attain higher levels of certification. An example of this is the "Fair Flowers, Fair Plants" label, a familiar consumer icon that was created recently. The label assures the implementation and sustenance of good practices of corporate eco-social responsibility.

Energy is an important production factor in glasshouse horticulture. In recent years, glasshouse growers are faced with increasing energy prices, which force them to think intensely about energy-saving opportunities. Today, average energy costs account for 25–30% of the overall costs of greenhouse production in most Central and Northern European countries. The production of greenhouse cut flowers is hit the hardest by rising energy prices, followed by flowering plants and houseplants. Energy saving installations such as the use of energy saving screens is nowadays already widely spread. New approaches as the application of solar energy, bio heating fuels and geothermal heat are currently important research items. A next step forward is the development of greenhouses in which fossil fuel dependency is largely reduced. Instead, these greenhouses are transformed into sources of sustainable energy. In various locations mainly located in the Netherlands, the possibility is now considered to use such greenhouses to provide green energy (warm water and electricity) to the suburban neighbourhood (http://www.energiek2020.nu). This innovated design is based on captures of the excess heat from solar radiation during summer. The heat is stored in underground natural water reservoirs (aquifers) and used for warming the greenhouse at night or during winter. Energy balances show that there is sufficient energy left to heat a large number of houses. Two ha greenhouse can heat up to 200 houses. In addition, the greenhouse supplies tap water, treats wastewater, and produces electricity. The whole complex is self-sufficient in energy and water and recycles nutrients and carbon.

Besides investments in new concepts of greenhouses, an increased input in new technology is also observed i.e. energy-efficient lamps, long-term use of LED lighting, camera sorting systems, etc. Intensification of production is taking place partly as a result of increased automation and robotization. In the Netherlands, there is increased investment in transport systems and mobile production areas, the aim being to augment productivity per surface unit. Capital investment is horticulture's life-line for European mid- and large-scale holdings. Also, investments in know-how and the development of new floricultural products with new characteristics and added value is one of the strengths of European floriculture.

Finally, changes in market and supply chains will have implications for producers. Consumer demands for higher quality products in bulk and niche products, the increasing role of supermarkets and garden centres, the need for year-round delivering of the same product, just in time delivery, ability to supply big quantities, new business alliances and vertical integrations in the chain are only some of the challenges producers have to deal with.

5. Conclusions

Although today most of the growth in production of ornamental products is being realised outside Europe, Europe itself remains one of the world's leaders in floricultural products. The European market is very diverse. While West-European countries have a more saturated market, opportunities and growing markets emerge in Eastern Europe. Both in floricultural production as well as flower trade, the Netherlands play the most important role. Increased competition in the international scene, changes in consumer demand, trade and markets, together with the rise in costs of energy and other inputs form the major challenges for growers in the near future.

References

1. AIPH (2007) International statistics flowers and plants, vol 55. Institut für Gartenbauökonomie der Leibniz Universität Hannover, Hannover, 127 p
2. Van Uffelen RLM, de Groot NSP (2005) Floriculture world wide: production, trade and consumption patterns show market opportunities and challenges. Agricultural Economics Research Institute (LEI) Paper Series
3. Jaarverslag (2007) Vereniging van Bloemenveilingen in Nederland (VBN)

Chapter 32

Thin Cell Layers: Power-Tool for Organogenesis of Floricultural Crops

Jaime A. Teixeira da Silva

Abstract

Thin cell layers or TCLs have been a fundamental corner-stone of tissue culture of all plant species, including floricultural and ornamental plants (FOPs). In this review, the current status of the use of TCL technology to specific FOPs will be outlined while certain successes and difficulties will be focused. The ability to control developmental and morphogenetic processes in vitro through the use of TCLs, especially where conventional methods have not always produced ideal results, is the key in taking lab-based theory to business-based applications. TCL technology, which focuses more on the size rather than the origin of the explant, is a gateway for FOP regeneration and transformation studies. Some general principles and guidelines for quality verification will be provided.

Key words: Thin cell layer, Organogenesis, Somatic embryogenesis, Callogenesis, Rhizogenesis

1. Introduction

Thin Cell Layers (TCLs) is a term coined by Khiem Tranh Than Van 35 years ago when she proved that it was possible to induce flowers, vegetative buds, and roots in vitro from thin, transverse slices of tissue from the pedicels of flowering *Nicotiana tabacum* (1). The "totipotency" concept, introduced by Haberlandt to plant science, existed for 75 years. Even though sterile cultures of plant cells and organs were well established by 1973, TCLs transformed plant tissue culture to control more strictly the outcome of an organogenesis "programme", primarily by controlling the size of the explants. Since 1973, TCLs have been used to control the organogenesis potential of a wide range of plants, including ornamentals (2, 3).

S.M. Jain and S.J. Ochatt (eds.), *Protocols for in Vitro Propagation of Ornamental Plants,* Methods in Molecular Biology, vol 589, DOI 10.1007/978-1-60327-114-1_32, © Humana Press, a part of Springer Science+Business Media, LLC 2010

This brief review chapter focuses on the application of TCL technology to floricultural and ornamental plants (FOPs), allowing for easier and more reproducible tissue culture of valuable germplasm without the need for expensive labor and technology, especially where conventional explants have not produced the desired result.

2. The Division of Cells and the Formation of Tissues and Organs

Higher plants develop from a single-celled zygote into a multicellular organism through co-ordinated cell divisions, and when this process occurs without patterning, disorganized callus tissue is produced (4). Body organization is generated by two distinct processes: the first, the primary organization of the body, represented by the seedling, including the shoot and root meristems, is laid down by embryonic pattern formation, while the meristems then take over to produce the adult plant during postembry onic development. Plant organs are composed of ordered collections of various cell types differing in their shape, size, position, function, and DNA content. Cell division, separation, and morphogenesis are intimately connected, and ontogenesis is determined by the genome and is influenced by external signals. Correct control of the cell-division cycle is required for the elaboration and execution of developmental programmes, while the patterning genes determine overall architecture of the plant. Cell division, a critical activity during the growth and development of a plant, together with cell signalling (5), provides the building blocks for the differentiation of *in vitro* TCLs or in planta tissues and organs and contributes to the overall size of the individual plant.

This review serves to show readers the application of the concept of the TCL model to FOPs. TCL technology is also a solution to many of the issues currently hindering the efficient progress of FOP improvement, since it resolves problems at the first stage, i.e. regeneration, by using the most basic developmental building blocks, cells, and tissues. Through the use of TCLs, the regeneration of specific organs may be effectively manipulated and, together with specific controlled in vitro conditions and exogenously applied plant growth regulators (PGRs), many problems hindering the improvement of in vitro plant systems are potentially removed, as has been reviewed by Nhut et al. (6). The possibility of TCLs as a developmental tool for molecular and genetic studies is put into perspective. The TCL system also allows for the mass propagation of a species of interest as has been demonstrated for *Cymbidium* hybrid orchids (7), and thus has profound potential economic benefits. The TCL system could in theory provide a simple but efficient micropropagation system for developing countries with limited resources and facilities and it could be used as a high-tech method for the production of perfect, uniform clonal FOPs in an ever-increasingly demanding FOP market.

The initial concept of a TCL was applied to thin sections of *N. tabacum* pedicels (1). One mm-thick layer of cells with variable area dimensions were defined as a longitudinal TCL (lTCL), while a transverse slice, a few mm thick, was termed a transverse TCL or tTCL. In a recent paper, I contested this terminology originally used and now widely adapted and suggest that the term be adjusted to Thin Tissue Layer or TTL (8).

3. Model Floricultural and Ornamental Plants

3.1. Snapdragon, Antirrhinum majus

Antirrhinum majus, whose many genes have homology with those of *Arabidopsis* and which is a model plant for the study of flowering genetics, is also an ornamental species in some countries. To date, it has been difficult to obtain de novo flower formation from snapdragon TCLs than from those of *Arabidopsis* (*see* refs. 3, 9 for more discussion). In their studies, Altamura and co-workers obtained flower neoformation from snapdragon TCLs, but the response was sporadic: lTCLs consisting of epidermis and cortical parenchyma were excised from floral pedicels and cultured under various hormonal and environmental conditions, using both a mono- and a two-phase method (*see* ref. 3). The macroscopic formation of flowers was poor at the end of the culture period (day 30), and the flower developmental process was arrested at the dome stage, possibly hindered by the co-formation of callus.

3.2. Chrysanthemum, Dendranthema grandiflora

Chrysanthemum is one of the top three global, economically-important FOPs. The biotechnology of this plant has been covered in detail in ref. 9. TCLs have been used to study the effect of numerous media additives (carbon source, antibiotics, PGRs, inter alia) and conditions on regeneration and morphogenesis in chrysanthemum.

tTCLs of cvs. "Shuhou-no-Chikara" and "Lineker" (standard and spray buds, respectively) placed on various media resulted in the controlled production of roots, shoots, or somatic embryos could be achieved when stem internode (10, 11). More comprehensive reviews on chrysanthemum also highlights these studies in which the use of TCLs (primarily stem tTCLs) as the choice explant for tissue culture and micropropagation has also been used for the genetic transformation of this species (9, 12). Despite the regeneration of whole plants (primarily adventitious shoots without an intermittent callus phase, and a prerequisite for any transformation system) from the tissue culture of various explant sources, few histologically confirmed reports on somatic embryogenesis in

Dendranthema exist. A mixed organogenic response (i.e., shoots, roots, callus, and embryos forming together) or multiple-organ formation is a result of the cellular heterogeneity present in the initial explant source, but this can be overcome by the use of TCLs in conjunction with a single PGR application (10). The de novo formation of roots from tTCLs, with or without wounding (11), may allow this procedure to be used for cryopreservation, artificial seed production (root "synseeds"), secondary metabolite production (root-specific), and the improvement of the genus through genetic engineering and transformation (12–14). Neither the choice of carbohydrate source or concentration had an effect on somatic embryogenesis from tTCLs, although somatic embryogenic cultures developed better in the dark (13). Numerous studies have been conducted on the effect that a number of factors and media additives have on chrysanthemum TCL morphogenesis. To further enhance the medium-dependence of explants, tTCLs, composed of several tissue types, but which are normally too small to separate as in the case of chrysanthemum, were used in the experiments. In most, if not all, in vitro morphogenetic studies as described below, flow cytometric analyses certified that even though low levels of endopolyploidy (8C cells) could be detected in embryogenic callus cultures, resulting shoot and root cultures, in vitro and greenhouse plantlets did not exhibit any endopolyploidy, i.e., 2C and 4C cells only.

Only one report truly addresses the direct effect of filter paper on plant tissue culture and in vitro growth and morphogenesis (15). In that study, the effect of filter paper on in vitro growth and morphogenesis of chrysanthemum tTCLs was examined, focusing on the buffering effect it has on the phytotoxicity of antibiotics and on in vitro growth and morphogenesis (callus, shoot, root, somatic embryo formation). Whatman #1 and Advantec filter paper positively stimulated organogenesis and buffered the phytotoxic activity of aminoglycoside antibiotics (AAs), but Whatman #3 filter paper negatively impacted chrysanthemum organogenesis.

The reaction of stem TCL morphogenesis to various carbon sources, selective agents for positive selection systems, was studied (13). The rationale behind those experiments lay in the fact that the plant cannot use or metabolize all carbon sources effectively, and thus those could be used as limiting factors to regeneration, growth, and development. Sucrose, glucose, and fructose were shown to support metabolic growth in cvs. "Lineker" and "Shuhou-no-chikara." Using this principle, it was possible to establish threshold survival levels (TSLs; the level at which a TCL does not differentiate, independent of the increase in carbon source) in response to varying concentrations of different carbon sources, since when using a nonantibiotic marker gene, a low level of nutrient medium is utilized, making the TCL highly dependent (more heterotrophic) on the carbon source. Since the shoot regeneration capacity of chrysanthemum is severely hampered by the presence of antibiotics in the selective caulogenic medium (16), which despite optimization of shoot production in the caulogenic program is disturbed at higher antibiotic concentrations, the search for alternative (positive) selection systems using other carbon sources may benefit the outcome and efficiency of chrysanthemum genetic transformation. TSL levels could be achieved in 64% of the carbon sources tested, enhancing their potential, provided that genes coding for their respective degrading enzymes can be cloned into a vector system.

Polyamines (PAs), which are regarded as PGRs implicated in the control of cell division cycle, growth, differentiation, and developmental processes such as flowering and adventitious rooting (17, 18) affected chrysanthemum tTCL development in vitro (19). The involvement of PAs in in vitro morphogenesis, in particular de novo vegetative bud, floral bud, and root formation from TCLs has been extensively investigated using several approaches (covered more extensively in (2)): (a) analysis of changes in endogenous PA titres and metabolic pathways; (b) treatment with PA biosynthetic inhibitors; and (c) application of exogenous PAs. When using putrescine, spermine, spermidine, and cadaverine on chrysantheumum tTCLs, depending on the choice of PA, either a growth-supporting or growth-inhibiting activity was observed in different in vitro morphogenic programs (callogenic, caulogenic, rhizogenic, somatic embryogenic (19)). Even though all PAs detrimentally affected caulogenesis and somatic embryogenesis, shoots that were harvested from the former were not significantly different from controls, not showing altered morphology and flowering when greenhouse-acclimatized. The use of PAs (spermine, spermidine, and putrescine), however, positively stimulated rhizogenesis, while cadaverine reduced the rhizogenic response.

Aminoglycoside antibiotics (AAs) are still the most common selective agents in genetic transformation experiments both in chrysanthemum and almost every other FOP, the most common being the use of kanamycin A, gentamycin, or G-418 as a selective agent for transgenic plants harboring the *nptII* gene (20). Most AAs had a negative effect on in vitro growth and morphogenesis (shoot and root formation) of chrysanthemum TCLs. The effect of the AA concentration on plant morphogenesis and explant survival depended on the size of the explant, the choice of explant source, the timing of infection by *A. tumefaciens*, and selection pressure in genetic transformation. In separate experiments on the effect of other antibiotics on TSL values, a gradient of phytotoxicity was shown: bialaphos® > chloramphenicol > rifampicin > streptomycin > minomycin > ampicillin > penicillin G = penicillin V (21). Other studies (16, 22) showed the importance that *Agrobacterium* selective agent (carbenicillin, cefotaxime, or vancomycin) has on maximizing *Dendranthema* SRC, while minimizing phytotoxicity and explant mortality. When 250 mg/L of cefotaxime was added, shoot regeneration capacity increased (but not significantly).

3.3. Easter lily, *Lilium longiflorum*

Lilies are a popular bulbous crop species globally. The traditional asexual propagation of *Lilium* spp. by bulb scales as well as the lack of efficient micropropagation systems for species within the *Lilium* genus prompted extensive studies of TCL as a tool and solution for these shortcomings (23). TCLs have been used extensively to study lily differentiation, successfully manipulating all morphogenic programs, and thus it has been considered as a model system. In an attempt to test the effect of tTCL explant source (receptacle, stem node and internode, pseudo-bulblet, leaf) together with different factors such as sucrose concentration, explant position, activated charcoal (AC) and PGRs on the mass propagation of *Lilium* was studied (24). For all the studies on bulblet formation using TCL methods, a MS/2 medium supplemented with 1–2.7 μM α-napthaleneacetic acid (NAA) or 10 μM indole-3-butyric acid (IBA) and sucrose at concentrations from 20 to 30 g/L were used for the rooting of shoots, bulblets, and pseudo-bulblets. Plantlets obtained in the light developed well on this medium and were subsequently transferred to the greenhouse, with a 90–100% survival rate. When young leaf explant tTCLs (0.3 mm) were excised and cultured on MS medium supplemented with 3% sucrose, 2 μM 6-benzyladenine (BA) and 6 μM NAA (25), pseudo-bulblets formed mainly on the adaxial surface. Shoots regenerated from pseudo-bulblet tTCLs using forchlofenuron (CPPU), while a maximum of 15 bulblets can be obtained from one pseudo-bulblet tTCL. At low sucrose concentrations (2–4%) shoots formed, but at high concentrations (6–9%) bulblets formed (26–28). When receptacle tTCLs were used, buds appeared within 3 weeks of culture, but not in the ovary or flower stalks (23). These buds continued to develop into bulblets. Stem node (2–3 mm) tTCL sections from plantlets derived from shoot tips of dormant bulbs formed pseudo-bulblets, which developed into the flowering stage without dormancy being observed (24). tTCL square epidermal layers from young stems that could form an average of four bulblets were formed per tTCL after 4-weeks culture (23). Shoots did not form in activated charcoal (AC)-free medium, indicating that AC has the same effect as a cytokinin-like hormone on the development of tTCL explants. In other studies (29, 30), similarly prepared transverse young stem sections of *L. longiflorum* were shown to form different organs (bulblets, roots, shoots, plantlets, PLBs, somatic embryos) when explants were exposed to different PGRs. Somatic embryogenesis has been achieved in tTCLs of in vitro *L. longiflorum* pseudo-bulblet explants (23).

4. Other Floricultural and Ornamental Plants

4.1. African violet, *Saintpaulia ionantha*

Direct bud organogenesis (vegetative bud, callus, roots, somatic embryo, polyembryonic-like structure) was achieved when the auxin/cytokinin (NAA: BA/TDZ) ratio in the TCL (derived from leaf petioles to the central nerve (leaf vein) and lamina) culture medium was strictly controlled. An average of 100–200 shoots per tTCL explant were obtained from 0.3–0.5 mm petiole or 3 × 3 mm lamina sections, within 4 weeks culture. Over 70,000 plants were produced from a single leaf within 3–4 months (31).

4.2. Amaranth, *Amaranthus edulis*

Shoots and embryo-like structures (ELSs) were obtained when 0.2–0.4-mm tTCLs excised from apical and subapical zones of *Amaranthus* seedlings formed directly on the epidermal cells after one week on MS with 3 μM TDZ (32).

4.3. Begonia, *Begonia rex*

One to six epidermal cell layers and subjacent collenchyma cells excised from the main vein of leaves. These were cultured on a mineral solution containing 1% sucrose and 10 g/l agar, to which 1 μM BA was added for bud formation, or 0.5 μM NAA for root formation, or 0.1 μM NAA for unicellular hair formation and expressed after 6 days of culture (33). TDZ induced a similarly high frequency of shoot regeneration in petiole tTCLs (34).

4.4. Gentian, *Gentiana spp.*

Gentians, relatively recalcitrant species for shoot regeneration, could be mass propagated when 0.3–0.5-mm tTCLs, excised from floral stalk receptacles, were cultured on 50 μM thidiazuron (TDZ) and 1 μM NAA (35). In TDZ-less medium, the percentage of buds per tTCL is low. Many buds developed after 2–4 weeks directly on the surface of receptacle tTCLs.

4.5. Gerbera, *Gerbera jamesonii*

A unique procedure for the mass shoot propagation of *Gerbera* using receptacle tTCLs was developed (36). Genotype, flower bud age, explant size, position of receptacle tTCLs, and culture media were found to affect the success of culture. Ten interspecific crosses of *Gerbera* showed different shoot regeneration rates and callus induction via receptacle tTCL culture, all of which had shoot regeneration rates higher than 57%. Flower buds collected on the 10th day resulted in 91% shoot regeneration after 6 weeks of culture on basal MS medium supplemented with 0.02 mg/L TDZ, 0.8 mg/L adenine, and 10% (v/v) coconut water. This was significantly higher than those from flower buds on the 7th and 14th days (22% and 54%, respectively). Shoot regeneration rate was the highest (94–100%) in the middle layers of the receptacle. For mass shoot propagation, shoot clusters were subcultured on half-strength MS medium supplemented with 0.5 mg/L IBA, 0.5 mg/L BA, and 2.0 mg/L kinetin after every 4 weeks. Plantlets formed when single shoots were cultured on half-strength MS medium containing 1 mg/l IBA. All plantlets acclimatized well in the greenhouse.

4.6. Geranium hybrid, Pelargonium × hortorum

In tTCL hypocotyl explants (1 × 10 mm) of 1-week-old geranium hybrid seedlings, somatic embryos could form in response to TDZ or a combination of IAA and BA, but the number of embryos was much less in the latter than with 1–1.5 μM TDZ (37). The development of somatic embryos was rapid and the number of embryos was about eightfold higher than in the culture of whole hypocotyl explants. In tTCLs, globular or early heart-shaped somatic embryos were formed within one week of culture, whereas in the whole hypocotyl explants they were visible only after 2 weeks. Hypocotyls from a 1-week-old plant can be the source of at least five and as many as ten tTCLs, each of which can develop into about 40–60 embryos, giving a total of 400–600 embryos per hypocotyl, as compared to approximately 50 embryos which can form directly from a whole hypocotyl of similar size and age. tTCLs of 1-week-old seedlings can produce a higher number of somatic embryos than those obtained from older seedlings, and regenerated somatic embryos develop into complete plantlets within 6–8 weeks of culture initiation.

4.7. Gladiola, Gladiolus spp.

The genus *Gladiolus* from the family *Iridaceae* comprises about 180 species and numerous cultivars and hybrids that are of ornamental value as cut flowers. For the mass propagation of *Gladiolus* within a short time interval, *Gladiolus* in vitro cormel tTCLs were aseptically cultured from corms on medium containing 10 μM BA (37). After one month, cormels developed on the surface of originally cultured cormels, near the axillary buds. In vitro cormel tTCL explants (0.3–0.5 mm) were excised and cultured on medium containing different concentrations of TDZ (1–10 μM). After two weeks, direct bud primordia without an intermediate callus phase were observed on the surface of tTCL explants at an optimal concentration of 1–3 μM TDZ. More than 50 buds per tTCL could be recovered after 3-weeks culture.

4.8. Heliconia, Heliconia psittacorum

Stem tTCLs from the shoot apex (0.8–1 mm) of *Heliconia psittacorum* L. "Choconiana" were cultured in vitro on MS with 80 μM 2,4-D, forming callus and PLBs, which developed and grew into plantlets after two 6-week subcultures on basal MS (39). The TCL system was used for both mass propagation and germplasm conservation.

4.9. Iris, Iris pallida

Thin sections (0.3–0.5 mm) of *I. pallida* made across a mature shoot (comprising 5–6 leaves) and from the base toward the apex were cultured. Somatic embryogenesis occurred on young leaf base tTCLs (40). Protocols have been extensively utilized using thin sections for the plant regeneration of Irises (41).

4.10. Petunia, Petunia hybrida

In vitro shoots and flowers are formed from thin epidermal cells excised from the first five internodes of basal flowering branches in *Petunia hybrida* (42). Explants (1×10 mm^2) consisted of 3–6 layers of subepidermal and epidermal cells, and when placed on basal MS with 1 µM each of IAA or kinetin, vegetative buds formed after 2 weeks and developed into vegetative shoots in all genotypes tested. In contrast, in vitro floral buds were obtained when all stages of flowering (floral buds to faded flowers) were present on basal flowering branches of the mother plant. Other morphogenic programs were observed when medium PGRs were modified, such as the substitution of 1 µM kinetin by 10 µM BAP, resulting in vegetative bud formation, or when 1 µM IAA was replaced by 10 µM IBA or when kinetin was used at 0.1 µM instead of 1 µM, resulting in root formation . A combination of 100 µM 2, 4-D and 0.1 µM kinetin induced callus.

4.11. Rose, Rosa spp.

lTCLs were excised longitudinally from dormant bud floral stalks and cultured on a full- or half-strength MS medium, supplemented with 0.05–5 mg/L 2,4-D or NAA usually in combination with BAP, zeatin, or kinetin. Preincubation at a high (100 µM) 2, 4-D concentration increased the frequency of both organogenic and embryogenic callus from *Rosa hybrida* "Baccara" leaf explants (43). Sucrose at 2–3% was used as the sole carbon source in most cases. However, replacement of sucrose by galactose or fructose increased somatic embryogenesis from leaf explants of some *R. hybrida* cultivars. Breaking of bud dormancy in roses is important for rapid multiplication of roses using grafting. lTCLs cultured on medium with 10 µM BAP and 3 µM GA_3 resulted in more than seven buds per lTCL after 4-weeks culture.

4.12. Sunflower, Helianthus annuus

Different hypocotyl *Helianthus annuus* tTCLs were compared for their embryogenic and callogenic capacities: (a) 2-cm-long segments of hypocotyls, (b) hypocotyls without epidermis, (c) monolayer of epidermis, (d) subepidermal layer, and (e) epidermis plus parenchyma layers, cultured on medium containing 1 mg/L NAA, 1 mg/L BA, and 20 ml/L coconut water (44). The epidermal monolayer, the subepidermal layers, and the hypocotyls without epidermis were not embryogenic. Only the tTCLs comprising the epidermis plus parenchyma layer and the hypocotyls segments were embryonic. The primary somatic embryos that differentiated on *Helianthus* TCLs gave rise to secondary embryos, which developed into normal fertile plants.

5. Orchids

Orchids occupy a special position among FOPs and are thus dealt with separately here. To mass produce a monopodial orchid hybrid *Aranda* "Deborah", 0.6–0.7-mm-thick tTCLs from a single shoot tip (6–7 mm) were used (45). After 45 days of culture, neoformation of protocorm-like bodies (PBLs) occurred on the same culture medium, 13.6 PBLs per TCL and 2.7 PBLs per shoot tip. The addition of 2.75 μM NAA to the same medium further increased PLB production (19.2 PBLs per TCL). The advantage of the tTCL system is to produce a high frequency of shoot regeneration and to reduce the time interval required, with potentially more than 80,000 plantlets produced from a single tTCL in a year as compared to the 11,000 plantlets produced by the conventional shoot tip method. Among monocot orchid species such as *Phalaenopsis*, young leaf lamina TCL explant (4 mm^2) or floral stalks can be induced to form protocorms directly along the wounded edges of the lamina and on the surface of the TCL (46). An efficient micropropagation protocol was developed for *Phalaenopsis amabilis* cv. "Cool Breeze" using inflorescence axis-derived protocorms through in vitro culture (47). TCLs of young inflorescence axes of *Phal. amabilis* hybrid, white cultivar, were cultured on half-strength MS medium amended with BA (2.0 mg/L), NAA (0.5 mg/L), 2% (w/v) sucrose, 10% (v/v) CW, 2 g/L peptone, and 1 g/L activated charcoal. Each cross section produced an average of 20 PLBs after 12 weeks. Within the first 32 weeks after initiation of culture, 960 plantlets as well as a huge amount of PLBs were achieved from a single explant section. Repeating the subculture of PLBs on proliferation medium and culturing leafy shoots on plantlet regeneration medium could produce 2,000,000 plantlets every 32 weeks. CPPU (10 μM) together with a low sucrose concentration (1%) resulted in the highest percentage of explant (95%) shoot growth and rooting in *Rhynchostylis gigantean* (48). The culture of PLB epidermal longitudinal TCLs or lTCLs results in the controlled development of *Cymbidium* Twilight Moon "Day Light", a hybrid, shoots, secondary PLBs, or somatic embryogenic callus (49, 50). *Dendrobium malones* "Victory" could be regenerated by direct differentiation and through callus formation following the use of Thin Section (TS) (syn. TCL) explants from a single leaf to form ten times more PLBs than from a single whole leaf, in a shorter time period. These authors used full MS media supplemented with auxins/cytokines either in single or in combination along with peptone, yeast extract, casein hydrolysate (each at 100 mg/L), banana powder (40 g/L), 15% coconut water, and 2% sucrose (51). An efficient in vitro propagation protocol for *Dendrobium candidum*

Wall ex Lindl. using tTCLs culture system was established (52). The frequency of shoot regeneration and the number of adventitious buds produced from the regenerated shoots significantly relied on the concentration of PGRs and the position and orientation of the explant. MS medium with half-strength macronutrients and 2% sucrose, supplemented with 1.2 mg/L NAA and 1.2 mg/L 6-BA, was optimal for shoot regeneration. Upon this medium, the youngest explant inoculated in the upright orientation exhibited a high frequency of shoot regeneration (92%) and the highest number of adventitious buds (an average of 24.5) per explant. Ontogenetic studies revealed that the shoots originated from the stem vascular bundles.

6. Conclusions and Perspectives

Almost two dozen FOPs *sensu strictu* have successfully been tissue cultured using this simple but efficient system, TCL technology, that enhances the prospects of using it for a much wider range of plants. If one looks at the broader sense of FOPs and begins to consider plants used in landscaping and gardening, which range from herbs and medicinal plants through to trees, then the application of TCLs extends even further and to an even larger group of plants (2). The direct application of TCL technology to plant genetic engineering (14), as a tool to study in vitro flowering (Table 32.1) (53) or plant physiology, e.g. the reaction of plants to CO_2 super-enrichment (54), brings great promise to the use of this technology for basic and applied FOP research.

Table 32.1
FOPs induced to flower in vitro using TCL explants

Plant (family)	Explant source	PGR	Other conditions	% Flower[a]	Var[b]	Reference
Begonia rex (Begoniaceae)	Leaf vein	BA/Z	Epidermis	100	No	(33)
Cichorium intybus (Asteraceae)	Flower stalk	NAA/BA/IAA	–	100	No	(55)
Cymbidium sp. (Orchidaceae)	SAM	BA/Kin	Light:dark (1:1)	7	No	(56)
Dianthus caryophyllus (Caryophyllaceae)	Petal/sepal	NAA/TDZ/BA	Light	100	No	(57)

(continued)

Table 32.1 (continued)

Plant (family)	Explant source	PGR	Other conditions	% Flower[a]	Var[b]	Reference
Geum urbanum (Rosaceae)	SAM	IBA/Kin	Oligosaccharins	100	No	(46)
Melia azedarach (Meliaceae)	Hypocotyl	None	–	5	No	(58)
Odontoglossum sp. (Orchidaceae)	SAM	None	Low light intensity	100	No	(46)
Odontonia sp. (Orchidaceae)	SAM	None	Low light intensity	100	No	(46)
Petunia hybrida (Solanaceae)	Internode/ pedicel	IAA/Kin	–	100	No	(42)
Silene coeli-rosa (Caryophyllaceae)	0.5 mm shoot tip	–	*Myo*-inositol	50	No	(59)
Torenia fournieri (Caryophyllaceae)	Stem epidermis	IAA/Kin	Epidermis only	unspecified	No	(60)

[a]Represents the maximum percentage recorded
[b]Variability observed in either anthesis, or total flower morphology

References

1. Van Tran Thanh M (1973) Direct flower neoformation from superficial tissue of small explant of *Nicotiana tabacum*. Planta 115: 87–92
2. Nhut DT, Van Le B, Tran Thanh Van K, Thorpe T (2003) Thin cell layer culture system: regeneration and transformation applications. Kluwer Academic, Dordrecht, The Netherlands
3. Teixeira da Silva JA, Van Tran Thanh K, Biondi S, Nhut DT, Altamura MM (2007) Thin cell layers: developmental building blocks in ornamental biotechnology. Floricult Ornam Biotechnol 1:1–13
4. Teixeira da Silva JA, Nhut DT (2003) Cells: functional units of TCLs. In: Nhut DT, Van Le B, Tran Thanh Van K, Thorpe T (eds) Thin cell layer culture system: regeneration and transformation applications. Kluwer Academic, Dordrecht, The Netherlands, pp 65–134
5. Wu X-L, Weigel D, Wigge PA (2002) Signaling in plants by intercellular RNA and protein movement. Genes Dev 16:151–158
6. Nhut DT, Hai NT, Don NT, Teixeira da Silva JA, Tran Thanh Van K (2006) Latest applications of thin cell layer (TCL) culture systems in plant regeneration and morphogenesis. In: Teixeira da Silva JA (ed) Floriculture, ornamental and plant biotechnology: advances and topical issues, 1st edn, vol II. Global Science Books, Isleworth, UK, pp 465–471
7. Teixeira da Silva JA, Tanaka M (2010) Thin cell layers: the technique. In: Davey M, Anthony P (eds) Plant cell culture: methods express. Scion, Bloxham, UK (in press)
8. Teixeira da Silva JA (2008) Plant thin cell layers: challenging the concept. Int J Plant Dev Biol 2:79–81
9. Shinoyama H, Anderson N, Furuta H, Mochizuki A, Nomura Y, Singh RP, Datta SK, Wang B-C, Teixeira da Silva JA (2006)

Chrysanthemum biotechnology. In: Teixeira da Silva JA (ed) Floriculture, ornamental and plant biotechnology: advances and topical issues, 1st edn, vol II. Global Science Books, Isleworth, UK, pp 140–163
10. Teixeira da Silva JA, Fukai S (2003) Chrysanthemum organogenesis through thin cell layer technology and plant growth regulator control. Asian J Plant Sci 2:505–514
11. Teixeira da Silva JA (2003) Thin cell layer technology for induced response and control of rhizogenesis in chrysanthemum. Plant Growth Regul 39(1):67–76
12. Teixeira da Silva JA (2003) Chrysanthemum: advances in tissue culture, postharvest technology, genetics and transgenic biotechnology. Biotechnol Adv 21:715–766
13. Teixeira da Silva JA (2004) Evaluation of carbon sources as positive selection agents for chrysanthemum (*Dendranthema* X *grandiflorum* (Ramat.) Kitamura) transformation. New Zealand J Crop Hortic Sci 32:55–67
14. Teixeira da Silva JA (2005) Simple multiplication and effective genetic transformation (4 methods) of *in vitro*-grown tobacco by stem thin cell layers. Plant Sci 169:1046–1058
15. Teixeira da Silva JA (2003) Filter paper significantly affects the morphogenic programmes, and buffers the phytotoxic effect of antibiotics in chrysanthemum and tobacco thin cell layer *in vitro* culture. Hortic Sci 38:1403–1407
16. Teixeira da Silva JA, Fukai S (2001) The impact of carbenicillin, cefotaxime and vancomycin on chrysanthemum and tobacco TCL morphogenesis and *Agrobacterium* growth. J Appl Hortic 3:18–27
17. Kuznetsov VV, Shevyakova NI (2007) Polyamines and stress tolerance of plants. Plant Stress 1:50–71
18. Pang X-M, Zhang Z-Y, Wen X-P, Ban Y, Moriguchi T (2007) Polyamines, all-purpose players in response to environment stresses in plants. Plant Stress 1:173–188
19. Teixeira da Silva JA (2002) Polyamines in the regulation of chrysanthemum and tobacco *in vitro* morphogenic pathways. Propagatn Ornam Plants 2:9–15
20. Teixeira da Silva JA (2002) Antibiotics in successful transformation and in the war against plant tissue culture infection. Newsletter Int Soc Chemother 6:13
21. Teixeira da Silva JA, Nhut DT, Tanaka M, Fukai S (2003) The effect of antibiotics on the *in vitro* growth response of chrysanthemum and tobacco stem transverse thin cell layers (tTCLs). Sci Hortic 97:397–410
22. Teixeira da Silva JA, Fukai S (2003) Effect of aminoglycoside antibiotics on *in vitro* morphogenesis from cultured cells of chrysanthemum and tobacco. J Plant Biol 46: 71–82
23. Nhut DT, Bui VL, Teixeira da Silva JA, Aswath CR (2001) Thin cell layer culture system in *Lilium*: regeneration and transformation perspectives. In Vitro Cell Dev Biol Plant 37:516–523
24. Nhut DT, Bui VL, Fukai S, Tanaka M, Van Tran Thanh K (2001) Effects of activated charcoal, explant size, explant position and sucrose concentration on plant and shoot regeneration of *Lilium longiflorum* via young stem culture. Plant Growth Regul 33:59–65
25. Nhut DT (1998) Micropropagation of lily (*Lilium longiflorum*) via *in vitro* stem node and pseudo-bulblet culture. Plant Cell Rep 117:913–916
26. Bui VL, Nhut DT, Van Tran Thanh K (1999) Plant production via shoot regeneration from thin cell layer pseudo-bulblet explants of *Lilium longiflorum in vitro.* C R AcadSci Paris 322:303–310
27. Nhut DT, Bui VL, Tanaka M, Van Tran Thanh K (2001) Shoot induction and plant regeneration from receptacle tissue of *Lilium longiflorum.* Sci Hortic 87:131–138
28. Nhut DT, Bui VL, Minh NT, Teixeira da Silva JA, Fukai S, Tanaka M, Van Tran Thanh K (2002) Somatic embryogenesis through pseudo-bulblet transverse thin cell layers of *Lilium longiflorum.* Plant Growth Regul 37:193–198
29. Nhut DT, Bui VL, Tran Thanh Van K (2001) Manipulation of the morphogenetic pathways of *Lilium longiflorum* transverse thin cell layer explants by auxin and cytokinin. In Vitro Cell Dev Biol Plant 37:44–49
30. Nhut DT, Huong NTD, Bui VL, Teixeira da Silva JA, Fukai S, Tanaka M (2002) The changes in shoot regeneration potential of protocorm-like bodies derived from *Lilium longiflorum* young stem explants exposed to medium volume, pH, light intensity and sucrose concentration pretreatment. J Hortic Sci Biotechnol 77:79–82
31. Ohki S (1994) Scanning electron microscopy of shoot differentiation *in vitro* from leaf explants of the African violet. Plant Cell Tissue Organ Cult 36:157–162
32. Bui VL, Nghieng Thao DM, Jeanneau M, Sadik S, Tu S, Vidal J, Van Tran Thanh K (1998) Rapid plant regeneration of a C_4 dicot species: *Amaranthus edulis.* Plant Sci 132: 45–54

33. Chlyah A, Van Tran Thanh M (1975) Differential reactivity in epidermal cells of *Begonia rex* excised and growth *in vitro*. Physiol Plant 35:16–20
34. Nhut DT, Hai NT, Huyen PX, Huong DTQ, Hang NTT, Teixeira da Silva JA (2005) Thidiazuron induces high frequency shoot regeneration in *Begonia* petiole transverse thin cell layer culture. Propagat Ornam Plants 5:151–157
35. Hosokawa K, Nakano M, Oikawa Y, Yamamura S (1996) Adventitious shoot regeneration from leaf, stem and root explants of commercial *Gentiana* cultivars. Plant Cell Rep 15: 578–581
36. Nhut DT, An TTT, Huong NTD, Don NT, Hai NT, Thien NQ, Vu NH (2007) Effect of genotype, explant size, position, and culture medium on shoot generation of *Gerbera jamesonii* by receptacle transverse thin cell layer culture. Sci Hortic 111:146–151
37. Gill R, Gerrath J, Saxena PK (1992) High-frequency direct embryogenesis in thin layer cultures of hybrid seed geranium (*Pelargonium*). Can J Bot 71:408–413
38. Stefaniak B (1994) Somatic embryogenesis and plant regeneration of gladiolus (*Gladiolus hort*). Plant Cell Rep 13:386–389
39. Goh CJ, Nathan MJ, Kumar PP (1995) Direct organogenesis and induction of morphogenic callus through thin section culture of *Heliconia psittacorum*. Sci Hortic 62:113–120
40. Tran Thanh Van K, Bui VL (2000) Curent status of thin cell layer method for the induction of organogenesis or somatic embryogenesis. In: Mohan SJ, Gupta PK, Newton RJ (eds) Somatic embryogenesis in woody plants, vol 6. Kluwer Academic, Dordrecht, The Netherlands, pp 51–92
41. Gozu Y, Yokoyama M, Nakamura M, Namba R, Yomogida K, Yanagi M, Nakamura S (1993) *In vitro* propagation of *Iris pallida*. Plant Cell Rep 13:12–16
42. Mulin M, Van Tran Thanh K (1989) Obtention of *in vitro* flowers from thin epidermal cell layers of *Petunia hybrida* (Hort.). Plant Sci 62:113–121
43. Hsia CN, Korban SS (1996) Factors affecting *in vitro* establishment and shoot proliferation of *Rosa hybrida* L. and *Rosa chinensis minima*. In Vitro Cell Dev Biol Plant 32:217–222
44. Pélissier B, Bouchefra O, Pépin R, Freyssinet G (1990) Production of isolated somatic embryos from sunflower thin cell layers. Plant Cell Rep 9:47–50
45. Lakshmanan P, Loh CS, Goh CJ (1995) An *in vitro* method for rapid regeneration of a monopodial orchid hybrid *Aranda* Deborah using thin section culture. Plant Cell Rep 14:510–514
46. Van Tran Thanh M (1974) Methods of acceleration of growth and flowering in a few species of orchids. Am Orchid Soc Bull 43:699–707
47. Sinha P, Hakim ML, Alam MF (2007) Efficient micropropagation of *Phalaenopsis amabilis* (L.) BL. cv. 'Cool Breeze' using inflorescence axis thin sections as explants. Propagat Ornam Plants 7:9–15
48. Bui VL, Hang Phuong NT, Anh Hong LT, Van Tran Thanh K (1999) High frequency shoot regeneration from *Rhynchostylis gigantea* (Orchidaceae) using thin cell layer. Plant Growth Regul 28:179–185
49. Teixeira da Silva JA, Singh N, Tanaka M (2005) Priming biotic factors for optimal protocorm-like body and callus induction in hybrid *Cymbidium* (Orchidaceae), and assessment of cytogenetic stability in regenerated plantlets. Plant Cell Tissue Organ Cult 84:119–128
50. Teixeira da Silva JA, Tanaka M (2006) Embryogenic callus, PLB and TCL paths to regeneration in hybrid *Cymbidium* (Orchidaceae). J Plant Growth Regul 25:203–210
51. Anjum S, Zia M, Chaudhary MF (2006) Investigations of different strategies for high frequency regeneration of *Dendrobium malones* 'Victory'. African J Biotechnol 5:1738–1743
52. Zhao P, Wei W, Feng F-S, Wu F, Yag Z-Q, Wang W-J (2007) High-frequency shoot regeneration through transverse thin cell layer culture in *Dendrobium Candidum* Wall Ex Lindl. Plant Cell Tissue Organ Cult 90: 131–139
53. Teixeira da Silva JA, Nhut DT (2003) Thin cell layers and floral morphogenesis, floral genetics and in vitro flowering. In: Nhut DT, Van Le B, Tran Thanh Van K, Thorpe T (eds) Thin cell layer culture system: regeneration and transformation applications. Kluwer Academic, Dordrecht, The Netherlands, pp 285–342
54. Teixeira da Silva JA, Giang DTT, Chan M-T, Sanjaya Norikane A, Chai M-L, Chico-Ruíz J, Penna S, Granström T, Tanaka M (2007) The influence of different carbon sources, photohetero-, photoauto- and photomixotrophic conditions on protocorm-like body organogenesis and callus formation in thin cell layer culture of hybrid *Cymbidium* (Orchidaceae). Orchid Sci Biotechnol 1:15–23
55. Li Y, Zhang H, Han B-W (1996) Changes of endogenous hormone contents during floral bud and vegetative bud differentiation in thin cell layer culture of *Cichorium intybus* L. explant. *Acta Bot Sin* 38:131–135

56. Van Tran Thanh M (1974) Growth and flowering of *Cymbidium* buds normally inhibited by apical dominance. J Am Soc Hort Sci 99:450–453

57. Sankhla D, Davis TD, Sankhla N, Upadhaa A (1995) *In vitro* regeneration of the heat-tolerant 'German Red' carnation through organogenesis and somatic embryogenesis. Gartenbauwissenschaft 60:228–233

58. Handro W, Flo EIS (2000) Neo-formation of flower buds and other morphogenetic responses in tissue cultures of *Melia azedarach*. Plant Cell Tissue Organ Cult 64: 73–76

59. Donnison IS, Francis D (1993) Determination of floral organ type in cultured *Silene* shoot apices. Physiol Plant 89:315–322

60. Chlyah A (1974) Inter tissue correlations in organ fragments. Organogenic capacity of tissues excised from stem segments of Torenia fournieri Lind. cultured separately in vitro. Plant Physiol 54:341–348

INDEX

R

S

T

V

W

MIX
Papier aus verantwortungsvollen Quellen
Paper from responsible sources
FSC® C105338

If you have any concerns about our products,
you can contact us on
ProductSafety@springernature.com

In case Publisher is established outside the EU,
the EU authorized representative is:
Springer Nature Customer Service Center GmbH
Europaplatz 3, 69115 Heidelberg, Germany

Printed by Libri Plureos GmbH
in Hamburg, Germany